AF213994

State Estimation of Multi-Agent Vehicle-Road Interaction Systems

IEEE Press
445 Hoes Lane
Piscataway, NJ 08854

IEEE Press Editorial Board
Sarah Spurgeon, *Editor-in-Chief*

Moeness Amin
Jón Atli Benediktsson
Adam Drobot
James Duncan
Hugo Enrique Hernandez Figueroa
Ekram Hossain
Brian Johnson
Hai Li
James Lyke
Joydeep Mitra
Albert Wang
Desineni Subbaram Naidu
Yi Qian
Tony Quek
Behzad Razavi
Thomas Robertazzi
Patrick Chik Yue

State Estimation of Multi-Agent Vehicle-Road Interaction Systems

Yan Wang
The Hong Kong Polytechnic University, Hong Kong

Guodong Yin
Southeast University, China

Chao Huang
The University of Adelaide, Adelaide, South Australia, Australia

IEEE Press
WILEY

Copyright © 2026 by The Institute of Electrical and Electronics Engineers, Inc.
All rights reserved.

Published by John Wiley & Sons, Inc., Hoboken, New Jersey.
Published simultaneously in Canada.

No part of this publication may be reproduced, stored in a retrieval system, or transmitted in any form or by any means, electronic, mechanical, photocopying, recording, scanning, or otherwise, except as permitted under Section 107 or 108 of the 1976 United States Copyright Act, without either the prior written permission of the Publisher, or authorization through payment of the appropriate per-copy fee to the Copyright Clearance Center, Inc., 222 Rosewood Drive, Danvers, MA 01923, (978) 750-8400, fax (978) 750-4470, or on the web at www.copyright.com. Requests to the Publisher for permission should be addressed to the Permissions Department, John Wiley & Sons, Inc., 111 River Street, Hoboken, NJ 07030, (201) 748-6011, fax (201) 748-6008, or online at http://www.wiley.com/go/permission.

The manufacturer's authorized representative according to the EU General Product Safety Regulation is Wiley-VCH GmbH, Boschstr. 12, 69469 Weinheim, Germany,
e-mail: Product_Safety@wiley.com.

Trademarks: Wiley and the Wiley logo are trademarks or registered trademarks of John Wiley & Sons, Inc. and/or its affiliates in the United States and other countries and may not be used without written permission. All other trademarks are the property of their respective owners. John Wiley & Sons, Inc. is not associated with any product or vendor mentioned in this book.

Limit of Liability/Disclaimer of Warranty: While the publisher and the authors have used their best efforts in preparing this work, including a review of the content of the work, neither the publisher nor the authors make any representations or warranties with respect to the accuracy or completeness of the contents of this work and specifically disclaim all warranties, including without limitation any implied warranties of merchantability or fitness for a particular purpose. No warranty may be created or extended by sales representatives, written sales materials or promotional statements for this work. The fact that an organization, website, or product is referred to in this work as a citation and/or potential source of further information does not mean that the publisher and authors endorse the information or services the organization, website, or product may provide or recommendations it may make. This work is sold with the understanding that the publisher is not engaged in rendering professional services. The advice and strategies contained herein may not be suitable for your situation. You should consult with a specialist where appropriate. Further, readers should be aware that websites listed in this work may have changed or disappeared between when this work was written and when it is read. Neither the publisher nor authors shall be liable for any loss of profit or any other commercial damages, including but not limited to special, incidental, consequential, or other damages.

For general information on our other products and services or for technical support, please contact our Customer Care Department within the United States at (800) 762-2974, outside the United States at (317) 572-3993 or fax (317) 572-4002.

Wiley also publishes its books in a variety of electronic formats. Some content that appears in print may not be available in electronic formats. For more information about Wiley products, visit our web site at www.wiley.com.

Library of Congress Cataloging-in-Publication Data Applied for

Hardback ISBN: 9781394293377

Cover Design: Wiley
Cover Image: © ArtemisDiana/Shutterstock

Set in 9.5/12.5pt STIXTwoText by Straive, Chennai, India

Contents

About the Authors

Yan Wang, PhD, is a Research Fellow at The Hong Kong Polytechnic University, Hong Kong. His current research interests include vehicle system dynamics and automotive active safety control, where he contributed over 50 papers and obtained over 10 patents. He is also an Associate Editor for the *Chinese Journal of Mechanical Engineering, Metrology and Measurement Systems, IEEE Open Journal of Signal Processing*, and an Editorial board member of the *Journal of Zhejiang University Science A, Measurement Science Review*. Dr. Wang was the recipient of the Best Paper of IEEE ICUS in 2024, the Distinguished Youth Paper of Chinese Journal of Mechanical Engineering in 2024, and the Excellent Doctoral Dissertation Award of Jiangsu Province in 2023.

Guodong Yin, PhD, is a Professor at the School of Mechanical Engineering, Southeast University, Nanjing, China. His current research interests include vehicle dynamics control and connected vehicles, where he has contributed three book chapters, over 200 papers, and obtained 70 patents. He is an Associate Editor for the *IEEE Transactions on Intelligent Vehicles, Journal of Intelligent and Connected Vehicles*, and an Editorial Board for *the Chinese Journal of Mechanical Engineering*.

Chao Huang, PhD, is a Senior Lecturer at The University of Adelaide, Adelaide, South Australia, Australia. Her research interests include human-machine collaboration, fault-tolerant control, mobile robot (EV, UAV), and path planning and control, where she has contributed two book chapters, and over 100 papers. She is an Associate Editor for the *IEEE Transactions on Transportation Electrification* and *IEEE Transactions on Intelligent Vehicles, IEEE Transactions on Consumer Electronics, Engineering Applications of Artificial Intelligence, International Journal of Control, Automation and Systems*, and *IEEE Women in Engineering Magazine*.

Preface

Globally, the automotive industry is undergoing a profound transformation. The transition from traditional vehicles to intelligent vehicles, and from human driving to autonomous driving, is fundamentally reliant on perception technologies that capture information about vehicles and roads. Information such as vehicle state and tire-road friction coefficient parameters is not only crucial for vehicle dynamics and control but also serves as core technologies in intelligent transportation systems and autonomous driving. With technological advancements, vehicle state estimation and tire-road friction coefficient identification have gradually transitioned from theoretical research to practical applications. This journey includes significant milestones such as estimating the motion state and tire-road friction coefficient of the host vehicle using onboard sensors and predicting the state of surrounding vehicles through vehicular communication networks. Additionally, the development of deep learning technologies has provided new solutions for acquiring this information.

To provide a unified description of the states of various traffic elements, the authors have proposed the concept of the multi-agent vehicle-road interaction system (MVRIS). This concept describes a system composed of a vehicle and its surrounding traffic elements, distinguishing it from traditional macro-traffic research. In our book, the MVRIS consists of three main traffic elements: the host vehicle, the preceding vehicle, and the road. The authors have compiled their latest research work in this area into this book. The main topics discussed regarding the state estimation of the MVRIS include:

1) Ego-vehicle state estimation considering sensor data loss.
2) Ego-vehicle state estimation with unknown noise and parameter perturbations
3) State estimation of the preceding vehicle with data loss and parameter perturbations
4) Tire-Road Friction Coefficient Estimation with parameters mismatch and data loss

With the development of 5G communication technology, edge computing, and cloud computing, MVRIS state estimation technology is set to play a crucial role in a broader range of applications. In the future, vehicles will not merely be means of transportation but will serve as intelligent mobile terminals and centers for data collection and processing. This evolution places higher demands and challenges on MVRIS state estimation technology. We believe that, with continuous technological advancements, the estimation of MVRIS states will play an increasingly important role in intelligent transportation and autonomous driving.

Professionals in the field of autonomous vehicles, as well as researchers, engineers, and scientists in related fields, can utilize ***State Estimation of Multi-Agent Vehicle-Road Interaction Systems*** to gain relevant knowledge. This book offers practical, precise, and validated algorithms that can be deployed in various real-world scenarios.

Yan Wang
Guodong Yin
Chao Huang

1

Introduction

1.1 The Definition of Vehicle–Road Interaction System

Traffic accidents are one of the main causes of human casualties [1]. Intelligent connected vehicles will provide a new possibility for the automotive industry to effectively solve safety and congestion problems due to their functions of intelligent decision-making and collaborative control. Some typical technologies include vehicle road coordination systems, advanced driver assistance systems (ADAS), etc. Some of the most representative technologies in ADAS include stability control systems [2, 3], braking control systems [4–7], local path planning systems [8, 9], active suspension control systems [10–12], etc. The prerequisite for these active safety systems to work effectively is to obtain accurate vehicle state and tire-road friction coefficient (TRFC) [1]. To describe these vehicle states and road surface information in a unified way, this book adopts the concept of "interaction system" and defines the set composed of the host vehicle, the preceding vehicle, and the current road as the vehicle–road interaction system. As shown in Fig. 1.1, the corresponding variables, such as vehicle sideslip angle, tire stiffness, and TRFC, constitute the key state parameters in the vehicle–road interaction system. However, onboard sensors fail to directly obtain this information. Therefore, estimating these states using only onboard sensors is a hot topic of current research.

In the context of vehicle-to-vehicle (V2V) communication, the host vehicle, the preceding vehicle, and the road form a multi-agent vehicle–road interaction system due to their dynamic interdependence and interactive roles. The host vehicle acts as an independent agent, constantly interacting with its environment by adjusting its control actions based on its own dynamic states and real-time information from the preceding vehicle and road conditions. The preceding vehicle, another independent agent, influences the host vehicle's behavior through its speed, acceleration, and position, impacting car-following decisions and safety measures. Additionally, the road, though not a vehicle, can be viewed as an agent due to its influence on vehicle dynamics via tire–road interactions,

State Estimation of Multi-Agent Vehicle-Road Interaction Systems, First Edition.
Yan Wang, Guodong Yin, and Chao Huang.
© 2026 The Institute of Electrical and Electronics Engineers, Inc. Published 2026 by John Wiley & Sons, Inc.

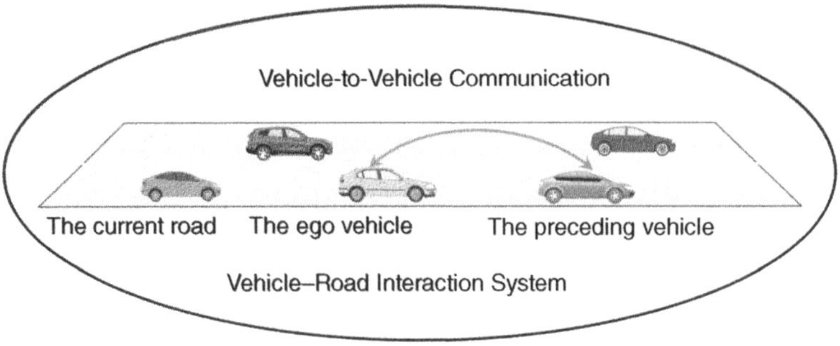

Figure 1.1 The vehicle–road interaction system.

such as road friction and surface conditions, which directly affect vehicle stability and performance. These three components—host vehicle, preceding vehicle, and road—communicate and interact within a shared environment, forming a tightly coupled multi-agent system. This framework allows for more accurate state estimation and decision-making, which is essential for the development of ADAS and autonomous driving technologies. In previous studies, host vehicle and preceding vehicle state estimation, as well as TRFC identification, are usually considered as two types of parameter identification problems. However, in this book, we try to define a new concept to describe a more macroscopic vehicle–road coupled system. This will provide a new perspective to researchers in this field. Therefore, these different states, or TRFC, will become the internal states of this macroscopic system.

1.2 The Importance of State Estimation for Vehicle–Road Interaction System

State estimation plays a crucial role in the development of vehicle–road interaction systems, directly impacting the safety, efficiency, and reliability of vehicles. This technology is fundamental for the operation of autonomous vehicles, as it allows for accurate sensing and understanding of both the vehicle's state and its surrounding environment. By doing so, state estimation enables intelligent vehicles to make autonomous decisions and execute complex driving maneuvers. Accurate state estimation supports ADAS such as adaptive cruise control (ACC), lane-keeping assist, and emergency braking. For ACC, state estimation helps maintain a safe distance from the vehicle ahead by continuously monitoring relative speed and distance. Lane-keeping assist relies on state estimation to ensure the vehicle stays centered in its lane by detecting lane markings and making necessary steering adjustments. Emergency braking systems use state estimation to detect potential collisions and apply brakes in time to avoid or mitigate the impact. Furthermore, state estimation enhances the efficiency and reliability of autonomous vehicles. By optimizing driving patterns based on accurate state information, vehicles can achieve smoother acceleration and braking, better fuel economy, and reduced emissions. Reliable state estimation ensures that autonomous vehicles can operate consistently in various conditions, from clear weather to rain or snow, thus building trust in autonomous vehicle technology. In summary, vehicle state estimation is a cornerstone of autonomous vehicle technology. It integrates sensor data, mathematical models, and advanced algorithms to provide a comprehensive understanding of a vehicle's dynamics and its immediate environment. This enables intelligent vehicles to make informed, autonomous decisions, ultimately improving the safety, efficiency, and reliability of modern transportation systems.

Vehicle state estimation involves the real-time determination of a vehicle's yaw rate, sideslip angle, velocity, and other pertinent parameters. These parameters are critical for assessing the vehicle's current status and predicting its future behavior, which is essential for facilitating safe and effective decision-making in autonomous vehicles.

The yaw rate of a vehicle, representing its rotational motion around the vertical axis, holds a central position in the realm of vehicle dynamics. It directly impacts the vehicle's stability during maneuvers such as turns and lane changes. A controlled and well-monitored yaw rate is critical for preventing oversteer or understeer conditions, both of which can lead to loss of control and compromise safety. In the context of vehicle dynamics, yaw rate plays a pivotal role in ensuring the vehicle's stability and responsiveness. During a turn, the yaw rate determines how quickly the vehicle rotates about its vertical axis. If the yaw rate is too high, it can result in oversteer, where the rear wheels lose traction and the vehicle turns more sharply than intended. Conversely, if the yaw rate is too low, it can lead to understeer, where the front wheels lose traction, causing the vehicle to turn less sharply than the driver intends. Both conditions can be dangerous, especially at high speeds or on slippery surfaces. By precisely managing the yaw rate, vehicles can navigate corners with optimal stability, reducing the risk of skidding or rollovers. This is achieved through advanced control systems such as electronic stability control (ESC), which continuously monitors the yaw rate and other parameters to make real-time adjustments. ESC systems apply brake force to individual wheels and adjust engine power to correct oversteer or understeer, helping the driver maintain control of the vehicle. The importance of yaw rate control becomes even more evident in emergency situations. Rapid changes in direction, such as during evasive maneuvers to avoid an obstacle, demand judicious control of the yaw rate to ensure the vehicle's response aligns with the driver's intentions. In such scenarios, the ability to swiftly and accurately adjust the yaw rate can make the difference between avoiding a collision and losing control. For instance, consider a situation where a driver must swerve to avoid a sudden obstacle on the road. The vehicle's stability control system, relying on yaw rate sensors, will detect the rapid change in direction and intervene to maintain stability. By modulating brake pressure on individual wheels and adjusting throttle input, the system helps the vehicle follow the desired path while preventing oversteer or understeer. This intervention occurs in a matter of milliseconds, often faster than a human driver can react, thereby enhancing safety. Moreover, maintaining an optimal yaw rate is crucial for ensuring passenger comfort. Sudden or excessive rotational movements can be unsettling for passengers, leading to discomfort and motion sickness. By managing the yaw rate effectively, the vehicle can provide a smoother ride, enhancing overall comfort and driving experience. In the realm of autonomous driving, yaw rate control is even more critical. Autonomous vehicles rely on precise control of all dynamic parameters, including yaw rate, to execute complex maneuvers safely and efficiently. Advanced algorithms and sensor fusion techniques are employed to continuously monitor and adjust the yaw rate, ensuring the vehicle remains stable and responsive under all conditions.

In conclusion, the yaw rate of a vehicle is a fundamental aspect of vehicle dynamics, and is crucial for maintaining stability and safety during various driving maneuvers. Whether it is preventing oversteer and understeer in everyday driving or ensuring precise control during emergency situations, effective yaw rate management is essential. Advanced stability control systems and autonomous driving technologies rely heavily on yaw rate data to enhance vehicle performance and passenger safety, underscoring its significance in modern automotive engineering.

The sideslip angle, indicating the angle between a vehicle's velocity vector and its heading angle, is a fundamental parameter influencing lateral stability. A controlled sideslip angle is integral to preventing uncontrollable skidding and maintaining the vehicle's trajectory during dynamic maneuvers. This parameter is critical for ensuring that the vehicle responds predictably to driver inputs, particularly during high-speed driving, abrupt steering inputs, or when navigating adverse road conditions. In the context of vehicle dynamics, the sideslip angle plays a pivotal role in maintaining the vehicle's stability. When a vehicle is in motion, its tires generate lateral forces to counteract any sideways motion. The sideslip angle quantifies the deviation between the vehicle's intended path and its actual path. If this angle becomes too large, it indicates that the tires are losing grip on the road surface, which can lead to a loss of control and potential skidding. Therefore, managing the sideslip angle is crucial for maintaining the vehicle's lateral stability and ensuring safe handling characteristics. Excessive sideslip angles can lead to loss of tire grip, compromising the vehicle's ability to respond predictably to driver commands. For instance, during a sharp turn or a sudden evasive maneuver, the sideslip angle increases as the lateral forces acting on the tires intensify. If the tires exceed their grip limit, they will start to slide sideways, resulting in a loss of control. This situation is particularly dangerous on slippery or uneven road surfaces, where the risk of skidding is higher. By controlling the sideslip angle, vehicles can maintain optimal tire grip, ensuring stable and predictable handling.

Advanced control systems play a crucial role in managing the sideslip angle to enhance vehicle safety. Modern vehicles are equipped with sophisticated systems such as ESC and traction control systems (TCS) that continuously monitor and adjust the sideslip angle. These systems use sensors to measure the vehicle's speed, steering angle, and yaw rate, among other parameters. By analyzing this data in real time, they can detect any deviation from the intended path and apply corrective measures. For example, if the ESC system detects that the sideslip angle is increasing beyond safe limits during a turn, it can selectively apply brake force to individual wheels and adjust engine power to counteract the sideways motion. This helps to bring the vehicle back on its intended trajectory, reducing the risk of skidding and enhancing overall stability. Similarly, the TCS system can modulate the power delivery to the wheels to prevent excessive wheel spin and maintain optimal traction, especially on slippery surfaces. The importance of controlling the sideslip angle is particularly evident in emergency situations. During sudden maneuvers to avoid obstacles or navigate sharp curves, the sideslip angle can change rapidly. Advanced control systems must react swiftly to these changes to maintain vehicle stability. By keeping the sideslip angle within stable operational limits, these systems enhance the vehicle's ability to respond effectively to driver commands, ensuring a safer driving experience. In autonomous vehicles, the management of the sideslip angle is even more critical. Autonomous driving algorithms rely on precise control of all vehicle dynamics to execute complex maneuvers safely. These algorithms use advanced sensor fusion techniques and predictive models to monitor and adjust the sideslip angle continuously. This ensures that the autonomous vehicle can navigate through various driving conditions with optimal stability and safety.

In conclusion, the sideslip angle is a vital parameter influencing the lateral stability of a vehicle. Effective management of the sideslip angle is essential to prevent uncontrollable skidding, maintain the vehicle's trajectory, and ensure predictable handling. Advanced control systems such as ESC and TCS are crucial in continuously monitoring and adjusting the sideslip angle to enhance vehicle safety. By maintaining stable operational limits, these systems contribute significantly to overall road safety, providing a safer and more reliable driving experience.

Longitudinal velocity, representing the rate of change of a vehicle's position along its direction of motion, and lateral velocity, depicting the rate of change of position perpendicular to the direction of motion, collectively play pivotal roles in determining a vehicle's stability, maneuverability, and response to various driving conditions. These two components of velocity are integral to understanding and managing a vehicle's dynamics. Longitudinal velocity, often associated with acceleration and deceleration, directly affects a vehicle's dynamics and braking performance. In emergency braking situations, the ability to modulate longitudinal velocity is crucial for avoiding collisions and ensuring the safety of occupants and pedestrians. Advanced antilock braking systems (ABS) and ESC mechanisms leverage longitudinal velocity data to optimize braking forces, preventing wheel lockup and skidding. ABS prevents the wheels from locking up during hard braking, allowing the driver to maintain steering control. ESC, on the other hand, helps to maintain vehicle stability by detecting and reducing the loss of traction. By continuously monitoring longitudinal velocity, these systems can adjust braking force distribution to ensure maximum efficiency and safety. Longitudinal velocity is also integral to the operation of ACC systems. ACC systems maintain a set following distance from the vehicle ahead by adjusting the throttle and brake based on longitudinal velocity. This enhances safety by providing a seamless response to changes in traffic conditions, reducing the risk of rear-end collisions. The ACC system uses sensors to monitor the speed and distance of the vehicle in front, adjusting the vehicle's speed accordingly. This not only ensures a safer driving experience but also enhances comfort by reducing the need for manual speed adjustments in varying traffic conditions. Moreover, collision avoidance systems utilize longitudinal velocity information to assess the risk of an impending collision and initiate pre-crash measures, such as autonomous emergency braking (AEB). AEB systems are designed to detect potential collisions and automatically apply the brakes if the driver does not respond in time. By analyzing longitudinal velocity along with other parameters like the distance to the obstacle and the relative speed, these systems can determine the likelihood of a collision and take preventive action. This significantly reduces the chances of accidents, protecting both the vehicle's occupants and other road users.

Lateral velocity, while less commonly discussed, is equally important for vehicle stability and maneuverability. Lateral velocity affects how the vehicle responds to steering inputs and how well it can maintain its intended path, especially during cornering or lane changes. High lateral velocities can lead to oversteer or understeer, where the vehicle either turns more sharply or less sharply than intended. Effective control of lateral velocity is essential for maintaining stability and preventing accidents, particularly in high-speed driving or adverse weather conditions. Advanced vehicle dynamics control systems, such as ESC and TCS, monitor and adjust both longitudinal and lateral velocities to enhance stability and safety. These systems use a network of sensors to gather real-time data on the vehicle's motion and the road conditions. By analyzing this data, they can make precise adjustments to the braking force, throttle, and steering inputs to maintain optimal stability. For instance, if the vehicle begins to oversteer, ESC can reduce engine power and apply braking to individual wheels to help regain control. Similarly, TCS can prevent wheel spin during acceleration by adjusting the throttle and brake. The integration of longitudinal and lateral velocity data is crucial for the development of autonomous driving technologies. Autonomous vehicles rely on accurate and continuous monitoring of these parameters to navigate safely and efficiently. Advanced algorithms and machine learning techniques are used to process the velocity data and make real-time decisions. For example, when navigating a sharp turn, the autonomous system must balance both longitudinal and lateral velocities to ensure a smooth and safe maneuver. This involves adjusting the speed and steering angle precisely to maintain stability and adhere to the intended path.

In summary, longitudinal and lateral velocities are fundamental to vehicle dynamics, playing critical roles in ensuring stability, maneuverability, and safety. Longitudinal velocity is crucial for acceleration, deceleration, and braking performance, impacting systems such as ABS, ESC, ACC, and collision avoidance. Lateral velocity, on the other hand, influences how well the vehicle maintains its path and responds to steering inputs. Advanced control systems continuously monitor and adjust both velocities to enhance overall vehicle performance and safety. As the automotive industry advances towards greater automation, the precise control and integration of longitudinal and lateral velocity data will be essential for developing safe and reliable autonomous vehicles.

As tires are the only components connecting the vehicle to the ground, motion control or stability control of vehicles ultimately translates into the control of motor torque and braking torque. The TRFC directly limits the maximum tire forces available for the vehicle. Furthermore, many ADAS or high-level autonomous vehicles require dynamic adjustments in longitudinal and lateral control to enhance vehicle safety based on the TRFC. Understanding and accurately assessing the TRFC are crucial for optimizing the performance and safety of these systems. The main function of ABS is to prevent the wheels from locking during heavy braking and to maintain the traction between the tires and the road at an optimal value. The magnitude of this optimal traction is usually determined based on the TRFC. ABS works by modulating the brake pressure to prevent wheel lockup, thereby maintaining steerability and stability during braking. When the TRFC is high, ABS can allow for more aggressive braking without the risk of wheel lockup. Conversely, when the TRFC is low, such as on icy or wet roads, the ABS adjusts to provide gentler braking to maintain control. ESC systems generate a yaw moment based on the desired yaw rate to ensure the lateral stability of the vehicle. The desired yaw rate normally shows a positive correlation with the TRFC. By continuously monitoring the TRFC, ESC systems can adjust the braking force applied to individual wheels to correct understeer or oversteer conditions. For instance, if the vehicle begins to oversteer, the ESC system can apply the brake to the outer front wheel, generating a counteracting force that helps stabilize the vehicle. Accurate TRFC information allows the ESC system to make precise adjustments, enhancing the vehicle's ability to maintain its intended path, especially in challenging driving conditions. Active collision-avoidance systems use a variety of sensors to obtain information about the surrounding environment of the vehicle to reduce the risk of accidents. These systems work by assessing the relative distance between the vehicle and potential obstacles and initiating preemptive actions when this distance falls below a safety threshold. This safety distance is negatively correlated with the TRFC. In other words, when the TRFC is low, the safety distance must be increased to account for the reduced traction and longer stopping distances. Conversely, when the TRFC is high, the vehicle can safely operate with a shorter safety distance. By integrating TRFC data, collision avoidance systems can more accurately determine when to initiate braking or evasive maneuvers, thereby reducing the likelihood of collisions. Additionally, TRFC plays a crucial role in the operation of ACC systems. ACC systems maintain a set following distance from the vehicle ahead by adjusting the throttle and brake based on the longitudinal velocity of the vehicle. When the TRFC is high, the ACC system can operate more aggressively, allowing for closer following distances and more responsive acceleration and deceleration. However, when the TRFC is low, the ACC system must adjust to maintain a greater following distance and smoother speed changes to ensure safety. The accurate assessment of TRFC is also essential for the performance of high-level autonomous vehicles. Autonomous driving algorithms rely on precise TRFC data to make real-time decisions about acceleration, braking, and steering. For example, when navigating a sharp turn, the autonomous system must balance both longitudinal and lateral forces to maintain stability. Accurate TRFC information allows the system to adjust the speed and steering angle precisely, ensuring a smooth and safe maneuver.

Moreover, ADAS such as TCS also depend on accurate TRFC data. TCS works by preventing wheel spin during acceleration by adjusting the throttle and brake. When the TRFC is high, TCS can allow for more aggressive acceleration without the risk of wheel spin. Conversely, when the TRFC is low, TCS must apply more conservative throttle control to maintain traction. By continuously monitoring and adjusting based on TRFC, TCS enhances the vehicle's ability to accelerate safely in various road conditions.

In summary, the TRFC is a critical parameter for vehicle dynamics and safety systems. It directly influences the performance of ABS, ESC, active collision avoidance, ACC, and TCS, among others. Accurate TRFC information allows these systems to make precise adjustments, enhancing vehicle stability, maneuverability, and safety. As the automotive industry continues to advance towards greater automation and improved safety features, the importance of accurate TRFC assessment will only grow. Ensuring that active safety systems have reliable TRFC data is essential for optimizing their performance and ultimately contributing to safer roads.

Although the yaw rate, sideslip angle, velocity of a vehicle, and TRFC are crucial dynamic parameters that significantly influence its handling and stability, these parameters cannot be directly measured and require estimation methods for determination. Directly measuring the yaw rate of a vehicle is challenging, as it represents the rotational speed around the vertical axis. Conventional vehicle sensors typically do not provide this precise information, necessitating estimation through alternative measured data and models. Direct measurement of the sideslip angle often requires specialized sensors, such as an inertial navigation system. However, in the case of most conventional vehicles, there is no sensor configuration designed for the direct measurement of the sideslip angle. Traditional vehicle sensors primarily focus on parameters like speed, angular velocity, and acceleration, lacking a dedicated sensor for sideslip angle measurement. Consequently, direct measurement of the sideslip angle is often impractical in many situations due to the absence of specific sensors. For longitudinal velocity, while some vehicles are equipped with speed-measuring sensors such as wheel speed sensors or GPS systems, there are situations where the measurements from these sensors may not be accurate or available. For lateral velocity, traditional vehicle sensors are effective for certain dynamic measurements; they do not provide a direct measurement of lateral velocity. Specific sensors designed solely for the direct measurement of lateral velocity are not commonly integrated into standard vehicle sensor setups. The absence of dedicated lateral velocity sensors limits the availability of direct measurement options. Similarly, TRFC cannot be measured by onboard sensors.

The process of vehicle state estimation relies on an intricate interplay between sensor data, mathematical models, and sophisticated algorithms. Sensors such as GPS, LiDAR, radar, and cameras provide raw data about the vehicle's position, speed, and surroundings. This sensor data is then processed using mathematical models that describe the vehicle's dynamics, including its mass, center of gravity, and aerodynamic properties. These models help in predicting how the vehicle will respond to different inputs, such as steering, acceleration, and braking. Sophisticated algorithms, such as Kalman filters, particle filters, and machine learning techniques, are employed to fuse the sensor data and refine the state estimates. Kalman filters, for example, are used to recursively estimate the state of the vehicle by combining predictions from the mathematical models with real-time sensor data. To this end, researchers have successively proposed various estimation approaches to address the challenge. Based on the above discussion, some key importance for vehicle–road interaction systems can be summarized.

1.2.1 Enhancing Safety

State estimation is one of the core technologies for ensuring the safe operation of intelligent vehicles. With accurate state estimation, vehicles can perceive their position, speed, and direction in real time, as well as detect other objects in the environment, such as pedestrians, other vehicles, and obstacles. This perception capability allows intelligent vehicles to react promptly and avoid collisions and other potential hazards. For instance, if an obstacle suddenly appears in front of the vehicle, state estimation can quickly identify and convey the information to the control system, enabling necessary evasive or emergency braking maneuvers.

1.2.2 Improving Driving Efficiency

Through state estimation, intelligent vehicles can optimize their driving paths and behaviors, thus improving overall driving efficiency. By monitoring vehicle states and traffic conditions in real time, intelligent vehicles can choose the optimal path, avoid congestion, and adjust speed and trajectory based on traffic signals and the behavior of other vehicles. This not only reduces travel time but also lowers fuel consumption and emissions, achieving more environmentally friendly travel.

1.2.3 Enhancing Autonomous Decision-making Capabilities

State estimation empowers intelligent vehicles with autonomous decision-making capabilities, allowing them to make independent judgments in complex and variable traffic environments. For example, when faced with changing traffic signals, pedestrian crossings, and emergency vehicle priority, intelligent vehicles need to quickly assess the current state and make corresponding decisions. Through accurate state estimation, vehicles can promptly obtain necessary information, perform risk assessments, and execute decisions to ensure a safe and efficient driving experience.

1.2.4 Supporting ADAS

State estimation is the foundation for realizing ADAS. ADAS features, such as ACC, lane-keeping assist, and automated parking, rely on accurate state estimation to perceive the vehicle and environmental states. These systems provide necessary driving assistance through real-time monitoring and data analysis, reducing driver burden and enhancing driving safety. For instance, ACC systems can accurately measure the distance and speed of the vehicle ahead through state estimation, adjusting their speed to maintain a safe following distance.

1.2.5 The Foundation of Future Traffic Systems

As technology advances, state estimation's role in future traffic systems will become more prominent. The development of intelligent transportation systems (ITS) relies on efficient communication and coordination between vehicles and between vehicles and infrastructure, and state estimation provides the foundation for achieving this goal. By sharing and integrating various state data, ITS can achieve more efficient traffic management, reduce congestion and accidents, and improve overall traffic flow.

1.2.6 Enhancing User Experience

State estimation not only improves the technical performance of intelligent vehicles but also significantly enhances the user experience. Through precise navigation and smooth driving behavior, passengers can enjoy a more comfortable ride. Additionally, various autonomous driving and driver assistance functions supported by state estimation make driving easier and safer, increasing user trust and satisfaction with intelligent vehicles.

In conclusion, the importance of state estimation for intelligent vehicles is undeniable. It is the key technology ensuring safe driving, precise navigation, improved driving efficiency, enhanced autonomous decision-making capabilities, and adaptation to complex environments. As intelligent vehicle technology continues to develop, state estimation will continue to play a central role in advancing autonomous driving and ITS. By continuously improving the accuracy and reliability of state estimation, we can look forward to a future with safer, more efficient, and more comfortable travel methods.

1.3 State Estimation Problems of Vehicle–Road Interaction System

According to the definition of vehicle–road interaction system, its key state estimation mainly includes ego-vehicle state estimation, preceding vehicle state estimation, and tire–road friction coefficient identification. Ego-vehicle state estimation usually includes vehicle body dynamic state estimation and model parameter identification. As shown in Fig. 1.2, some typical states include longitudinal and lateral speeds, tire forces, tire cornering stiffness, sideslip angle, the height of the center of gravity, vertical load, etc. Accurate state information is essential for ADAS [13–15]. However, onboard sensors fail to directly obtain this information. Therefore, estimating these vehicle states using only onboard sensors is a hot topic of current research.

Apart from obtaining the ego vehicle state, ADAS usually needs to obtain the state of the preceding vehicle. For example, the ACC system needs to get the preceding vehicle state to achieve better vehicle following. However, the preceding vehicle's state is usually difficult to measure directly via the in-vehicle sensors of the ego vehicle [16]. Therefore, it is a common method to use in-vehicle sensors of

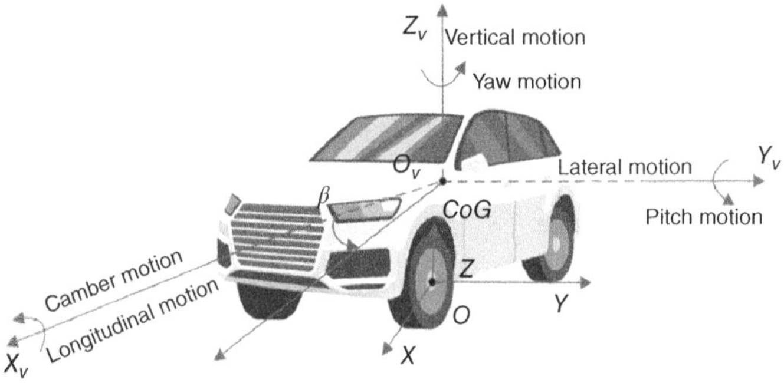

Figure 1.2 The vehicle coordinate systems.

the host vehicle to obtain some motion information about the preceding vehicle and combine it with a complex vehicle model to estimate the preceding vehicle state. In the vehicle–road interaction system, in addition to obtaining the vehicle state, we also need to get the TRFC. Studies have shown that traffic accidents are more likely to occur on roads with low friction coefficients. The magnitude of the longitudinal and lateral tire forces is related to the TRFC. Hence, the TRFC can indirectly affect vehicle stability [17]. However, the TRFC cannot be measured directly by onboard sensors. Thus, some estimation methods need to be utilized to solve the problem. As can be seen from the above description, the use of estimation methods to obtain the state of a vehicle–road interaction system is an effective solution due to factors such as sensor configuration and cost. The current related studies, despite the fruitful results, still have the following problems:

1) In terms of vehicle state estimation, existing studies rarely consider the impact of mass changes on the accuracy of vehicle models. This oversight is significant because the accuracy of estimating critical variables, such as the sideslip angle, is highly dependent on the precision of the vehicle model. Perturbations caused by vehicle mass variation must be accounted for during the estimation process to ensure accurate results. Changes in vehicle mass can occur due to various factors, such as passenger load, cargo, or fuel consumption, and these changes can significantly affect the dynamics of the vehicle. The sideslip angle and tire cornering stiffness are particularly sensitive to mass variations. As the vehicle's mass changes, the distribution of forces on the tires and the overall handling characteristics of the vehicle are altered. This, in turn, impacts the sideslip angle, which is a crucial parameter for maintaining lateral stability. Accurate estimation of the sideslip angle is essential for the effective functioning of stability control systems and other ADAS. Therefore, incorporating the effects of mass variation into the vehicle model is necessary for precise state estimation.

Additionally, the process noise of the vehicle system is often treated as a constant when estimating the vehicle state. However, due to the nonlinearity of the vehicle system and the complexity of the driving environment, the time-varying vehicle dynamics cause the estimated process noise to exhibit dynamic characteristics. The assumption of constant process noise can lead to inaccuracies in state estimation, particularly in scenarios involving rapid changes in driving conditions, such as sharp turns, sudden accelerations, or varying road surfaces. It is crucial to consider the dynamic nature of process noise and its impact on state estimation accuracy.

In practical applications, sensor data loss is a common phenomenon that further complicates vehicle state estimation. Most existing research focuses on the effect of sensor measurement noise on estimation accuracy, rarely considering the impact of sensor data loss. Sensor data loss can occur due to communication issues, hardware malfunctions, or environmental factors. When sensor data is lost, the state estimation algorithms must rely on incomplete information, leading to potential inaccuracies. Effective state estimation must account for the possibility of data loss and implement strategies to mitigate its impact. Furthermore, the noise from sensors is not always Gaussian white noise; it can also be colored noise. In real road driving scenarios, the type of noise from sensors is uncertain and can vary depending on the environment and sensor conditions. Gaussian white noise is a common assumption in many state estimation algorithms because of its simplicity and mathematical convenience. However, this assumption does not always hold true in real-world conditions. Colored noise, which has a correlation structure, can arise from various sources such as road texture, weather conditions, or electromagnetic interference. Ignoring the presence of colored noise can lead to significant errors in state estimation. The multifactorial nature of vehicle state estimation, involving mass variations, dynamic process noise, sensor data loss, and the presence of colored noise, makes it a challenging problem. Addressing these factors requires the development of robust and adaptive state estimation algorithms that can handle the uncertainties and complexities of real-world driving scenarios. For instance, advanced filtering techniques such as adaptive Kalman filters or particle filters can be employed to better account for dynamic process noise and sensor data loss. These filters can adjust their parameters in real time based on the observed data, improving the accuracy of state estimation. In conclusion, accurate vehicle state estimation is critical for the safe and efficient operation of intelligent vehicles. To achieve this, it is necessary to consider the influence of changes in vehicle mass, dynamic process noise, sensor data loss, and the presence of colored noise.

2) In terms of the prediction of the motion state of the preceding vehicle, the use of V2V communication for state prediction can effectively avoid the problem

that the in-vehicle sensor is disturbed by the environment and causes the estimation accuracy to decrease. Existing studies estimating the longitudinal and lateral states of the preceding vehicle require a combination of two different models and assume that they are uncorrelated. In addition, estimating the state of the preceding vehicle based on V2V communication is currently a time-triggered prediction method, which lacks an effective mechanism to achieve a balance between communication rate and estimation accuracy. This information transmission mechanism makes the vehicle regularly broadcast its movement information regardless of whether other vehicles need the data, which will inevitably occupy too much bandwidth. In addition, wireless communication usually has the problem of data packet loss. Data loss can occur for a variety of reasons, such as cyberattacks. Once some key sensor data is lost, it will inevitably affect the estimation accuracy of the vehicle state. Furthermore, unknown system dynamics usually have a direct impact on system state estimation and control. Therefore, the model parameters of the vehicle, such as the vehicle mass, can directly affect the accuracy of the model and consequently reduce the estimation performance of the vehicle state. Although many research efforts are devoted to the state estimation of connected vehicles, these studies seldom consider the effect of model parameters perturbation and packet loss simultaneously.

3) In terms of predicting the motion state of preceding vehicles, leveraging V2V communication for state prediction can effectively mitigate issues arising from in-vehicle sensor disturbances due to environmental factors, thereby preserving estimation accuracy. Current research focusing on estimating the longitudinal and lateral states of preceding vehicles typically involves combining two different models and assuming their independence. However, estimating the state of preceding vehicles using V2V communication currently relies on a time-triggered prediction method, which lacks an efficient mechanism to balance communication frequency and estimation precision. This transmission mechanism mandates vehicles to regularly broadcast their movement data, irrespective of whether other vehicles require this information, potentially monopolizing bandwidth resources unnecessarily. Furthermore, wireless communication commonly encounters challenges like data packet loss, stemming from causes such as cyberattacks. Loss of critical sensor data inevitably undermines the accuracy of vehicle state estimation. Moreover, unknown system dynamics significantly impact state estimation and control. Vehicle model parameters, such as mass, play a pivotal role in model accuracy and consequently influence vehicle state estimation performance.

Despite extensive research efforts devoted to connected vehicle state estimation, studies seldom address the concurrent effects of model parameter perturbations and packet loss. Addressing these challenges necessitates innovative approaches in vehicle state estimation methodologies. Enhanced utilization of V2V communication can mitigate reliance on potentially compromised in-vehicle sensors, ensuring more reliable estimation results. However, current methodologies rely on time-triggered transmission, lacking adaptive mechanisms to optimize communication rates based on real-time needs, which can lead to inefficient bandwidth usage. Additionally, managing data packet loss in wireless communication is crucial. Implementing robust error correction and redundancy techniques can help mitigate the impact of packet loss on state estimation accuracy. Furthermore, considering the influence of unknown system dynamics on model parameters is essential. Incorporating adaptive modeling techniques that account for parameter variations, such as vehicle mass, can improve the robustness and accuracy of state estimation algorithms. Despite advancements in connected vehicle technologies, the integrated consideration of model parameter uncertainties and communication challenges remains underexplored in state estimation research.

In the realm of TRFC identification, significant strides have been made in improving estimation techniques. However, the impact of parameter perturbations on estimation accuracy remains largely overlooked. For instance, inaccuracies in vehicle mass parameters can lead to deviations in estimated axle forces, subsequently affecting the precision of TRFC estimation. These deviations arise because variations in vehicle mass alter the distribution of forces on tires, influencing their frictional interaction with the road surface. Many hybrid estimation methods enhance TRFC estimation accuracy through weighted fusion of results from multiple models. Yet, a common limitation is that each model updates its estimates based solely on its previous predictions, without adapting to real-time conditions. This static approach can constrain accuracy in complex driving scenarios where dynamics rapidly change. To address this, dynamic adjustment of model weights based on current driving conditions is essential for achieving optimal TRFC estimation performance.

Moreover, existing TRFC estimation approaches often overlook the impact of missing sensor measurements on accuracy. Data loss from onboard sensors is prevalent, resulting from various factors such as communication disruptions or sensor malfunctions. The incomplete transmission of sensor signals to the estimation system can lead to incomplete or unreliable data inputs, thereby compromising the accuracy of TRFC estimates. Furthermore, traditional model-based methods typically assume precise knowledge of noise statistical characteristics for optimal results. This assumption is often unrealistic in practical scenarios where noise characteristics may vary or be inadequately understood. Data-driven approaches offer promise but face challenges related to data collection, quality, and the ability to generalize across diverse driving conditions. Both model-based and data-driven approaches require real-time access to accurate sensor data to function effectively. Addressing data loss and developing noise-adaptive algorithms are critical steps toward improving the robustness and applicability of TRFC estimation methods. Adaptive algorithms can dynamically adjust to fluctuations in sensor data quality and availability, thereby enhancing estimation accuracy under real-world conditions. In conclusion, advancing TRFC estimation methods requires addressing the complexities introduced by parameter perturbations, adapting to dynamic driving conditions through flexible model weighting, and mitigating the impact of sensor data loss with noise-adaptive algorithms.

1.4 Overview and Organization of the Book

To address the above problems, this book gives the corresponding technical solutions for ego-vehicle state estimation, the preceding vehicle state estimation, and TRFC identification, and carries out simulation and real-vehicle verification. The chapters are organized as follows.

In Chapter 2, the principles of the extended Kalman filter (EKF) are first introduced, and an adaptive fault-tolerant EKF is established for vehicle state estimation in cases of data loss. In Chapter 3, a fuzzy adaptive robust cubature Kalman filter is established to improve the vehicle state estimation accuracy by considering the effects of uncertainties in model parameters and noise parameters. Furthermore, in order to construct a vehicle state estimation method that adapts to multiple noises, a fusion of physical and data-driven estimation methods is introduced. In Chapter 4, a method for estimating the state of the preceding vehicle under event triggering is established by utilizing the V2V communication technology, which can effectively deal with the impact of data loss on the estimation accuracy and can effectively save communication resources at the same time. Furthermore, while considering model parameter uncertainty and data loss effects, a strong tracking event-triggered cubature Kalman filter is designed to enhance estimation accuracy. In Chapter 5, the problem of TRFC under complex operating conditions is investigated, and an interactive multi-model identification method is designed to estimate the TRFC. Furthermore, a fault-tolerant estimation scheme is established to realize the accurate estimation of TRFC considering the sensor data loss problem. Finally, in order to improve the applicability of the estimation method, such as multiple noise adaptation, a model-based learning method is designed to estimate the TRFC. Summaries and recommendations are presented in Chapter 6.

References

1 Wang, Y., Hu, J., Wang, F.A. et al. (2022). Tire road friction coefficient estimation: review and research perspectives. *Chinese Journal of Mechanical Engineering* 35 (1): 6.

2 Li, S.E., Chen, H., Li, R. et al. (2020). Predictive lateral control to stabilise highly automated vehicles at tire-road friction limits. *Vehicle System Dynamics* 58 (5): 768–786.

3 Hu, Z., Liao, Y., Liu, J., and Xu, H. (2022). Investigation of vehicle stability by integration of active suspension, torque vectoring, and direct yaw control. *SAE International Journal of Vehicle Dynamics, Stability, and NVH* 6 (4): 441–459.

4 He, L., Ye, W., He, Z. et al. (2020). A combining sliding mode control approach for electric motor anti-lock braking system of battery electric vehicle. *Control Engineering Practice* 102: 104520.

5 Chen, X., Wei, L., Wang, X. et al. (2021). Hierarchical cooperative control of anti-lock braking and energy regeneration for electromechanical brake-by-wire system. *Mechanical Systems and Signal Processing* 159: 107796.

6 Zhao, Q., Zheng, H., Kaku, C., and Cheng, F. (2023). Safety spacing control of truck platoon based on emergency braking under different road conditions. *SAE International Journal of Vehicle Dynamics, Stability, and NVH* 7 (1): 69–81.

7 Rajendran, S., Spurgeon, S.K., Tsampardoukas, G., and Hampson, R. (2019). Estimation of road frictional force and wheel slip for effective antilock braking system (ABS) control. *International Journal of Robust and Nonlinear Control* 29 (3): 736–765.

8 Wang, Z., Duan, Y., Wu, J., and Zhang, Y. (2022). Cooperative game approach to merging sequence and optimal trajectory planning of connected and automated vehicles at unsignalized intersections. *SAE Technical Paper, 2022-01-0295.*

9 Wang, Y., Yin, G., Li, Y. et al. (2021). Self-learning control for coordinated collision avoidance of automated vehicles. *Proceedings of the Institution of Mechanical Engineers, Part D: Journal of Automobile Engineering* 235 (4): 1149–1163.

10 Sande, T.P., Merks, M.H.M., Lindeman, E., and Nijmeijer, H. (2022). Rule-based control of a semi-active suspension system for road holding using limited sensor information: design and experiments. *Vehicle System Dynamics* 60 (12): 4226–4244.

11 He, X., Chen, J., Tang, D. et al. (2023). Using an inerter-based suspension to reduce carbody flexible vibration and improve riding-comfort. *SAE International Journal of Vehicle Dynamics, Stability, and NVH* 7 (2): 137–151.

12 Aghasizade, S., Mirzaei, M., and Rafatnia, S. (2019). The effect of road quality on integrated control of active suspension and anti-lock braking systems. *AUT Journal of Mechanical Engineering* 3 (1): 123–135.

13 Alquran, M. and Mayyas, A. (2021). Design of a nonlinear stability controller for ground vehicles subjected to a tire blowout using double-integral sliding-mode controller. *SAE International Journal of Vehicle Dynamics, Stability, and NVH* 5 (3): 291–305.

14 Hu, Y. and Wang, H. (2020). Robust tracking control for vehicle electronic throttle using adaptive dynamic sliding mode and extended state observer. *Mechanical Systems and Signal Processing* 135: 106375.

15 Lin, F., Wang, S., Zhao, Y., and Cai, Y. (2021). Research on autonomous vehicle path tracking control considering roll stability. *Proceedings of the Institution of Mechanical Engineers, Part D: Journal of Automobile Engineering* 235 (1): 199–210.

16 Wang, Y., Zhou, Z., Wei, C. et al. (2018). Host–target vehicle model-based lateral state estimation for preceding target vehicles considering measurement delay. *IEEE Transactions on Industrial Informatics* 14 (9): 4190–4199.

17 Ahn, C., Peng, H., and Tseng, H.E. (2012). Robust estimation of road friction coefficient using lateral and longitudinal vehicle dynamics. *Vehicle System Dynamics* 50 (6): 961–985.

2

Ego-vehicle State Estimation Considering Sensor Data Loss

2.1 Introduction

Certain vehicle state information crucial for active safety control in intelligent vehicles cannot always be directly measured by in-vehicle sensors. To acquire these critical data in real time, numerous advanced estimation algorithms have been proposed. However, current research predominantly examines the influence of sensor measurement noise on estimation accuracy, often overlooking the significant impact of sensor data loss. In this chapter, we focus on developing and evaluating extended Kalman filter (EKF) and adaptive fault-tolerant extended Kalman filter (AFTEKF) algorithms designed to estimate vehicle states even in scenarios where sensor data is partially lost. Unlike traditional approaches that assume continuous and complete sensor data availability, the AFTEKF is tailored to maintain accurate state estimation despite intermittent sensor failures or data losses. The EKF is a well-established method used to estimate the state of nonlinear dynamic systems based on noisy sensor measurements. It linearizes the system model around the current state estimate and updates predictions based on sensor inputs. However, it can be susceptible to inaccuracies and divergences when faced with abrupt changes in sensor data availability. In contrast, the AFTEKF extends traditional EKF capabilities by integrating adaptive mechanisms that adjust estimation parameters in response to varying levels of sensor data availability and quality. By dynamically adapting to sensor failures or partial data losses, the AFTEKF algorithm aims to maintain robust and accurate state estimation under challenging conditions.

Experimental validation of these algorithms under diverse operating conditions is crucial to assess their effectiveness and reliability in real-world applications. By subjecting the algorithms to scenarios involving partial sensor data loss, varying environmental conditions, and different driving dynamics, we aim to demonstrate their ability to provide reliable vehicle state estimates despite imperfect sensor inputs. Through these experiments, we seek to validate the practical utility of the

State Estimation of Multi-Agent Vehicle-Road Interaction Systems, First Edition.
Yan Wang, Guodong Yin, and Chao Huang.
© 2026 The Institute of Electrical and Electronics Engineers, Inc. Published 2026 by John Wiley & Sons, Inc.

AFTEKF in enhancing the active safety control capabilities of intelligent vehicles. By bridging the gap between theoretical advancements and practical implementation challenges, this research contributes to the development of more resilient and reliable vehicle state estimation technologies, paving the way for safer and more efficient autonomous driving systems.

2.2 Related Works

In recent years, the field of vehicle dynamics has seen a proliferation of research exploring various observer designs aimed at estimating critical parameters, such as the sideslip angle. A state observer, also known as an estimator or observer, is a crucial component in control systems engineering designed to estimate the internal state variables of a dynamic system based on available input–output measurements. Its primary function is to provide real-time estimates of the system's states that may not be directly measurable by sensors. This is achieved through mathematical models, typically based on differential equations or transfer functions, that describe the system dynamics and the relationship between inputs, outputs, and states. State observers play a pivotal role in various applications, including aerospace, automotive, industrial control, and robotics, enabling feedback control strategies even when direct state measurements are impractical or expensive. Advanced observer designs, such as sliding mode observers (SMOs), aim to minimize estimation errors, account for uncertainties, and improve robustness against disturbances and noise, thereby enhancing the overall performance and reliability of control systems. These observers encompass a spectrum of methodologies, including linear, nonlinear, and sliding mode variants, each contributing distinctively to advancing the accuracy and reliability of vehicle state estimation. A state observer, which provides estimates of the internal state of a system based on input and output measurements, is integral to achieving precise estimations crucial for vehicle control and safety. Linear observers have been extensively studied for their effectiveness in estimating sideslip angle. Stephant et al. [1] proposed the application of the Luenberger observer tailored specifically for sideslip angle estimation. Their approach demonstrated robust performance in real-world vehicle tests, achieving high accuracy in sideslip angle estimation under varying driving conditions. This method leverages input and output measurements to predict the sideslip angle dynamics, highlighting its applicability and reliability in practical scenarios. Furthermore, the H-infinity observer is a sophisticated estimation technique used in control systems to accurately estimate the states of a dynamic system while minimizing the effect of disturbances and uncertainties. Unlike traditional observers, which aim to minimize error based on a specific criterion (such as least squares or Kalman filtering (KF)), the H-infinity observer seeks to minimize the worst-case estimation error across all possible disturbances within a specified frequency range. This robustness criterion is achieved by formulating the observer design as an optimization problem, where the objective is to maximize the attenuation of disturbances while ensuring that the estimation error remains within acceptable bounds. The H-infinity observer is particularly valuable in applications where uncertainties in system parameters or external disturbances are significant and need to be actively managed to ensure accurate state estimation and stable system performance. Its theoretical foundation lies in optimal control theory and robust control, making it a powerful tool for high-performance control systems in aerospace, automotive, and industrial applications. Zhang et al. [2] explored the development of an H-infinity observer to estimate sideslip angle using information from front wheel angles. The H-infinity observer framework is known for its robustness against disturbances and uncertainties, making it suitable for handling variations in vehicle dynamics and road conditions. By integrating front wheel angle data into the estimation process, this approach enhances the accuracy and responsiveness of sideslip angle estimation, crucial for precise vehicle control and stability management. Beyond sideslip angle estimation, the exploration of controller output observers [3] has also proven beneficial, particularly in estimating tire forces. These observers adaptively estimate tire forces by analyzing controller outputs, accommodating uncertainties in model parameters and dynamics, especially within linear operational regimes.

The introduction of nonlinear observers has been instrumental in addressing the complexities inherent in the nonlinear dynamics of vehicles. Zhao et al. [4] applied a nonlinear observer to estimate vehicle velocity, demonstrating its effectiveness through comparative tests that highlighted its superiority over linear counterparts [5]. Expanding on these findings, Grip et al. [6] further advanced the field by developing a nonlinear observer tailored specifically for sideslip angle estimation. Subsequently, a high-gain nonlinear observer was proposed for additional adaptability [7]. The application of nonlinear observers extended beyond vehicle dynamics estimation. Hashemi et al. [8] utilized a nonlinear observer for road bank identification, showcasing the versatility of such methodologies. This broader application emphasizes the potential of nonlinear observers to address various challenges within the domain of vehicle dynamics. In summary, the integration of nonlinear observers signifies a noteworthy advancement in the pursuit of more accurate and versatile estimation techniques. The work of Zhao et al. and Grip et al. objectively illustrates the efficacy of nonlinear observers in enhancing vehicle velocity and sideslip angle estimation, as well as their potential application in broader aspects of vehicle dynamics, such as road bank identification. This pragmatic approach underscores the role of nonlinear observers in contributing to the ongoing refinement of methods for understanding and estimating the intricate dynamics of vehicles.

As technology evolves, the integration of sophisticated filtering methods continues to play a pivotal role in advancing the accuracy and reliability of vehicle state estimation. These methods not only contribute to real-time monitoring and control of vehicle dynamics but also address challenges posed by uncertainties, nonlinearities, and practical limitations in obtaining precise vehicle models. In the realm of observer-based methods for vehicle dynamics, the SMO has risen to prominence as a preferred and efficient choice. An SMO is an advanced estimation technique used in control systems to accurately estimate the states of a dynamic system, especially in the presence of uncertainties, disturbances, and nonlinearities. It operates by generating a sliding surface in the state space, where the dynamics of the observer force the system state to slide along this surface toward the true state trajectory. This approach ensures robustness against model uncertainties and disturbances by actively driving the estimation error to zero along the sliding surface. The design of an SMO involves choosing a sliding surface and designing the observer dynamics to ensure robust convergence and rejection of disturbances within a specified frequency range. SMOs find applications in diverse fields such as robotics, automotive systems, aerospace, and power systems, where accurate state estimation and robust performance are critical. They are particularly valued for their ability to handle nonlinearities and uncertainties that challenge traditional observers. Renowned for its computational efficiency and resilience to parameter variations and modeling uncertainties, the SMO has been widely applied to estimate critical parameters such as vehicle speed, tire cornering stiffness, tire forces, and sideslip angle. This methodological versatility has spurred the development of enhanced SMO versions, including the second-order SMO [9], reduced-order SMO [10], and higher-order SMO [11]. Each of these iterations contributes to the advancement of refined vehicle state estimations, catering to the intricate dynamics of modern vehicles. Beyond the SMO family, observers optimized using alternative techniques have surfaced, broadening the spectrum of estimation methodologies. Boada et al. [12], for example, proposed an observer based on an adaptive neuro-fuzzy inference system. This specialized observer is meticulously tailored for sideslip angle estimation, demonstrating the diversification of observer-based methods. While observer-based methods showcase substantial potential, their effectiveness relies on accurate vehicle models. Practical scenarios, however, often present challenges in obtaining precise models, leading to potential shortcomings in estimation accuracy. Recognizing these challenges, a more robust solution emerges through model-based estimation methods designed to withstand model mismatches.

The realm of state estimation in vehicle dynamics has witnessed a paradigm shift with the advent of KF and its advanced modifications, marking a transformative journey toward more accurate and robust estimation methodologies. KF is a recursive mathematical technique used extensively in control systems and signal processing for estimating the state of a dynamic system from a series of noisy measurements. It operates by combining predictions from a mathematical model of the system with real-time measurement data, optimizing the estimation by minimizing the mean squared error. KFs are particularly effective in environments where measurements are prone to noise and uncertainty, providing optimal estimates of the state variables even when the underlying system dynamics are nonlinear. They have wide-ranging applications across fields such as navigation, autonomous systems, financial forecasting, and biomedical engineering, where accurate real-time estimation of states or parameters is crucial for decision-making and control. The KF's adaptability to varying noise characteristics and its ability to handle uncertain initial conditions make it a cornerstone in modern estimation theory and practical implementation in complex, real-world systems. These approaches are widely acclaimed for their proficiency in obtaining minimum mean squared error estimates, especially in the presence of Gaussian noise, and their resilience to model mismatches. The extensive utilization of Kalman-based methods in the state estimation of vehicle systems is a testament to their efficacy. The foundational work of Venhovens et al. [13] in 1999 represents a milestone in the integration of KF into vehicle state estimation. Their pioneering efforts laid the groundwork for subsequent advancements that have significantly shaped the landscape of state estimation techniques in vehicle dynamics. Subsequent research, such as the work by Cho et al. [14], demonstrated the versatility of KF by employing a random-walk KF for predicting tire forces. This application showcased the adaptability of KF to diverse parameters, contributing to the refinement of tire force predictions. A method in the realm of sideslip angle estimation was achieved by Zhang et al. [15], who proposed a Kalman-based estimator specifically tailored for this critical parameter. This tailored approach exemplifies the adaptability of Kalman-based methods to address specific challenges within the intricate dynamics of vehicle systems, showcasing their potential for targeted applications. To further enhance estimation accuracy, multisensor fusion techniques rooted in KF have emerged as instrumental tools for predicting vehicle states. These techniques incorporate data from diverse sources, including Global Positioning System (GPS) [16, 17] and grade inertial sensors [18]. The synergy of information from multiple sensors contributes to a more comprehensive and accurate representation of the vehicle's state, enabling effective real-time monitoring and control.

However, the effectiveness of linear model-based methods, such as traditional KF, predominantly prevails under linear working conditions. Challenges surface when vehicles transition into nonlinear working regions, where the inherent nonlinearities in the vehicle model and tires compromise the accuracy of traditional KF, potentially leading to divergence. Recognizing this limitation, researchers have delved into advanced techniques, among which the EKF stands out. The EKF is an extension of the traditional KF, designed to handle nonlinear dynamics in dynamic systems. It approximates the state estimation process by linearizing the system model at each time step, using Taylor series expansion around the current estimated state and input values. This allows the EKF to propagate mean and covariance estimates of the state through nonlinear functions. Despite its linearization approach, the EKF can effectively estimate the state variables of systems where the process and measurement models are nonlinear. It has widespread applications in fields such as robotics, aerospace engineering, and econometrics, where nonlinearities are common and accurate state estimation is critical for decision-making and control purposes. The EKF's computational efficiency and relative simplicity compared to more complex nonlinear filters make it a popular choice for practical implementations requiring real-time estimation in dynamic and uncertain environments. Wenzel et al. [19] spearheaded the exploration of EKF for simultaneous vehicle state and model parameter estimation, offering a more robust solution to the challenges posed by nonlinear dynamics. Recognizing the pivotal role of tire forces in ensuring vehicle stability [20], Baffet et al. [21] leveraged EKF to predict sideslip angle and tire forces, accounting for variations in tire-road friction coefficient. This nuanced consideration of tire–road interaction dynamics contributes to a more accurate representation of the vehicle's state. Moreover, a dual EKF estimation method, as employed by researchers [22], adds a layer of comprehensiveness to vehicle state estimation. In this approach, one EKF is dedicated to estimating the overall vehicle state, while another integrates the Highway Safety Research Institute tire model to estimate tire–road friction using EKF. The real-time communication between these two EKFs enhances estimation accuracy, underscoring the potential of advanced filtering techniques in addressing the challenges posed by nonlinear working conditions and model uncertainties.

To augment the versatility of the estimation algorithm, a combined approach of the interacting multiple model and EKF (IMM-EKF) has been employed for predicting vehicle states [23]. The IMM-EKF improves state estimation in systems with varying dynamics. It maintains multiple modes or models representing different possible behaviors of the system and dynamically switches between them based on observed data. Each model has its own EKF, which processes measurements independently, and a mixing algorithm integrates their estimates to produce a final state estimation. This approach enhances robustness by adapting to changes in system behavior and uncertainty, making it suitable for applications such as tracking maneuvering targets, navigation in complex environments, and autonomous systems where the dynamics are not fully known or may change over time. Furthermore, the exploration of various vehicle dynamics models has been pivotal in refining sideslip angle estimation using EKF. Notable models include the single-track vehicle model [24–30], the seven-degree-of-freedom (DOF) model [31], the four-DOF model [32], the eight-DOF model [33], and others, each contributing to a nuanced understanding of vehicle dynamics. Recognizing the need for improved noise adaptivity in EKF, a fuzzy EKF has been introduced to enhance the accuracy of vehicle state estimation [34]. Moreover, the dynamic nature of vehicle inertial parameters during driving introduces complexities. To mitigate the impact of model parameter perturbations on estimation performance, the extended H-infinity KF (H-inf KF) has been specifically designed [35]. The utilization of IMM-EKF showcases a commitment to adaptability, offering a dynamic approach to predicting vehicle states. Furthermore, the incorporation of diverse vehicle dynamics models reveals a comprehensive exploration of sideslip angle estimation methodologies, contributing to a deeper understanding of the intricacies of vehicle behavior. The introduction of a fuzzy EKF not only refines noise adaptivity but also highlights the role of advanced filtering techniques in overcoming challenges inherent in real-world scenarios. Additionally, the introduction of the Extended H-inf KF demonstrates a forward-looking approach to addressing challenges associated with changing inertial parameters during driving. These advancements collectively underscore the ongoing efforts to refine and expand the applicability of EKF-based methods in the dynamic field of vehicle state estimation.

2.3 State Estimation Based on EKF

The EKF stands as a prominent recursive estimation algorithm, originally developed as an extension of the classic KF. It has become a cornerstone in state estimation applications, particularly in dynamic systems characterized by non-linear dynamics and uncertainties. The choice of EKF for vehicle state estimation is driven by its adaptability to nonlinear systems, a characteristic well-suited for capturing the complexities of vehicle dynamics. Traditional linear models often fall short of accurately representing the intricate relationships between vehicle state variables, especially in scenarios involving significant nonlinearity.

2.3.1 Preliminary Knowledge

Wiener filtering, a classical method, operates in the frequency domain and is suitable for processing single-channel, stationary random signals. However, its applicability is limited to scenarios with stored historical data, making it challenging to extend to nonstationary and multidimensional situations, rendering it less practical in engineering applications. In response to these limitations, experts and scholars sought a method to directly design optimal filters in the time domain, leading to the emergence of KF. In the 1960s, Hungarian mathematician Rudolf Emil Kalman introduced the classical KF method [36]. This method employs state-space equations to describe the input–output relationship of a system. By utilizing the system's state and observation equations along with noise excitation, and through statistical characteristics and recursive algorithms, KF significantly reduces the storage requirements for data. Since the information used is temporal variables, KF is not only applicable to one-dimensional and stationary random processes but also extends to multi-dimensional and non-stationary random processes. As described earlier, KF is a recursive algorithm that utilizes the system's state and observation equations and noise excitation for estimation. Therefore, before using the KF reasonably, knowledge of three fundamental characteristics of the estimated system is required: the state equation, observation equation, and noise characteristics (generally including state and observation noise characteristics). Classical KF is limited to linear systems. Taking the example of a vehicle traveling on a road, a discrete state system is introduced. The simple state of the vehicle can be represented by the body position and vehicle speed. Assuming the body position at time, k is represented by p_k, the vehicle speed at time k is represented by v_k, and the vehicle acceleration at time k is represented by a_k, and the state vector is represented by $\boldsymbol{x}$. For a vehicle moving linearly on the road, the position and speed of the vehicle at time k can be expressed as

$$\begin{cases} p_k = p_{k-1} + T_0 v_{k-1} + \frac{T_0^2}{2} a_{k-1} \\ v_k = v_{k-1} + T_0 a_{k-1} \end{cases} \tag{2.1}$$

where p_{k-1} represents the body's position at time $k-1$, T_0 is the sampling time of the discrete system, v_{k-1} represents the vehicle's speed at time $k-1$, and a_{k-1} represents the vehicle's acceleration at time $k-1$. From Eq. (2.1), it can be seen that the current body position and vehicle speed at time k are linear combinations of the position and speed at time $k-1$, represented in matrix form as

$$\begin{bmatrix} p_k \\ v_k \end{bmatrix} = \begin{bmatrix} 1 & T_0 \\ 0 & 1 \end{bmatrix} \begin{bmatrix} p_{k-1} \\ v_{k-1} \end{bmatrix} + \begin{bmatrix} \frac{T_0^2}{2} \\ T_0 \end{bmatrix} a_{k-1} \tag{2.2}$$

Then the state equation of the estimated system can be expressed as

$$\boldsymbol{x}_k = \boldsymbol{A}\boldsymbol{x}_{k-1} + \boldsymbol{B}\boldsymbol{u}_{k-1} \tag{2.3}$$

$$\boldsymbol{x}_k = \begin{bmatrix} p_k \\ v_k \end{bmatrix} \tag{2.4}$$

Assuming that the state transition matrix and control matrix are represented by A and B, respectively, then

$$A = \begin{bmatrix} 1 & T_0 \\ 0 & 1 \end{bmatrix} \tag{2.5}$$

$$B = \begin{bmatrix} \frac{T_0^2}{2} \\ T_0 \end{bmatrix} \tag{2.6}$$

In the equations, x_k and x_{k-1}, respectively, represent the state variables of the target state equation at time k and $k-1$. u_{k-1} represents the vector composed of external control variables at time $k-1$, that is, the control vector. Thus, we have obtained the first fundamental characteristic equation of the estimated system, namely the state equation.

As an observer, assuming that only an observation device capable of observing the vehicle displacement is currently provided, the observed variable z_k at time k can be represented as

$$z_k = p_k \tag{2.7}$$

Then, the observation equation of the estimated system at time k can be expressed as

$$z_k = Hx_k \tag{2.8}$$

It can be expressed as in matrix form

$$z_k = [1 \;\; 0] \begin{bmatrix} p_k \\ v_k \end{bmatrix} \tag{2.9}$$

Assuming z represents the observation vector and H represents the observation matrix, it is known that

$$H = [1 \;\; 0] \tag{2.10}$$

Thus, another basic characteristic equation of the estimated system can be obtained, namely the observation equation.

However, due to the absence of "perfect" sensors and "perfect" models, the data obtained through sensors often contain some noise, defined as observation noise v. Process noise ω is used to represent the model's inaccuracy resulting from model simplification and external disturbances. Generally, the process noise and observation noise in the system model satisfy the Gaussian white noise characteristics with mean 0 and covariance matrices Q and R respectively, and are uncorrelated. Therefore, the state equation and observation equation of the system can be expressed as:

$$\begin{cases} x_k = Ax_{k-1} + Bu_{k-1} + \omega_{k-1} \\ z_k = Hx_k + v_k \end{cases} \tag{2.11}$$

In the classical KF algorithm, state estimation refers to the calculation of the optimal quantity for a research object with a known mathematical model within a specified period based on the current measurement value, the previous predicted value, and the prediction error. It mainly includes two parts: time update and measurement update.

Time Update:
State Prediction Equation:

$$\hat{\boldsymbol{x}}_k^- = \boldsymbol{A}\boldsymbol{x}_k^- + \boldsymbol{B}\boldsymbol{u}_{k-1} \tag{2.12}$$

State Covariance Prediction Equation:

$$\hat{\boldsymbol{P}}_k^- = \boldsymbol{A}\hat{\boldsymbol{P}}_k^-\boldsymbol{A}^T + \boldsymbol{Q} \tag{2.13}$$

Measurement Update:
Gain Matrix:

$$\boldsymbol{K}_k = \hat{\boldsymbol{P}}_k^-\boldsymbol{H}^T\left(\boldsymbol{H}\hat{\boldsymbol{P}}_k^-\boldsymbol{H}^T + \boldsymbol{R}\right)^{-1} \tag{2.14}$$

Filter Equation:

$$\hat{\boldsymbol{x}}_k = \hat{\boldsymbol{x}}_k^- + \boldsymbol{K}_k\left(\boldsymbol{z}_k - \boldsymbol{H}\hat{\boldsymbol{x}}_k^-\right) \tag{2.15}$$

Error Covariance Update Equation:

$$\hat{\boldsymbol{P}}_k = (\boldsymbol{I} - \boldsymbol{K}_k\boldsymbol{H})\hat{\boldsymbol{P}}_k^- \tag{2.16}$$

Equations (2.12) to (2.16) form the foundation of the KF algorithm and are the core of KF theory. Here, $\hat{\boldsymbol{x}}_k^-$ represents the priori estimate vectors at time k. $\hat{\boldsymbol{x}}_k$ represents the posterior estimate vectors at time k, $\hat{\boldsymbol{P}}_k^-$, and $\hat{\boldsymbol{P}}_k$, respectively, represent the prior and posterior estimate covariance matrices at time k. $\boldsymbol{K}_k$ is the Kalman gain matrix at time k, $\boldsymbol{z}_k$ represents the observation vector at time k, H is the observation matrix, and I is the identity matrix.

2.3.2 Vehicle Model and Problem Statement

To estimate vehicle state, the single-track model (see Fig. 2.1) is employed. The single-track model, also known as the bicycle model, simplifies vehicle dynamics by representing a four-wheeled vehicle as having characteristics akin to a bicycle, with a focus on lateral and yaw dynamics. It assumes a fixed wheelbase, no tire slip, and a point mass at the vehicle's center, making it a practical tool for studying steering control, stability analysis, path planning, and performance evaluation in automotive engineering and autonomous vehicle development. Despite its simplicity, the model provides valuable insights into vehicle behavior under various conditions, contributing to the design and optimization of vehicle control systems and enhancing the overall understanding of vehicle dynamics. It is important to note that the choice of model depends on the needs of the research problem, and

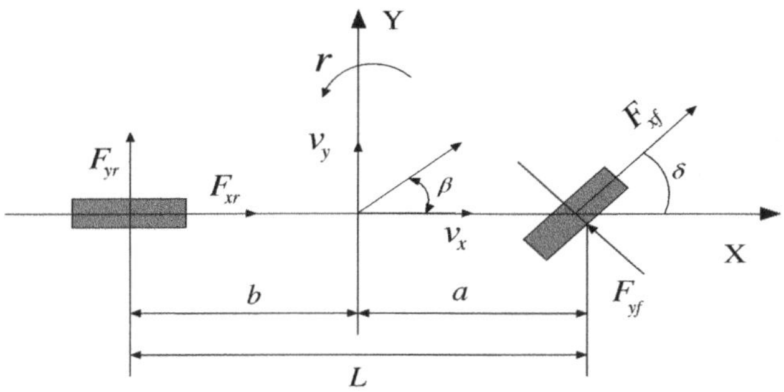

Figure 2.1 The single-track model.

the authors need to choose different vehicle models depending on the different vehicle states.

$$\dot{r} = \frac{a^2k_f + b^2k_r}{I_z v_x} r + \frac{ak_f - bk_r}{I_z}\beta - \frac{ak_f}{I_z}\delta \tag{2.17}$$

$$\dot{\beta} = \left(\frac{ak_f - bk_r}{m{v_x}^2} - 1\right) r + \frac{k_f + k_r}{mv_x}\beta - \frac{k_f}{mv_x}\delta \tag{2.18}$$

$$\dot{v}_x = r\beta v_x + a_x \tag{2.19}$$

$$a_y = \frac{ak_f - bk_r}{mv_x} r + \frac{k_f + k_r}{m}\beta - \frac{k_r}{m}\delta \tag{2.20}$$

where β, r, and δ are sideslip angle, yaw rate, and front wheel angle; a and b are the distances from the center of gravity to the front and rear axles; L is the wheelbase; k_f and k_r are cornering stiffnesses of the front axle and the rear axle; I_z is the moment of inertia along with the Z axis; v_x, a_x, a_y, and m are vehicle speed, longitudinal acceleration, lateral acceleration, and vehicle mass; and F_x and F_y represent the longitudinal and lateral tire forces, respectively.

Equations (2.17) to (2.20) need to be rewritten as a state space model to estimate the vehicle state using discrete measurement signals.

$$\begin{cases} \mathbf{x}_k = \boldsymbol{f}(\mathbf{x}_{k-1}, \mathbf{u}_{k-1}) + \boldsymbol{\omega}_{k-1} \\ \mathbf{y}_k = \boldsymbol{h}(\mathbf{x}_k, \mathbf{u}_k) + \boldsymbol{v}_k \end{cases} \tag{2.21}$$

$$\mathbf{x}_k = [r_k, \beta_k, v_{x,k}]^T \quad \mathbf{y}_k = [a_{y,k}]^T \quad \mathbf{u} = [\delta_k, a_{x,k}]^T$$

Some parameters in Eq. (2.21) are given in Table 2.1.

2.3.3 Methodology

This section describes the use of EKF for real-time and accurate estimation of vehicle states based on a nonlinear vehicle model in combination with conventional onboard sensors. The design of the scheme is illustrated in Fig. 2.2. The onboard

Table 2.1 Parameters of Eq. (2.21).

Symbol	Quantity
$\mathbf{y}$	Measurement vector
$\boldsymbol{\omega}$	Process noise
$\boldsymbol{\nu}$	Measurement noise
$\mathbf{x}$	State vector
$h(\cdot)$	Output function
$\mathbf{u}$	Input vector
$f(\cdot)$	State transition function
$\mathbf{Q}$	Covariance matrix of the process noise
$\mathbf{R}$	Covariance matrix of the measurement noise

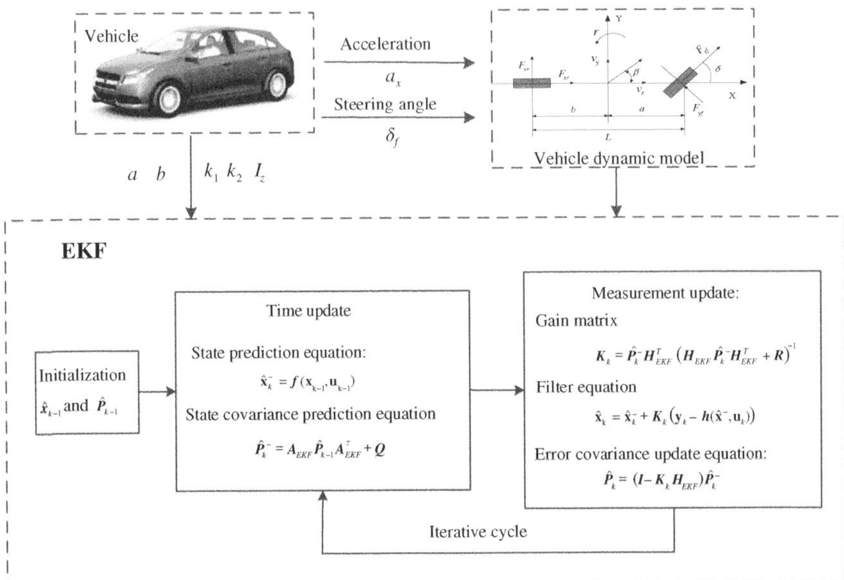

Figure 2.2 The estimation framework using EKF.

sensors provide acceleration and angular signals to the algorithm, and the vehicle dynamic model supplies the calculation equations for state prediction. Additionally, some inertial parameters of the vehicle are assumed to be known and input to the algorithm module. The subsequent iterations of the EKF algorithm are then carried out to accomplish the estimation of the vehicle state.

The core concept of the EKF algorithm revolves around approximating the nonlinear functions that describe the state equation and observation equation of a general nonlinear system model. This approximation is crucial for making the system amenable to the KF process, which traditionally applies to linear systems. The EKF achieves this approximation by employing a first-order Taylor series expansion around its reference point, neglecting higher-order terms beyond the first order. By linearizing the system dynamics and observation functions using this approach, the EKF effectively transforms the nonlinear system into a linear one for the purpose of filtering and state estimation. This transformation simplifies the computation involved in predicting the state of the system based on noisy sensor measurements and updates. The process starts with initializing the system state and covariance matrix at a known reference point. As new sensor measurements become available, the EKF predicts the next state of the system using the linearized state transition model and then adjusts this prediction based on the observed measurements. The filter iteratively refines its estimates of the system state and its uncertainty, incorporating new measurements while maintaining an optimal balance between prediction and observation. The key advantage of the EKF lies in its ability to handle nonlinearities in system dynamics and observation functions without requiring a complete transformation into a linear model. This makes it suitable for a wide range of nonlinear systems encountered in practical applications, such as robotics, aerospace, and autonomous vehicles. By approximating the nonlinear relationships with linear ones locally around the current state estimate, the EKF provides robust and computationally efficient state estimation, essential for real-time applications where rapid decision-making is critical.

Considering the nonlinear equation (2.21), during the iteration loop of the EKF, it is necessary to solve the Jacobian matrix to obtain the equivalent of matrices A and B in the classical KF.

$$\boldsymbol{A}_{EKF} = \frac{\partial \boldsymbol{f}}{\partial \boldsymbol{x}} = \begin{bmatrix} \frac{\partial f_1}{\partial x_1} & \frac{\partial f_1}{\partial x_2} & \cdots & \frac{\partial f_1}{\partial x_n} \\ \frac{\partial f_2}{\partial x_1} & \frac{\partial f_2}{\partial x_2} & \cdots & \frac{\partial f_2}{\partial x_n} \\ \vdots & \vdots & \vdots & \vdots \\ \frac{\partial f_n}{\partial x_1} & \frac{\partial f_n}{\partial x_2} & \cdots & \frac{\partial f_n}{\partial x_n} \end{bmatrix} \tag{2.22}$$

$$\boldsymbol{H}_{EKF} = \frac{\partial \boldsymbol{h}}{\partial \boldsymbol{x}} = \begin{bmatrix} \frac{\partial \boldsymbol{h}_1}{\partial \boldsymbol{x}_1} & \frac{\partial \boldsymbol{h}_1}{\partial \boldsymbol{x}_2} & \cdots & \frac{\partial \boldsymbol{h}_1}{\partial \boldsymbol{x}_n} \\ \frac{\partial \boldsymbol{h}_2}{\partial \boldsymbol{x}_1} & \frac{\partial \boldsymbol{h}_2}{\partial \boldsymbol{x}_2} & \cdots & \frac{\partial \boldsymbol{h}_2}{\partial \boldsymbol{x}_n} \\ \vdots & \vdots & \vdots & \vdots \\ \frac{\partial \boldsymbol{h}_i}{\partial \boldsymbol{x}_1} & \frac{\partial \boldsymbol{h}_i}{\partial \boldsymbol{x}_2} & \cdots & \frac{\partial \boldsymbol{h}_i}{\partial \boldsymbol{x}_n} \end{bmatrix} \tag{2.23}$$

The iterative process of the EKF for vehicle state estimation is illustrated in Fig. 2.3, and the specific steps are as follows:

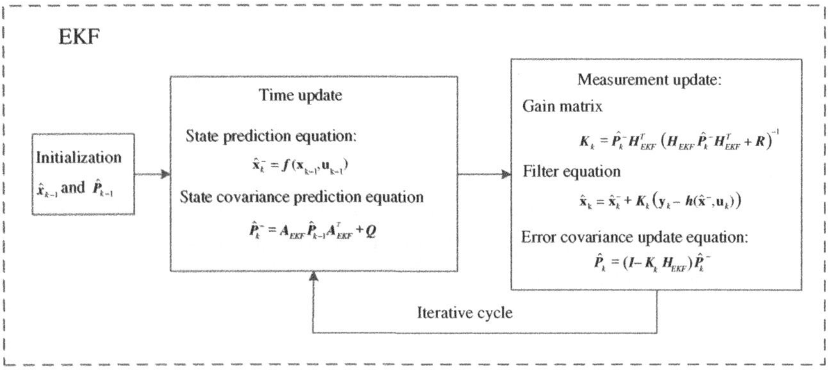

Figure 2.3 The overall iteration of the EKF.

Initialization:

Given the initial value of the state $\hat{x}_0$ and the initial value of the covariance $\hat{P}_0^-$

Time Update:

State Prediction Equation:

$$\hat{x}_k^- = f(x_{k-1}, u_{k-1}) \tag{2.24}$$

State Covariance Prediction Equation:

$$\hat{P}_k^- = A_{EKF}\hat{P}_{k-1}A_{EKF}^T + Q \tag{2.25}$$

Measurement Update:

Gain Matrix:

$$K_k = \hat{P}_k^- H_{EKF}^T s\left(H_{EKF}\hat{P}_k^- H_{EKF}^T + R\right)^{-1} \tag{2.26}$$

Filter Equation:

$$\hat{x}_k = \hat{x}_k^- + K_k\left(y_k - h\left(\hat{x}_k^-, u_k\right)\right) \tag{2.27}$$

Error Covariance Update Equation:

$$\hat{P}_k = (I - K_k H_{EKF})\hat{P}_k^- \tag{2.28}$$

The overall iteration block diagram is as follows:

The EKF begins with an initial estimate of the state vector and its associated covariance. This initial guess provides the starting point for the recursive estimation process. In the prediction step, the current state and covariance are projected forward in time based on the vehicle dynamics model. This involves applying the nonlinear motion model to estimate the expected state at the next time step. The core of the EKF lies in its ability to fuse predictions with real measurements.

In the measurement update step, in-vehicle sensor measurements are incorporated, adjusting the predicted state based on the innovation covariance and accounting for measurement noise characteristics. The update involves calculating the Kalman gain, which determines the weight given to the prediction and measurement information. The EKF cycle is repeated iteratively, continuously updating the state estimate as new sensor measurements become available. This recursive process ensures that the estimate evolves over time, providing a refined and accurate representation of the vehicle state.

2.3.4 Simulation Tests

To validate the effectiveness of the EKF, virtual simulations were conducted using a sophisticated testing setup. The chosen test scenario involved executing a double lane change (DLC) maneuver under typical operating conditions, providing a realistic and challenging environment to assess the algorithm's performance. The simulations aimed to evaluate the EKF's ability to predict critical vehicle dynamics parameters such as longitudinal velocity, yaw rate, and sideslip angle. For the test, specific inputs were fed into the estimation algorithms, including the front wheel angle and longitudinal acceleration. These inputs were essential for the EKF to generate predictions of the aforementioned vehicle dynamics parameters. The accuracy of these predictions was then meticulously compared with reference values derived from a highly detailed vehicle model within the CarSim software, which is known for its precise and reliable vehicle dynamics simulation capabilities. The entire simulation testing was executed using MATLAB R2021b in conjunction with CarSim 2016, running on a robust computing platform featuring an AMD Ryzen 7 5800HS CPU. This hardware setup ensured that the simulations could be conducted efficiently and effectively, providing a solid foundation for accurate and high-resolution data processing. The sampling time for the simulations was set at 0.01 seconds, enabling fine-grained temporal resolution and ensuring that the dynamic responses of the vehicle could be captured and analyzed with high precision.

2.3.4.1 The Test on the Asphalt Road

To establish a realistic simulation environment, the vehicle's initial speed is configured at 70 km/h, ensuring that the test conditions closely mimic typical real-world driving scenarios. It is assumed that we possess prior knowledge of the statistical properties of process noise and its covariance, as well as the statistical characteristics of observation noise. This information is crucial for the accurate implementation of the EKF algorithm, as it relies on these statistical properties to refine its estimations. The dynamic changes in lateral acceleration and front wheel steering angle are meticulously depicted in Figs. 2.4 and 2.5. These figures provide a

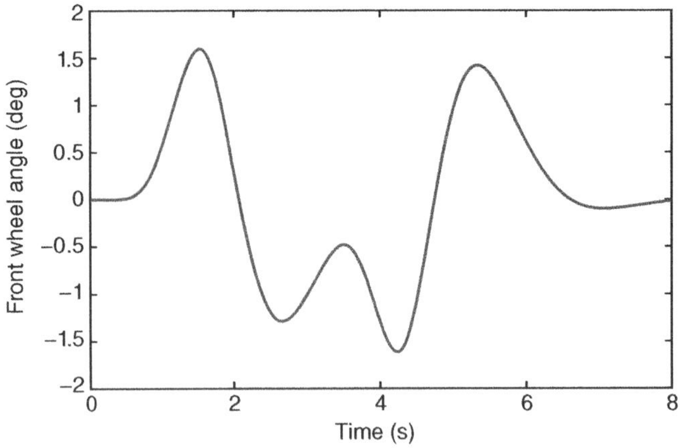

Figure 2.4 The front wheel angle on the asphalt road.

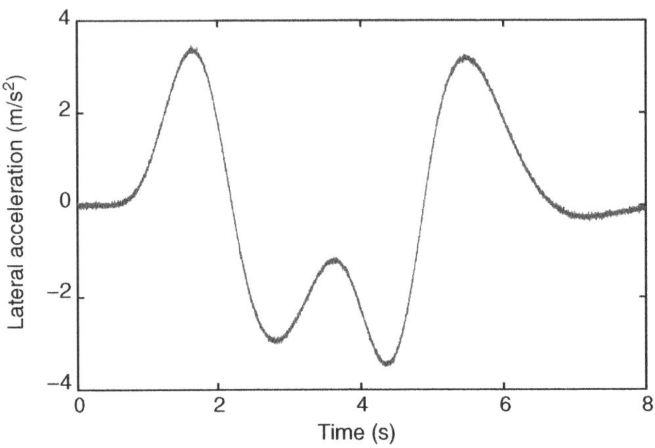

Figure 2.5 The lateral acceleration on the asphalt road.

comprehensive representation of the simulation scenarios, capturing the nuanced interactions between vehicle dynamics and control inputs. In the simulation, the vehicle undertakes a steering maneuver to execute a lane change, a common but critical driving task that tests the vehicle's stability and the EKF's estimation accuracy. During the lane change maneuver, the changes in lateral acceleration are synchronized with adjustments in the front wheel steering angle. This alignment between lateral acceleration and steering inputs is essential for the realistic simulation of vehicle dynamics, as it mirrors the real-time response of the vehicle to driver commands. Additionally, to further enhance the realism of the simulation, noise is introduced into the lateral acceleration curve. This noise simulates the

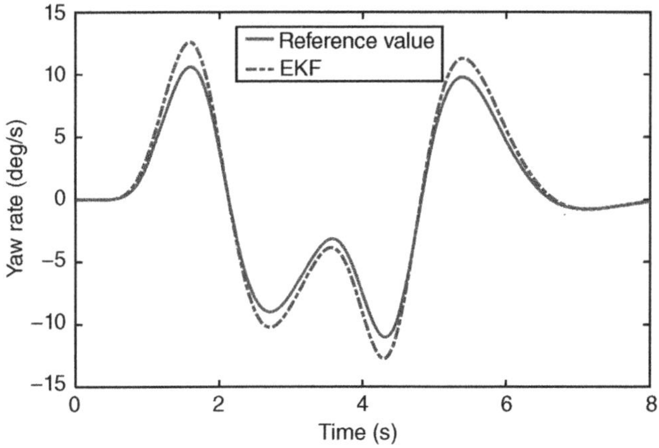

Figure 2.6 The yaw rate on the asphalt road.

measurement noise that is typically encountered in real-world sensor data, providing a more rigorous test of the EKF's noise-handling capabilities.

As shown in Fig. 2.6, the estimation results of the yaw rate using the EKF are presented. In the figure, the red solid line represents the reference values output by the CarSim software, while the blue dotted line corresponds to the EKF estimation results. From this comparison, it is evident that the EKF estimation closely tracks the trend of the reference values, capturing the dynamic behavior of the yaw rate with remarkable accuracy. Notably, the estimated values exhibit a slight increase at the inflection points, reflecting the algorithm's responsiveness to changes in vehicle dynamics. Figure 2.7 illustrates the estimation results for the sideslip angle, with the color coding of the curves consistent with that of Fig. 2.6. The sideslip angle estimation curve generated by the EKF closely aligns with the reference values from CarSim, indicating that the EKF performs effectively in estimating this critical parameter. The proximity of the estimation curve to the reference values demonstrates the EKF's capability to accurately capture the lateral dynamics of the vehicle, essential for ensuring vehicle stability and safety during maneuvers. In Fig. 2.8, the estimation results for the longitudinal velocity are depicted. Initially, the estimation curve aligns very well with the reference values, demonstrating the EKF's strong performance in predicting the vehicle's forward motion. Over time, there is a slight divergence, but the estimation deviation remains within acceptable limits, with a maximum error of no more than 0.3 m/s. This minor deviation indicates that while the EKF maintains high accuracy throughout most of the simulation, there are some challenges in perfectly tracking the longitudinal velocity over extended periods or under varying conditions. Nevertheless, the overall estimation error is minimal and deemed acceptable for practical applications.

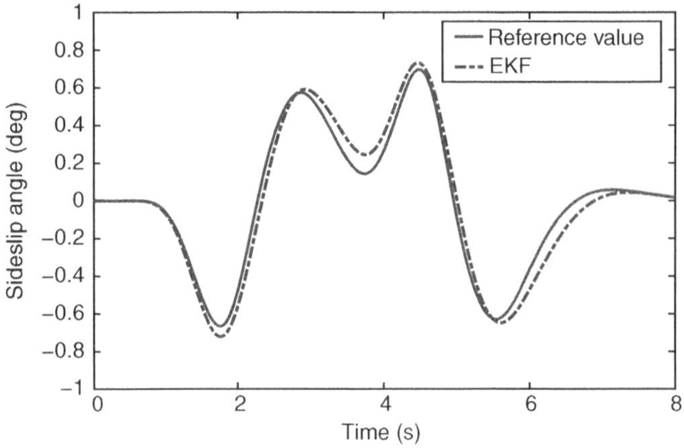

Figure 2.7 The sideslip angle on the asphalt road.

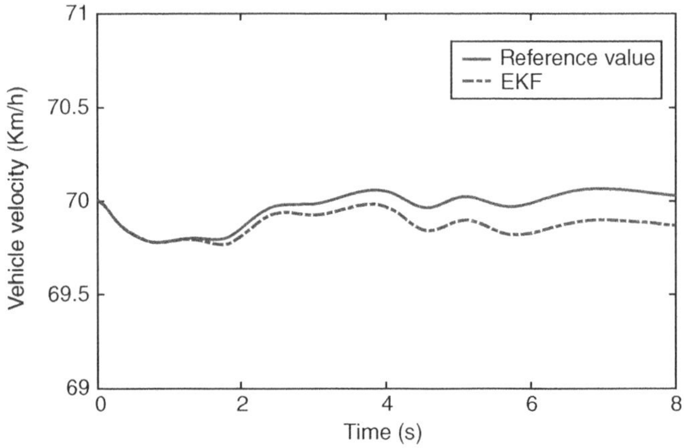

Figure 2.8 The longitudinal velocity on the asphalt road.

2.3.4.2 The Test on the Ice Road

In this simulation scenario, the tire-road friction coefficient is set at 0.2, and the vehicle's initial speed is configured at 30 km/h to create a realistic and controlled test environment. It is also assumed that there is prior knowledge of the statistical properties of both process noise and observation noise, which is crucial for the accurate implementation and performance of the EKF. Figures 2.9 and 2.10 illustrate the dynamic changes in lateral acceleration and front wheel steering angle, respectively. These figures provide a comprehensive visual representation of the vehicle's behavior during the simulation. In this scenario, the vehicle executes a steering maneuver to perform a lane change. This maneuver involves a

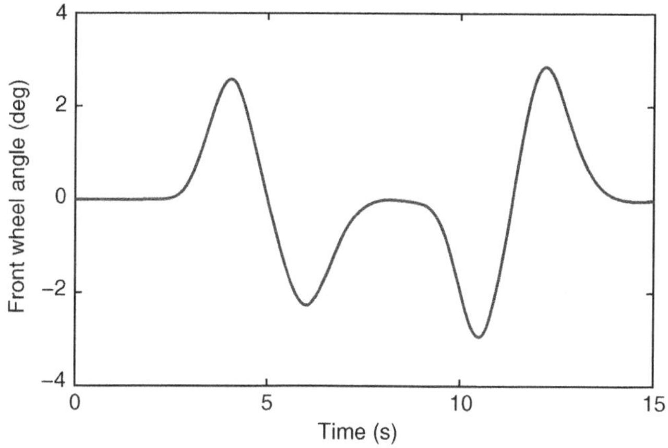

Figure 2.9 The front wheel angle on the ice road.

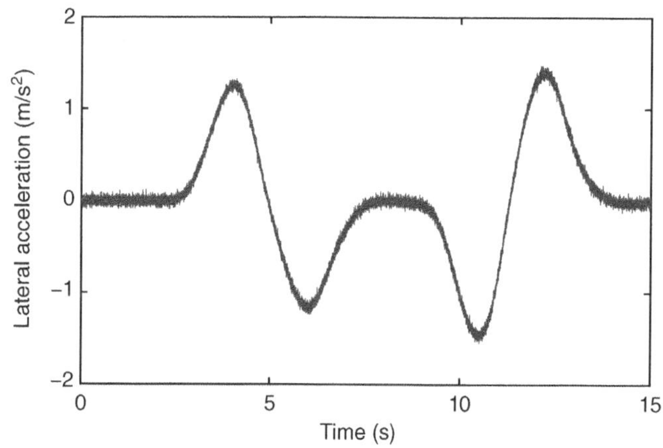

Figure 2.10 The lateral acceleration on the ice road.

series of coordinated adjustments in the front wheel steering angle, which in turn affect the lateral acceleration of the vehicle. To further enhance the realism of the simulation and to test the robustness of the EKF, irregularities are intentionally introduced into the lateral acceleration curve. These irregularities simulate measurement noise, reflecting the real-world imperfections and variability that are typically encountered in sensor data. By incorporating this simulated measurement noise, the study aims to evaluate the EKF's performance in conditions that closely mimic actual driving scenarios, where sensor data is often noisy and subject to fluctuations.

As illustrated in Fig. 2.11, the EKF-based estimation results for the yaw rate are presented. The red solid line represents the reference value output by CarSim, while the blue dotted line signifies the EKF estimation result. Examining Fig. 2.11 reveals a close alignment between the estimated curve and the reference curve, particularly evident in the larger estimated value at the inflection point. Figure 2.12 displays the estimation outcomes for the sideslip angle, with the color-coded curves holding the same significance as in Fig. 2.11. Notably, the estimation curves for the sideslip angle closely approximate the reference values. Moving on to Fig. 2.13, which portrays the estimation results for longitudinal

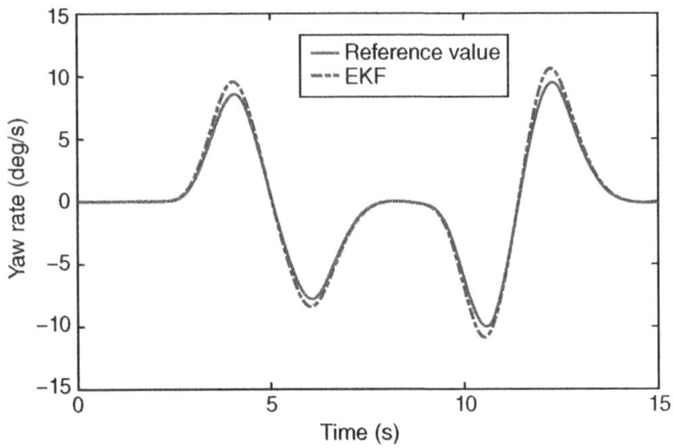

Figure 2.11 The yaw rate on the ice road.

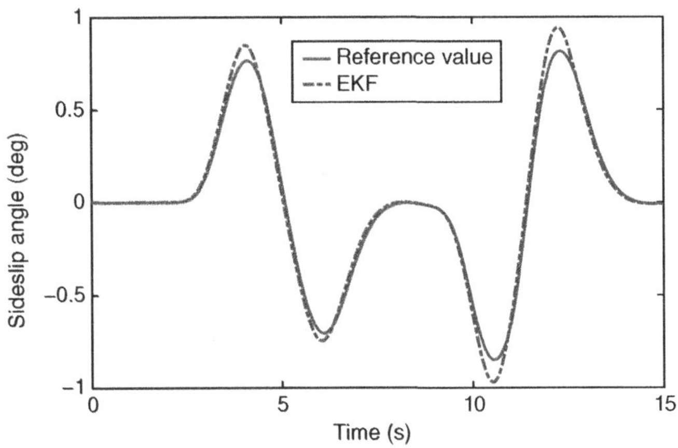

Figure 2.12 The sideslip angle on the ice road.

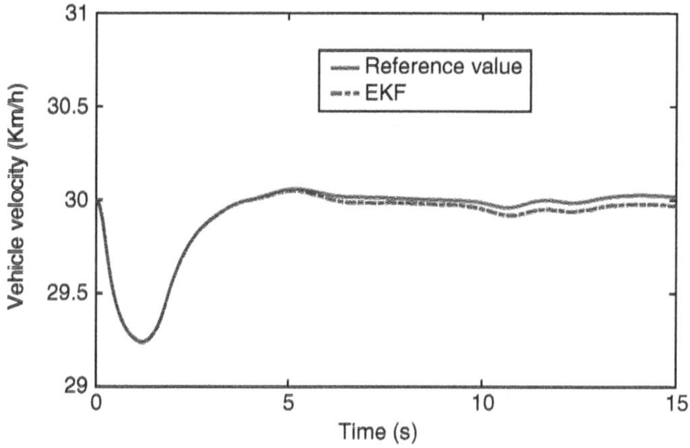

Figure 2.13 The longitudinal velocity on the ice road.

velocity, initial fitting is observed with a gradual tendency to diverge. However, the maximum estimation deviation remains below 0.15 m/s, rendering the overall estimation error acceptable.

2.4 AFTEKF for Estimating Vehicle State with Data Loss

Many existing research papers predominantly concentrate on exploring the influence of sensor measurement noise on estimation accuracy, often neglecting the equally significant impact of sensor data loss. In practical scenarios, sensor data loss is a pervasive occurrence that warrants careful consideration. As highlighted in [37], the data received from sensors may not always be accurate, often containing corrupted signals stemming from sensor malfunctions. In response to this challenge, several scholars have undertaken meaningful research endeavors. For instance, certain studies [38, 39] assumed a binary state for the sensor—either it operates correctly or fails. These studies employed the Bernoulli distribution to model and analyze missing measurements caused by sensor failures. However, it's essential to acknowledge that the probability of experiencing a complete loss of sensor data is minimal. More commonly, partial data loss occurs, which introduces an added layer of complexity to the estimation process. It is noteworthy that the field of state estimation for nonlinear systems has been extensively explored over the years. However, there is a discernible gap in the literature, as only a limited number of papers [40, 41] have delved into the challenges of state estimation in the presence of partial missing measurements.

To the best of the authors' knowledge, the specific intricacies and challenges of vehicle state estimation in the context of partial missing measurements have yet to be thoroughly investigated. This represents a significant research gap and an opportunity for further exploration and understanding within the realm of sensor-based estimation techniques. Additionally, it is essential to acknowledge that the EKF algorithm exhibits suboptimal performance when confronted with model parameter perturbations and abrupt changes in the system state [42]. Consequently, the vehicle state estimates derived from the EKF algorithm in specific driving conditions may manifest reduced accuracy. The literature, as documented in [43], has illuminated a viable solution to this challenge by integrating the EKF with a fading factor based on orthogonal theory. The integration of the EKF with a fading factor, rooted in orthogonal theory, has proven to be a robust strategy for mitigating the limitations associated with model parameter perturbations and sudden variations in system states. This innovative approach has found successful applications across diverse domains, including navigation [44] and vehicle state estimation [45]. The noteworthy success stories documented in these applications underscore the efficacy of incorporating a fading factor into the estimation process. Furthermore, the successful applications of this integrated approach in navigation and vehicle state estimation highlight its versatility and efficacy across diverse domains. The orthogonal theory-based fading factor serves as a dynamic adjustment mechanism, ensuring that the estimation algorithm remains robust and responsive in the face of evolving conditions.

Inspired by the insights gleaned from the preceding discussion, a solution has been introduced to address the challenges posed by the diminishing accuracy of estimations resulting from sensor data loss. This new approach is a fault-tolerant EKF (FTEKF). The FTEKF strategically incorporates a discrete distribution within the interval [0,1], serving as a method to present the stochastic nature of sensor data loss. Furthermore, the AFTEKF emerges as an evolution of the FTEKF, designed to bolster its resilience against not only sensor data loss but also the intricate dynamics of parameter perturbations and state mutations. At the core of this augmentation lies the integration of a fading factor, a dynamic mechanism from the orthogonal theory. This dynamic component imbues the AFTEKF with a heightened adaptability to the evolving conditions.

2.4.1 Vehicle Model and Problem Statement

As illustrated in Fig. 2.14, a comprehensive approach to vehicle state estimation is undertaken by leveraging a sophisticated four-wheel vehicle model, intricately coupled with the highly regarded Magic Formula (MF) tire model [46]. The four-wheel vehicle model offers a holistic representation of the vehicle's dynamics, encompassing key parameters that play a pivotal role in determining

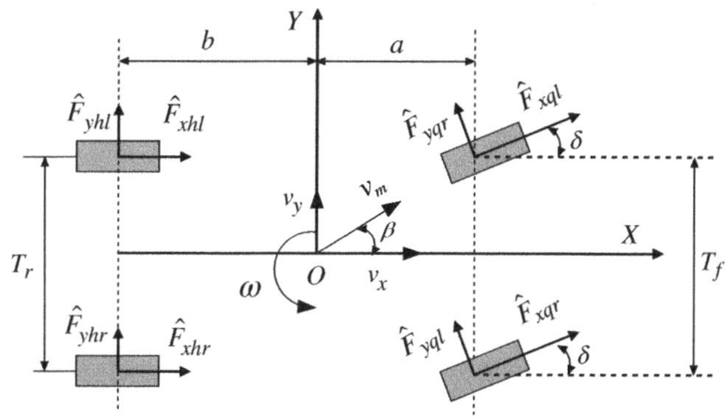

Figure 2.14 The four-wheel vehicle model.

its state. Complemented by the MF tire model, renowned for its precision in capturing tire-road interactions, this integrated model elevates the fidelity of the vehicle state estimation, ensuring a nuanced understanding of the complex dynamics governing vehicular motion. The four-wheel vehicle model is a comprehensive representation used in vehicle dynamics to simulate the movement and behavior of automobiles under various driving conditions. Unlike simpler models, it incorporates the dynamics of all four wheels individually, accounting for their interactions with the road surface and vehicle mass distribution. This model enables detailed analysis of longitudinal and lateral motions, making it essential for designing and optimizing vehicle handling, stability control systems, and performance characteristics such as acceleration, braking, and cornering capabilities in both conventional and autonomous vehicles. The model provides a detailed representation of vehicle dynamics, capturing real-world behaviors more accurately than simpler models. It accommodates various vehicle types, from passenger cars to commercial vehicles, allowing engineers to tailor simulations to specific design requirements. By examining the interactions between all four wheels and the road, the model offers deeper insights into factors influencing vehicle performance and safety. It serves as a foundation for developing advanced stability control, traction control, and autonomous driving algorithms by predicting vehicle responses to control inputs and external disturbances. In summary, the four-wheel vehicle model stands as a cornerstone in automotive engineering, enabling detailed analysis and optimization of vehicle dynamics across a wide range of operational scenarios. Its ability to simulate complex interactions between wheels, tires, and vehicle dynamics parameters contributes significantly to advancing vehicle performance, safety, and overall driving experience in modern automotive design and development.

The four-wheel vehicle model is expressed as follows

$$\dot{v}_m = \frac{1}{M}\begin{bmatrix}(\hat{F}_{xql} + \hat{F}_{xqr})\cos(\beta - \delta) + (\hat{F}_{xhl} + \hat{F}_{xhr})\cos\beta \\ +(\hat{F}_{yql} + \hat{F}_{yqr})\sin(\beta - \delta) + (\hat{F}_{yhl} + \hat{F}_{yhr})\sin\beta\end{bmatrix} \tag{2.29}$$

$$\dot{\beta} = \frac{1}{Mv_m}\begin{bmatrix}(\hat{F}_{yql} + \hat{F}_{yqr})\cos(\beta - \delta) + (\hat{F}_{yhl} + \hat{F}_{yhr})\cos\beta \\ -(\hat{F}_{xql} + \hat{F}_{xqr})\sin(\beta - \delta) - (\hat{F}_{xhl} + \hat{F}_{xhr})\sin\beta\end{bmatrix} - \omega \tag{2.30}$$

$$a_y = [(\hat{F}_{xql} + \hat{F}_{xqr})\sin\delta + (\hat{F}_{yql} + \hat{F}_{yqr})\cos\delta + \hat{F}_{yhl} + \hat{F}_{yhr}]/M \tag{2.31}$$

where the essential parameters are listed in Table 2.2.

The MF model is semi-empirical and provides more accurate information than other tire models such as those developed by Burckhardt [47] and Dugoff [48]. The MF tire model, conceived by Hans B. Pacejka in the late 1980s, serves as a crucial tool in vehicle dynamics simulations by providing a semi-empirical mathematical representation of tire behavior. It captures the complex, nonlinear relationship between tire forces and slip variables through a sophisticated equation involving empirically derived coefficients. This formula, often referred to as Pacejka's MF, is versatile and can be adapted to various types of tires by adjusting its coefficients, allowing it to accurately reflect tire behavior under different operational conditions. The principal advantages of the MF tire model include its high accuracy in representing real-world tire dynamics, flexibility to accommodate different tire characteristics, and comprehensive capability to model intricate behaviors such as

Table 2.2 Parameters of the four-wheel model.

Symbol	Quantity
β	Sideslip angle
ω	Yaw rate
M	Vehicle mass
$\hat{F}_{xij}$	Longitudinal forces of the wheels
$\hat{F}_{yij}$	Lateral forces of the wheels
δ	Front wheel steering angle
a_x	Longitudinal acceleration
a_y	Lateral acceleration
T_f	Front track width
T_r	Rear track width
I_Z	Moment of inertia about the vehicle vertical axis
a	Distance from the center of gravity to the front axle
b	Distance from the center of gravity to the rear axle

load sensitivity, lateral and longitudinal force interactions, and combined slip conditions. These attributes make the model invaluable for a wide range of automotive applications, including vehicle design, where it helps engineers predict how tires will interact with the road under various scenarios; control system development, enabling the creation of more responsive and safer vehicle control algorithms; and performance optimization, where it aids in fine-tuning tire and vehicle parameters to achieve the best possible performance. Additionally, its widespread adoption in the automotive industry attests to its reliability and robustness. By providing an accurate and adaptable representation of tire forces and moments, the MF tire model plays an essential role in enhancing vehicle safety, performance, and overall design efficiency, underscoring its significance in modern automotive engineering. Therefore, the MF tire model is adopted in this study, which is expressed as follows:

$$F(x) = D\sin[C\arctan\{Bx - E(Bx - \arctan(Bx))\}] \tag{2.32}$$

where $F(x)$ represents longitudinal tire forces or lateral tire forces; B, C, D, and E are functions of the vertical load of the tire, which are taken from [46]; x refers to the longitudinal slip ratio λ or the wheel sideslip angle α, as shown in Eqs. (2.33)–(2.37). Note that the specific parameters in these functions are usually obtained by experiments, and the parameters of our tire model are collected from the automobile companies that we cooperate with.

$$\lambda_{ij} = sgn(v_x - \hat{R}_{ij}w_{ij})\frac{|v_x - \hat{R}_{ij}w_{ij}|}{\max(\hat{R}_{ij}w_{ij}, v_x)} \quad i = q, h; j = l, r \tag{2.33}$$

$$\alpha_{ql} = \delta_{ql} - \arctan\left(\frac{v_y + a\omega}{v_x - T_f\omega/2}\right) \tag{2.34}$$

$$\alpha_{qr} = \delta_{qr} - \arctan\left(\frac{v_y + a\omega}{v_x + T_f\omega/2}\right) \tag{2.35}$$

$$\alpha_{hl} = -\arctan\left(\frac{v_y - b\omega}{v_x - T_r\omega/2}\right) \tag{2.36}$$

$$\alpha_{hr} = -\arctan\left(\frac{v_y - b\omega}{v_x + T_r\omega/2}\right) \tag{2.37}$$

where $\hat{R}$ and w are the radius and angular velocity of the wheel, respectively.

In addition, the MF tire model also needs to satisfy the friction ellipse constraint, and the improved MF tire model is expressed as

$$\begin{cases} \hat{F}_{xij} = \frac{|\xi_{xij}|}{\xi_{ij}}F(\lambda_{ij}), \ \hat{F}_{yij} = \frac{|\xi_{yij}|}{\xi_{ij}}F(\alpha_{ij}) \\ \xi_{xij} = \frac{\lambda_{ij}}{1+\lambda_{ij}}, \ \xi_{yij} = \frac{\tan\alpha_{ij}}{1+\lambda_{ij}}, \ \xi_{ij} = \sqrt{\xi_{xij}^2 + \xi_{yij}^2} \end{cases} i = q, \ h; j = l, r \tag{2.38}$$

Based on the nonlinear vehicle model described by Eqs. (2.29)–(2.31), the discrete vehicle state-space model with partial missing measurements can be given by

$$\begin{cases} x_{\theta+1} = f(x_\theta, u_\theta) + v_\theta \\ z_\theta = \begin{pmatrix} \gamma_\theta^1 \Gamma^1(x_\theta, u_\theta) + \varsigma_\theta^1 \\ \gamma_\theta^2 \Gamma^2(x_\theta, u_\theta) + \varsigma_\theta^2 \\ \vdots \\ \gamma_\theta^m \Gamma^m(x_\theta, u_\theta) + \varsigma_\theta^m \end{pmatrix} = \Xi_\theta \Gamma(x_\theta, u_\theta) + \varsigma_\theta \end{cases} \tag{2.39}$$

$x_\theta = [v_m, \beta]^T, z_\theta = [a_y]^T$ $u_\theta = [\delta, \omega, \hat{F}_{xij}, \hat{F}_{yij}]^T$ $i = q, h; j = l, r$

where θ is the sampling instant, x_θ is the state variables, u_θ is the control input, z_θ is the measurement output with partial data loss, m is the number of measured variables, f and Γ are the process and measurement functions, respectively. Let $diag\{\cdot\}$ denote the diagonal matrix. $\Xi_\theta = diag\left\{\gamma_\theta^1, \gamma_\theta^2, \dots \gamma_\theta^m\right\}$ and $\gamma_\theta^i (i = 1, 2 \dots m)$ are m independent random variables in θ and i, which are independent of all noise signals. In addition, $\Gamma(x_\theta, u_\theta) = diag\{\Gamma^1(x_\theta, u_\theta), \Gamma^2(x_\theta, u_\theta), \dots \Gamma^m(x_\theta, u_\theta)\}$. Furthermore, we assume that the probability density function of γ_θ^i on the interval [0,1] with expectation $\overline{\mu}_\theta^i$ and variance $\overline{\sigma}_\theta^i (i = 1, 2 \dots m)$. v_θ is the system process noise, whose covariance is expressed by Q_θ. ς_θ refers to the system measurement noises, whose covariance is R_θ. v_θ and ς_θ are zero-mean Gaussian white noises, which are uncorrelated with each other. The initial state x_0 is also independent of all noise signals.

2.4.2 Methodology

In this section, we design an AFTEKF to estimate vehicle state with partial missing measurements. A recursive filter and its linearization with first-order approximation are employed to derive the FTEKF. Then, a fading factor based on orthogonal theory is used to reduce the effect of parameter perturbations and state mutation on estimation accuracy.

To derive FTEKF, some necessary lemmas are introduced first.

Lemma 2.1 [49] Given a matrix $A_{m\times n}$ and $B_{n\times n} = B_{n\times n}^T$, then taking the partial derivative of $tr(ABA^T)$ with respect to A can be stated as

$$\frac{\partial tr(ABA^T)}{\partial A} = 2AB \tag{2.40}$$

where $tr(\cdot)$ is the trace of the matrix.

Lemma 2.2 [50] Let matrix $A = [a_{ij}]_{p\times p}$ be a real-valued matrix and $B = diag(b_1, b_2, \dots b_p)$ be a diagonal random matrix, then

$$E(BAB^T) = \begin{bmatrix} E\left(b_1^2\right) & E(b_1b_2) & \dots & E(b_1b_p) \\ E(b_2b_1) & E\left(b_2^2\right) & \dots & E(b_2b_p) \\ \vdots & \vdots & & \vdots \\ E(b_pb_1) & E(b_pb_2) & \dots & E\left(b_p^2\right) \end{bmatrix} \otimes A \tag{2.41}$$

where $\otimes$ is the Hadamard product and $E(\cdot)$ means the mathematical expectation.

To express convenience, we define x_θ^+ and x_θ^- as the posterior and prior estimators of x_θ, respectively, as shown in Eq. (2.42). Note that x_θ^+ is the conditional expected value of x_θ based on all measured values, including the measured value at time θ, and x_θ^- is the conditional expected value of x_θ based on all measured values excluding the measured value at time θ.

$$\begin{cases} x_\theta^+ = \mathrm{E}[x_\theta \mid z_1, z_2 \dots, z_\theta,] \\ x_\theta^- = \mathrm{E}[x_\theta \mid z_1, z_2 \dots, z_{\theta-1},] \end{cases} \tag{2.42}$$

2.4.2.1 The FTEKF

Theorem 2.1 Consider the system in Eq. (2.39) with sensor data loss where the measurement missing probability distribution is discrete in the interval [0,1]. If the following conditions

$$\begin{cases} \mathrm{E}[v_\theta] = 0, \mathrm{E}[\varsigma_\theta] = 0, \mathrm{E}\left[v_\theta \varsigma_j^T\right] = 0, \\ \mathrm{E}\left[v_\theta v_j^T\right] = Q_\theta \Omega_{\theta-j}, \mathrm{E}\left[\varsigma_\theta \varsigma_j^T\right] = R_\theta \Omega_{\theta-j}, \\ \Omega_{\theta-j} = 1(\theta = j); \Omega_{\theta-j} = 0 \ \ (\theta \neq j) \\ \mathrm{E}\left[v_\theta x_0^T\right] = 0, \mathrm{E}\left[\varsigma_\theta x_0^T\right] = 0 \end{cases} \tag{2.43}$$

are satisfied, then the posterior state estimation error covariance $P_{\theta+1}^+$ can be expressed as

$$\begin{aligned} P_{\theta+1}^+ &= (I - K_{\theta+1}\Xi_{\theta+1}C_{\theta+1})P_{\theta+1}^- \times (I - K_{\theta+1}\Xi_{\theta+1}C_{\theta+1})^T + K_{\theta+1}R_{\theta+1}K_{\theta+1}^T \\ &\quad + K_{\theta+1}E\left[\tilde{\Xi}_{\theta+1}\Gamma\left(\hat{x}_{\theta+1}^-, u_{\theta+1}\right) \times \Gamma^T\left(\hat{x}_{\theta+1}^-, u_{\theta+1}\right)\tilde{\Xi}_{\theta+1}^T\right]K_{\theta+1}^T \end{aligned} \tag{2.44}$$

and the optimal filter gain $K_{\theta+1}$ is written by

$$\begin{aligned} K_{\theta+1} = &\left(P_{\theta+1}^- C_{\theta+1}^T \Xi_{\theta+1}^T\right) \times \left[\Xi_{\theta+1}C_{\theta+1}P_{\theta+1}^- C_{\theta+1}^T \Xi_{\theta+1}^T\right. \\ &\left. + \hat{\Xi}_{\theta+1} \otimes E\left(\Gamma\left(\hat{x}_{\theta+1}^-, u_{\theta+1}\right) \times \Gamma^T\left(\hat{x}_{\theta+1}^-, u_{\theta+1}\right)\right) + R_{\theta+1}\right]^{-1} \end{aligned} \tag{2.45}$$

where $(\cdot)^T$ is matrix transpose, $\Omega_{\theta-j}$ is the Kronecker delta function, $(\cdot)^{-1}$ is the inverse of a matrix, I is the identity matrix, $P_{\theta+1}^-$ is the prior state estimation error covariance, $\tilde{\Xi}_{\theta+1} = \Xi_{\theta+1} - \overline{\Xi}_{\theta+1}$, and $\overline{\Xi}_{\theta+1} = \mathrm{E}(\Xi_{\theta+1}) = diag\left(\overline{\mu}_{\theta+1}^1, \overline{\mu}_{\theta+1}^2 \dots \overline{\mu}_{\theta+1}^m\right)$.

Proof: The recursive filtering form can be formulated as

$$\hat{x}_{\theta+1}^{-} = f\left(\hat{x}_{\theta}^{+}, u_{\theta}\right) \tag{2.46}$$

$$\hat{x}_{\theta+1}^{+} = \hat{x}_{\theta+1}^{-} + K_{\theta+1}\left[z_{\theta+1} - \overline{\Xi}_{\theta+1}\Gamma\left(\hat{x}_{\theta+1}^{-}, u_{\theta+1}\right)\right] \tag{2.47}$$

$e_{\theta+1}^{+} = x_{\theta+1} - \hat{x}_{\theta+1}^{+}$ and $e_{\theta+1}^{-} = x_{\theta+1} - \hat{x}_{\theta+1}^{-}$ denote the system posterior state estimation error and the prior state estimation error, respectively. Based on Eqs. (2.39), (2.46), and (2.47), $e_{\theta+1}^{+}$ and $e_{\theta+1}^{-}$ can be rewritten as

$$e_{\theta+1}^{-} = f(x_{\theta}, u_{\theta}) + v_{\theta} - f\left(\hat{x}_{\theta}^{+}, u_{\theta}\right) \tag{2.48}$$

$$e_{\theta+1}^{+} = f(x_{\theta}, u_{\theta}) + v_{\theta} - \hat{x}_{\theta+1}^{-} - K_{\theta+1}\left[z_{\theta+1} - \overline{\Xi}_{\theta+1}\Gamma\left(\hat{x}_{\theta+1}^{-}, u_{\theta+1}\right)\right] \tag{2.49}$$

Linearizing $f(x_{\theta}, u_{\theta})$ and $\Gamma(x_{\theta+1}, u_{\theta+1})$ by using first-order Taylor series expansion for $\hat{x}^{+}{}_{\theta}$ and $\hat{x}^{-}{}_{\theta+1}$, respectively. After ignoring the higher-order terms, we can obtain

$$\begin{cases} f(x_{\theta}, u_{\theta}) = f\left(\hat{x}_{\theta}^{+}, u_{\theta}\right) + A_{\theta}e_{\theta}^{+} \\ A_{\theta} = \dfrac{\partial f(x_{\theta}, u_{\theta})}{\partial x_{\theta}}\Big|_{x_{\theta}=\hat{x}_{\theta}^{+}} \end{cases} \tag{2.50}$$

$$e_{\theta+1}^{-} = A_{\theta}e_{\theta}^{+} + v_{\theta} \tag{2.51}$$

$$\begin{cases} \Gamma(x_{\theta+1}, u_{\theta+1}) = \Gamma\left(\hat{x}_{\theta+1}^{-}, u_{\theta+1}\right) + C_{\theta+1}e_{\theta+1}^{-} \\ C_{\theta+1} = \dfrac{\partial \Gamma(x_{\theta+1}, u_{\theta+1})}{\partial x_{\theta+1}}\Big|_{x_{\theta+1}=\hat{x}_{\theta+1}^{-}} \end{cases} \tag{2.52}$$

$$\begin{aligned} e_{\theta+1}^{+} &= f\left(\hat{x}_{\theta}^{+}, u_{\theta}\right) + A_{\theta}e_{\theta}^{+} + v_{\theta} - f\left(\hat{x}_{\theta}^{+}, u_{\theta}\right) \\ &\quad - K_{\theta+1}\begin{bmatrix} (\Xi_{\theta+1} - \overline{\Xi}_{\theta+1})\Gamma\left(\hat{x}_{\theta+1}^{-}, u_{\theta+1}\right) \\ + \varsigma_{\theta+1} + \Xi_{\theta+1}C_{\theta+1}e_{\theta+1}^{-} \end{bmatrix} \\ &= (I - K_{\theta+1}\Xi_{\theta+1}C_{\theta+1})e_{\theta+1}^{-} - K_{\theta+1}\tilde{\Xi}_{\theta+1}\Gamma\left(\hat{x}_{\theta+1}^{-}, u_{\theta+1}\right) - K_{\theta+1}\varsigma_{\theta+1} \end{aligned} \tag{2.53}$$

According to Eq. (2.51), $P_{\theta+1}^{-}$ can be expressed as

$$P_{\theta+1}^{-} = E\left[e_{\theta+1}^{-}\left(e_{\theta+1}^{-}\right)^{T}\right] = A_{\theta}P_{\theta}^{+}A_{\theta}^{T} + Q_{\theta} \tag{2.54}$$

According to Eq. (2.53), $P_{\theta+1}^{+}$ can be expressed as

$$\begin{aligned} P_{\theta+1}^{+} &= \mathrm{E}\left[e_{\theta+1}^{+}\left(e_{\theta+1}^{+}\right)^{T}\right] \\ &= \mathrm{E}\Big\{\left[(I - K_{\theta+1}\Xi_{\theta+1}C_{\theta+1})e_{\theta+1}^{-} - K_{\theta+1}\varsigma_{\theta+1} - K_{\theta+1}\tilde{\Xi}_{\theta+1}\Gamma\left(\hat{x}_{\theta+1}^{-}, u_{\theta+1}\right)\right] \\ &\quad \times \left[(I - K_{\theta+1}\Xi_{\theta+1}C_{\theta+1})e_{\theta+1}^{-} - K_{\theta+1}\varsigma_{\theta+1} - K_{\theta+1}\tilde{\Xi}_{\theta+1}\Gamma\left(\hat{x}_{\theta+1}^{-}, u_{\theta+1}\right)\right]^{T}\Big\} \end{aligned}$$

$$= \sum_{\hat{m}=1}^{3}\sum_{\hat{n}=1}^{3} P_{\theta+1}^{+}(\hat{m}\hat{n}) \tag{2.55}$$

where $P_{\theta+1}^{+}(\hat{m}\hat{n})$ are equal to

$$P_{\theta+1}^{+}(11) = (I - K_{\theta+1}\Xi_{\theta+1}C_{\theta+1})P_{\theta+1}^{-} \times (I - K_{\theta+1}\Xi_{\theta+1}C_{\theta+1})^{T} \tag{2.56}$$

$$\begin{aligned} P_{\theta+1}^{+}(22) &= K_{\theta+1}E\left(\varsigma_{\theta+1}\varsigma_{\theta+1}^{T}\right)K_{\theta+1}^{T} \\ &= K_{\theta+1}R_{\theta+1}K_{\theta+1}^{T} \end{aligned} \tag{2.57}$$

$$P_{\theta+1}^{+}(33) = K_{\theta+1}E\begin{bmatrix} \tilde{\Xi}_{\theta+1}\Gamma\left(\hat{x}_{\theta+1}^{-}, u_{\theta+1}\right) \times \\ \Gamma^{T}\left(\hat{x}_{\theta+1}^{-}, u_{\theta+1}\right)\tilde{\Xi}^{T}{}_{\theta+1} \end{bmatrix} K_{\theta+1}^{T} \tag{2.58}$$

$$P_{\theta+1}^{+}(12) = -(I - K_{\theta+1}\Xi_{\theta+1}C_{\theta+1})E\left(e_{\theta+1}^{-}\varsigma_{\theta+1}\right)^{T}K_{\theta+1}^{T} \tag{2.59}$$

$$P_{\theta+1}^{+}(13) = -(I - K_{\theta+1}\Xi_{\theta+1}C_{\theta+1}) \times E\left[e_{\theta+1}^{-}\Gamma^{T}\left(\hat{x}_{\theta+1}^{-}, u_{\theta+1}\right)\tilde{\Xi}^{T}{}_{\theta+1}\right]K_{\theta+1}^{T} \tag{2.60}$$

$$P_{\theta+1}^{+}(23) = K_{\theta+1}E\left(\varsigma_{\theta+1}\Gamma\left(\hat{x}_{\theta+1}^{-}, u_{\theta+1}\right)^{T}\tilde{\Xi}^{T}{}_{\theta+1}\right)K_{\theta+1}^{T} \tag{2.61}$$

$$P_{\theta+1}^{+}(21) = \left(P_{\theta+1}^{+}(12)\right)^{T} \tag{2.62}$$

$$P_{\theta+1}^{+}(31) = \left(P_{\theta+1}^{+}(13)\right)^{T} \tag{2.63}$$

$$P_{\theta+1}^{+}(32) = \left(P_{\theta+1}^{+}(23)\right)^{T} \tag{2.64}$$

Because v_{θ}, e_{θ}, ς_{θ}, and $\tilde{\Xi}_{\theta}$ are mutually independent, the results of Eqs. (2.59)–(2.64) are zeros. Therefore, Eq. (2.55) can be reformulated as

$$\begin{aligned} P_{\theta+1}^{+} = {} & (I - K_{\theta+1}\Xi_{\theta+1}C_{\theta+1})P_{\theta+1}^{-} \times (I - K_{\theta+1}\Xi_{\theta+1}C_{\theta+1})^{T} + K_{\theta+1}R_{\theta+1}K_{\theta+1}^{T} \\ & + K_{\theta+1}E\left[\tilde{\Xi}_{\theta+1}\Gamma\left(\hat{x}_{\theta+1}^{-}, u_{\theta+1}\right) \times \Gamma^{T}\left(\hat{x}_{\theta+1}^{-}, u_{\theta+1}\right)\tilde{\Xi}^{T}{}_{\theta+1}\right]K_{\theta+1}^{T} \end{aligned} \tag{2.65}$$

Based on Lemma 1, the following equation can be obtained

$$\begin{aligned} \frac{\partial tr\left(P_{\theta+1}^{+}\right)}{\partial K_{\theta+1}} = {} & -2(I - K_{\theta+1}\Xi_{\theta+1}C_{\theta+1})P_{\theta+1}^{-}C_{\theta+1}^{T}\Xi_{\theta+1}^{T} + 2K_{\theta+1} \\ & \left\{E\left[\tilde{\Xi}_{\theta+1}\Gamma\left(\hat{x}_{\theta+1}^{-}, u_{\theta+1}\right) \times \Gamma^{T}\left(\hat{x}_{\theta+1}^{-}, u_{\theta+1}\right)\tilde{\Xi}^{T}{}_{\theta+1}\right]\right\} + 2K_{\theta+1}R_{\theta+1} \end{aligned} \tag{2.66}$$

Let Eq. (2.66) equal to zero, the optimal filter gain $K_{\theta+1}$ is expressed as

$$\begin{aligned} K_{\theta+1} = {} & \left(P_{\theta+1}^{-}C_{\theta+1}^{T}\Xi_{\theta+1}^{T}\right) \times \Big[\Xi_{\theta+1}C_{\theta+1}P_{\theta+1}^{-}C_{\theta+1}^{T}\Xi_{\theta+1}^{T} \\ & + E\left(\tilde{\Xi}_{\theta+1}\Gamma\left(\hat{x}_{\theta+1}^{-}, u_{\theta+1}\right) \times \Gamma^{T}\left(\hat{x}_{\theta+1}^{-}, u_{\theta+1}\right)\tilde{\Xi}^{T}{}_{\theta+1}\right) + R_{\theta+1}\Big]^{-1} \end{aligned} \tag{2.67}$$

According to Eq. (2.45), Eq. (2.67) can be reformulated as

$$K_{\theta+1} = \left(P^-_{\theta+1} C^T_{\theta+1} \Xi^T_{\theta+1}\right) \times \left[\Xi_{\theta+1} C_{\theta+1} P^-_{\theta+1} C^T_{\theta+1} \Xi^T_{\theta+1} + \hat{\Xi}_{\theta+1} \otimes E\left(\Gamma\left(\hat{x}^-_{\theta+1}, u_{\theta+1}\right) \times \Gamma^T\left(\hat{x}^-_{\theta+1}, u_{\theta+1}\right)\right) + R_{\theta+1}\right]^{-1} c \quad (2.68)$$

This completes the proof.

2.4.2.2 The AFTEKF Algorithm

In the above FTEKF, the effect of the system state mutation and parameter perturbations on the estimation accuracy is not considered. However, in some emergency cases, such as collision avoidance [51], the driver may make some extreme maneuvers that cause vehicle state mutation. Similarly, in some time-varying driving conditions, such as empty or full load, the model parameters may deviate from the nominal value. To deal with these problems, an EKF with a fading factor was proposed in [43], and the results indicated the method is effective. To sum up, the filter with fading factor has the following advantages: 1) strong adaptability to parameter perturbations and 2) strong tracking ability even when a state jump occurs. Thus, we combine the FTEKF and a fading factor to develop the AFTEKF. The innovation sequence $\eta_{\theta+1}$ of the FTEKF is

$$\eta_{\theta+1} = z_{\theta+1} - \overline{\Xi}_{\theta+1} h\left(\hat{x}^-_{\theta+1}, u_{\theta+1}\right) \quad (2.69)$$

The key of the AFTEKF algorithm is to choose a suitable fading factor and embed it into the prior state estimation error covariance $P^-_{\theta+1}$, which can adjust the filter gain $K_{\theta+1}$ and must satisfy the following orthogonal principle.

$$E[x_{\theta+1} - \hat{x}_{\theta+1}][x_{\theta+1} - \hat{x}_{\theta+1}]^T = \min \quad (2.70)$$

$$E\left[\eta_{\theta+1+k} \eta^T_{\theta+1}\right] = 0, k = 1, 2 \ldots \quad (2.71)$$

It can be seen from Eqs. (2.70) and (2.71) that filter gain $K_{\theta+1}$ can be determined only when the estimation mean square error is the smallest and the innovation sequence is orthogonal.

The prior state estimation error covariance can be updated using a fading factor Λ_θ is given by

$$P^-_{\theta+1} = \Lambda_{\theta+1} A_\theta P^+_\theta A^T_\theta + Q_\theta \quad (2.72)$$

where $\Lambda_{\theta+1} = diag\left[\tau^1_{\theta+1}, \tau^2_{\theta+1}, \ldots \tau^n_{\theta+1}\right]$

$$\tau^t_{\theta+1} = \begin{cases} \Upsilon_t d_{\theta+1} & \Upsilon_t d_{\theta+1} \geq 1 \\ 1 & \Upsilon_t d_{\theta+1} < 1 \end{cases} \quad t = 1, 2, \ldots n \quad (2.73)$$

$$d_{\theta+1} = \frac{tr[N_{\theta+1}]}{\sum_{t=1}^{n} \Upsilon_t M_{\theta+1}} \quad (2.74)$$

$$N_{\theta+1} = V_{\theta+1} - R_{\theta+1} - C_{\theta+1} Q_\theta C_{\theta+1}^T \tag{2.75}$$

$$M_{\theta+1} = C_{\theta+1} A_\theta P_\theta^+ A_\theta^T C_{\theta+1}^T \tag{2.76}$$

$$V_{\theta+1} = \begin{cases} \eta_1 \eta_1^T & \theta = 0 \\ \frac{\rho V_\theta + \eta_{\theta+1} \eta_{\theta+1}^T}{1+\rho} & \theta \geq 1 \end{cases} \tag{2.77}$$

Υ_t is a constant that is determined based on prior information; $\rho \in [0, 1]$ is the forgetting factor. The value of this forgetting factor is usually set as 0.95. $d_{\theta+1}$ is the undetermined parameter.

2.4.3 Simulation and Experiment Tests

To validate the robustness and efficacy of the AFTEKF, a comprehensive assessment is conducted through a combination of virtual simulations and real vehicle experiments. In the virtual domain, the AFTEKF algorithm undergoes rigorous testing using a high-fidelity vehicle model meticulously implemented within the CarSim software environment. This virtual validation serves as an initial benchmark to evaluate the AFTEKF's performance under controlled conditions. Subsequently, the AFTEKF is further scrutinized through real-world vehicle experiments, where professional-grade measurement data obtained from specialized equipment is employed to assess the algorithm's performance in authentic driving scenarios. This dual-layered validation approach, encompassing both virtual and real-world testing environments, ensures a comprehensive evaluation of the AFTEKF's capabilities across a spectrum of conditions, reinforcing its credibility and applicability in practical vehicular applications.

Within the simulation framework, a comprehensive set of scenarios, including the DLC and continuous steering (CS) working conditions, are meticulously executed. The aim is to emulate diverse and challenging driving situations to thoroughly assess the performance of various estimation methods, particularly the AFTEKF. To bridge the simulation and real-world contexts, a concerted effort is made to align the dynamic parameters of the high-fidelity vehicle model used in the virtual tests with those of the actual vehicle. This meticulous calibration ensures a reduction in discrepancies between the simulated and real-world test environments, enhancing the relevance and accuracy of the simulation outcomes. To gauge the effectiveness of different estimation methods, the output values generated by the high-fidelity vehicle model within the CarSim software serve as benchmark reference values. These reference values provide a standardized basis for evaluating and comparing the performance of the AFTEKF and other estimation techniques under consideration. This rigorous testing protocol facilitates a robust analysis of the AFTEKF's capabilities in capturing the intricacies of dynamic driving scenarios, validating its reliability and suitability for real-world applications.

2.4.3.1 The DLC Test in Simulation Systems

In the virtual simulation environment, the simulated driver undertakes lane-changing maneuvers under challenging conditions, specifically on roads characterized by a low friction coefficient, adding complexity and realism to the test. The scenario begins with the vehicle traveling at an initial speed of 30 km/h, setting the stage for a series of precise and demanding maneuvers. As the vehicle navigates these low-friction roads, a controlled braking force is strategically applied within the temporal window from 4 to 4.5 seconds. This application of braking force is deliberate and designed to introduce additional dynamic variations to the simulation. The controlled braking force adds an element of complexity, requiring the estimation algorithms to accurately track and respond to the resulting changes in the vehicle state. Throughout the duration of the test, the vehicle's behavior is meticulously monitored, focusing on how the braking force impacts its dynamics during the lane-changing maneuvers.

The steering wheel angle depicted in Fig. 2.15 serves as a crucial input parameter, significantly influencing the vehicle's trajectory during the simulation. The precise control of the steering wheel angle directly impacts the vehicle's ability to navigate through the simulated environment, making it a vital component for evaluating the effectiveness of the estimation algorithms. Simultaneously, Fig. 2.16 illustrates the lateral acceleration, showcasing another critical aspect of the vehicle's dynamic behavior. What sets this scenario apart is the intentional introduction of partial missing measurements within the lateral acceleration data. This

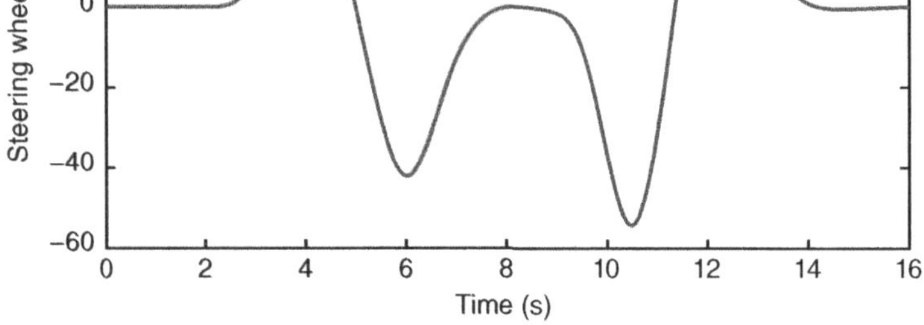

Figure 2.15 The steering wheel angle in the DLC.

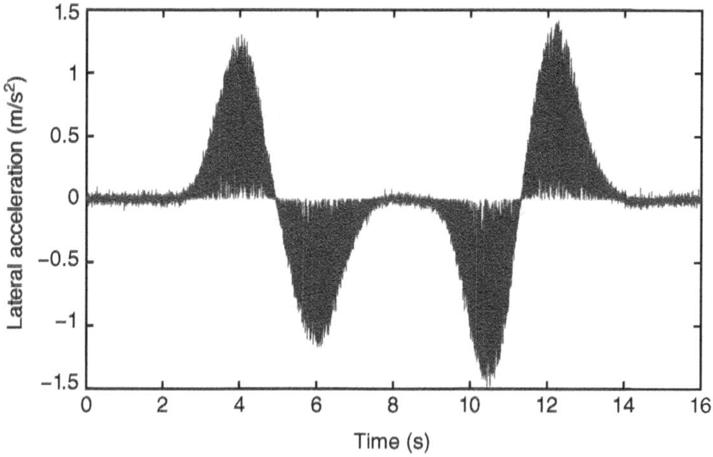

Figure 2.16 Lateral acceleration in the DLC.

deliberate inclusion of missing measurements is a key element designed to evaluate the robustness and effectiveness of the estimation algorithms under conditions of imperfect data availability. By incorporating this challenge, the simulation aims to mimic real-world scenarios where sensor data may be intermittently unavailable or compromised, thus testing the algorithms' ability to adapt and perform under such conditions.

In this context, the probability of missing measurements is assumed to adhere to a uniform distribution within the range of [0,1]. This probabilistic modeling introduces an element of uncertainty, further complicating the simulation and making it more representative of actual driving conditions. The random nature of the missing data means that the estimation algorithms must be capable of handling unpredictability and maintaining accuracy despite gaps in the data. The discrete nature of the simulation, with a sampling time of 0.01 seconds, adds another layer of complexity to the test conditions. This high-frequency sampling accentuates the dynamic nature of the vehicle's behavior, necessitating the estimation algorithms to rapidly process and adapt to the intermittent nature of the lateral acceleration data. The algorithms are challenged not only to estimate the vehicle's state accurately but also to do so in real-time, despite the presence of partial data loss. The intentional introduction of missing measurements and the probabilistic modeling of data loss are critical for evaluating the resilience and adaptability of the estimation algorithms. By simulating conditions sensor data is not always fully available, the test scenario provides a rigorous assessment of how well the algorithms can maintain performance and ensure vehicle safety and stability.

The results of the vehicle state estimation, along with corresponding reference values, are meticulously illustrated in Figs. 2.17 and 2.18, providing a detailed visual representation of the performance achieved by the estimation algorithms. Table 2.3 complements these visuals with a quantitative assessment using root mean square error calculations across three distinct estimation approaches. Figure 2.17 prominently showcases the accuracy of vehicle velocity estimation achieved by the AFTEKF algorithm, demonstrating its superior performance

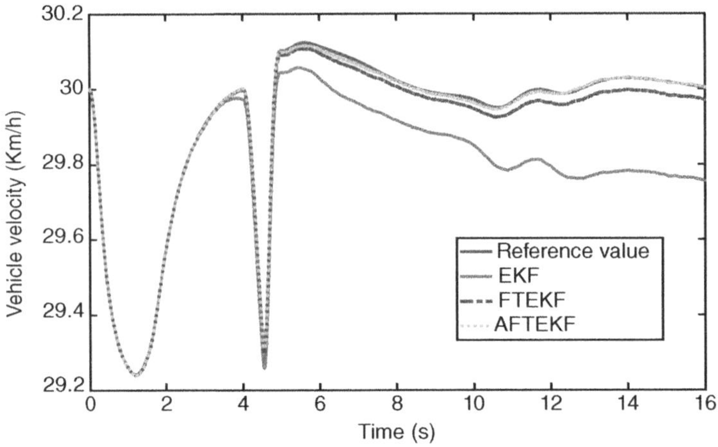

Figure 2.17 Vehicle velocity estimation in the double lane change.

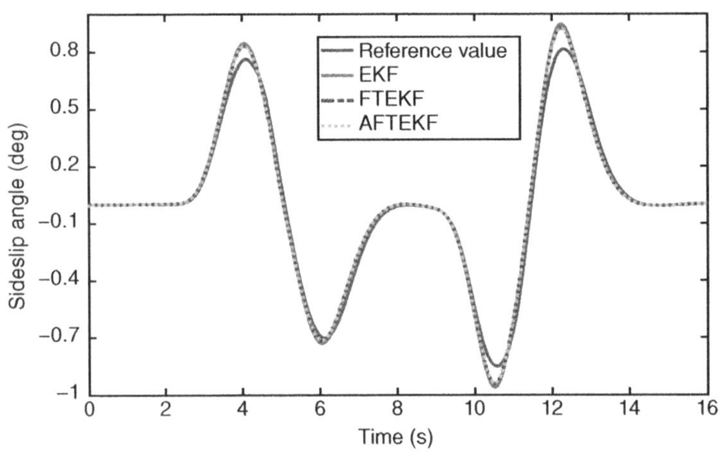

Figure 2.18 Sideslip angle estimation in the DLC.

compared to alternative methods. Both AFTEKF and FTEKF algorithms outperform the conventional EKF, primarily due to their effective handling of sensor data loss effects. A notable observation in Fig. 2.17 is the rapid decline in vehicle velocity between 4 and 4.5 seconds, attributed to the sudden application of braking force simulating dynamic changes in operational conditions. Turning to Fig. 2.18, estimates of sideslip angle by various methods are presented, with AFTEKF exhibiting the highest level of estimation accuracy. This finding is further supported by Table 2.3, which quantifies the error in vehicle state estimation. Remarkably, AFTEKF demonstrates the smallest error among the three methods, thanks to the incorporation of a fading factor that enhances its adaptability to rapid changes in operational conditions. These findings underscore the effectiveness of the AFTEKF algorithm in achieving precise and reliable vehicle state estimation, crucial for applications in autonomous driving, ADAS, and vehicle dynamics research. By integrating adaptive filtering techniques with robust error-handling mechanisms, AFTEKF not only improves estimation accuracy but also enhances the resilience of vehicle control systems against uncertainties and dynamic variations encountered in real-world driving scenarios.

2.4.3.2 The CS Test in Simulation Systems

The virtual driver undertakes a series of CS maneuvers on high-friction coefficient roads, beginning with an initial vehicle velocity of 80 km/h. The dynamic evolution of the steering wheel angle is meticulously illustrated in Fig. 2.19. This figure portrays a scenario where, for the initial 4 seconds, the frequency and amplitude of the steering wheel angle remain relatively modest. Subsequently, there is a sudden surge in both the frequency and amplitude of the steering wheel angle. This abrupt change is strategically introduced to simulate emergency driving situations, such as sudden obstacle avoidance or rapid lane changes. The simulation aims to test the vehicle's response and the robustness of the estimation algorithms under these high-stress conditions, which are critical for evaluating the safety and reliability of intelligent vehicle systems. Figure 2.20 provides an insightful depiction of the lateral acceleration during these maneuvers. This figure highlights the complexity of real-world driving scenarios by including partial missing measurements. The

Table 2.3 RMSE in DLC working condition.

Symbol	Sideslip angle	Vehicle velocity
EKF	0.0462	0.1466
FTEKF	0.0414	0.0215
AFTEKF	0.0408	0.0048

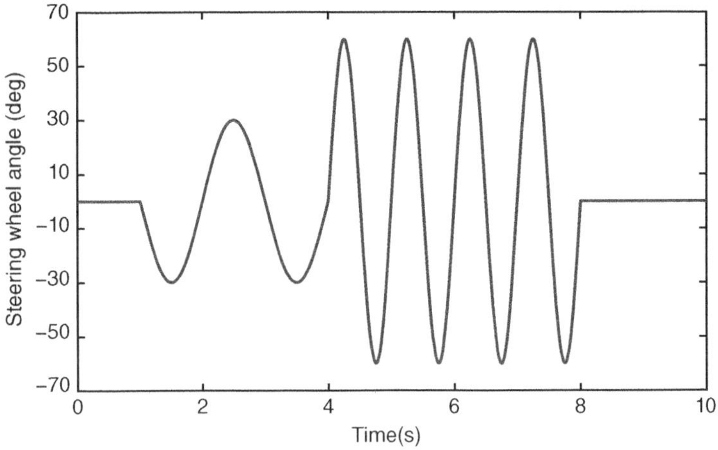

Figure 2.19 The steering wheel angle in CS.

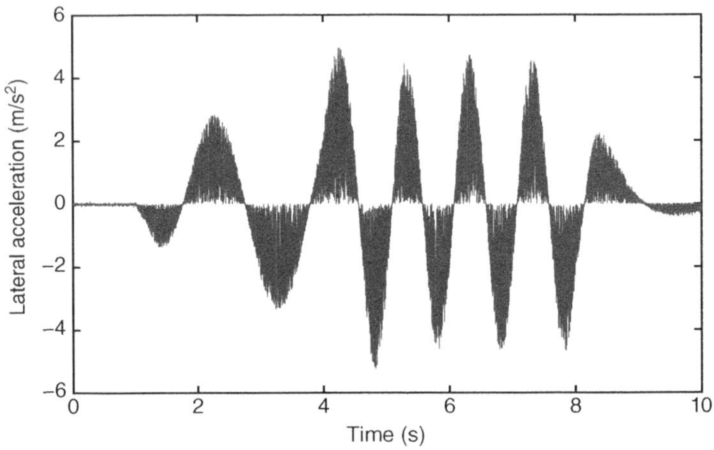

Figure 2.20 Lateral acceleration in CS.

presence of these missing measurements is intentional, designed to emulate situations where sensor data may be intermittently unavailable or compromised. This aspect of the simulation is crucial for assessing the estimation algorithms' ability to maintain accuracy and reliability despite imperfect data conditions. The probability distribution for missing measurements is modeled as a uniform distribution between [0,1], adding a layer of randomness to the data loss. This probabilistic approach reflects the unpredictable nature of sensor data availability in real-world environments. The consistent sampling time of 0.01 seconds further accentuates the dynamic nature of the test conditions, requiring the estimation algorithms to process and adapt to rapidly changing data. This rigorous simulation setup, with its controlled initial conditions followed by sudden, intense steering maneuvers and the introduction of missing measurements, creates a challenging environment for the estimation algorithms. The aim is to thoroughly evaluate their performance, robustness, and adaptability under a wide range of conditions, from normal driving to emergency situations.

Figures 2.21 and 2.22 clearly demonstrate that the AFTEKF consistently outperforms the other two methods in terms of estimation accuracy. The estimation accuracies achieved by both AFTEKF and FTEKF surpass those of the traditional EKF. This superior performance is quantitatively highlighted in Table 2.4, which presents the vehicle state estimation errors. The AFTEKF stands out with the smallest estimation errors, showcasing its effectiveness and reliability. The AFTEKF algorithm's adaptability to the dynamic nature of vehicle state mutation is a key factor in its success. This adaptability is achieved by dynamically adjusting the estimation error covariance of the prior state in real-time. By doing so, the AFTEKF can effectively track the reference value, maintaining high accuracy even as the vehicle's state changes rapidly. This ability to adapt in real-time underscores the robustness of the AFTEKF method, allowing it to accommodate various road friction conditions and diverse initial states. The robustness of the AFTEKF is particularly significant in the context of varying road friction conditions and different initial states. These factors can introduce significant variability in vehicle dynamics, challenging the estimation algorithms to maintain accuracy. The AFTEKF's capability to deliver optimal estimation results across

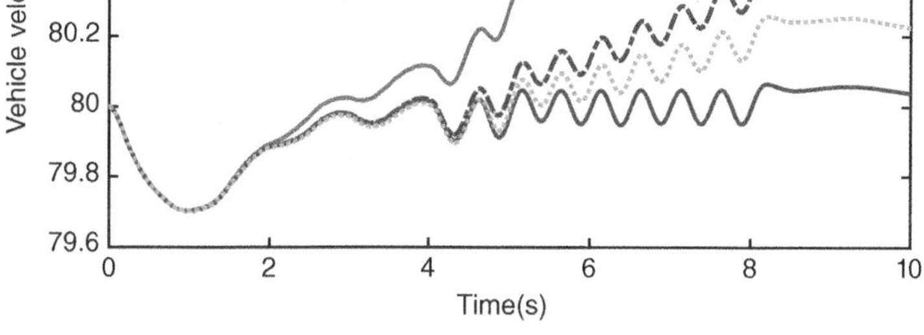

Figure 2.21 Vehicle velocity estimation in CS.

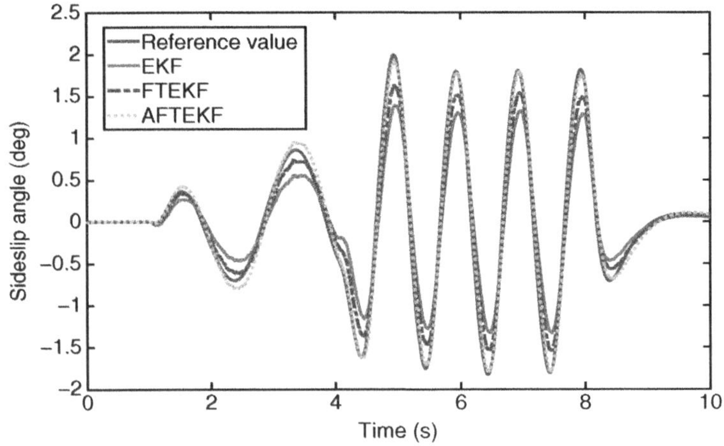

Figure 2.22 Sideslip angle estimation in CS.

Table 2.4 RMSE in CS working condition.

Symbol	Sideslip angle	Vehicle velocity
EKF	0.2970	0.5008
FTEKF	0.1835	0.1865
AFTEKF	**0.0987**	**0.1048**

a spectrum of challenging scenarios highlights its potential for application in real-world conditions, where such variability is commonplace.

Figure 2.23 illustrates the real-world validation of the AFTEKF through vehicle tests conducted on both a wet asphalt road (WAR) and a dry asphalt road (DAR). These tests are essential for verifying the practical applicability and effectiveness of the AFTEKF under diverse driving conditions. For obtaining reference values, the vehicle state data is sourced from differential GPS (DGPS), which provides high-accuracy data. The validation process involves a meticulous comparison between the estimated values derived from three different methods, including AFTEKF, and the reference values obtained from DGPS. By comparing these estimations with the highly accurate DGPS data, it becomes possible to evaluate the precision and reliability of each estimation method under real-world conditions. The vehicle tests on the WAR and DAR simulate different traction and friction scenarios, challenging the estimation algorithms to adapt to varying levels of road grip. Wet asphalt typically presents lower friction and greater variability in vehicle dynamics, while dry asphalt offers higher friction and more predictable vehicle

Figure 2.23 Test vehicle on WAR and DAR.

behavior. These contrasting conditions are crucial for a comprehensive evaluation of the estimation algorithms, as they reflect the range of environments that vehicles may encounter in everyday driving.

2.4.3.3 The Real Vehicle Test on the WAR

During the real-world vehicle experiments, an experienced driver performs the lane-changing maneuver, as illustrated in Fig. 2.24. Unlike in simulation tests, where the virtual driver can execute precise and controlled steering inputs, the real-world scenario presents a more dynamic and variable steering wheel angle. The driver's inability to perform as smooth and stable maneuvers as a virtual driver leads to more pronounced and erratic alterations in the steering wheel angle. This variability is a key distinction between virtual simulations and real-world tests. In simulations, the virtual driver follows a pre-programmed, ideal path with consistent and smooth steering inputs, resulting in a stable and predictable trajectory.

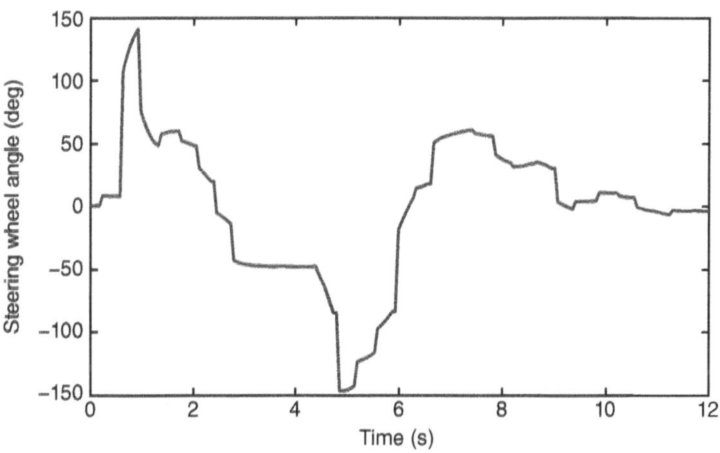

Figure 2.24 The steering wheel angle on WAR.

However, in real-world experiments, human drivers are subject to numerous external factors, such as road conditions, vehicle dynamics, and human reaction times. These factors contribute to the less predictable and more pronounced changes in the steering wheel angle during maneuvers. The pronounced steering wheel angle variations observed during the real-world experiments underscore the importance of robust estimation algorithms capable of handling real-world complexities. The ability to accurately estimate vehicle states under such conditions is critical for the development of reliable and effective ADAS and autonomous driving technologies.

The lateral acceleration, portrayed in Fig. 2.25, incorporates partial missing measurements, where the probability distribution for missing acceleration sensor data follows a uniform distribution between [0,1]. The sampling time remains consistent at 0.01 seconds. Notably, the lateral acceleration was obtained from the in-vehicle sensor. However, for the purpose of validating the AFTEKF, incomplete measurement data is essential. Consequently, we employ data processing techniques to transform normal data into incomplete data, acknowledging the offline nature of this test. Despite being offline, this approach effectively verifies the AFTEKF's efficacy and real-time performance. It is imperative to highlight that during this offline test, we exclusively constrain the distribution interval of the data without imposing limitations on the real-time performance of the data. Furthermore, our estimation algorithm operates on a rapid control prototype system in real time. The nominal value for the test vehicle mass stands at 995 kg, mirroring the mass parameter in the AFTEKF model. However, the actual total mass exceeds the nominal value due to the presence of two experimenters loaded

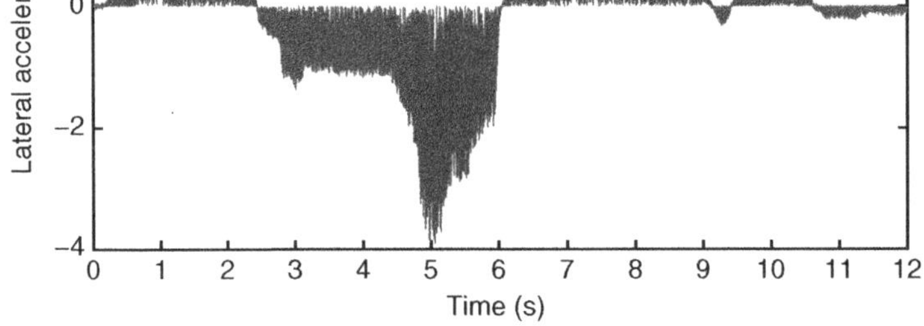

Figure 2.25 Lateral acceleration on WAR.

onto the test vehicle. This discrepancy indicates a deviation of the real vehicle mass parameter from its nominal value.

Figures 2.26 and 2.27 illustrate the estimated curves, which display slight fluctuations due to the experienced driver's steering maneuvers. These maneuvers, executed in a real-world setting, inherently lack the smoothness and precision that a virtual driver can achieve in a controlled simulation environment. Despite these variations and the natural inconsistencies introduced by human operation, the AFTEKF consistently demonstrates the highest estimation accuracy among

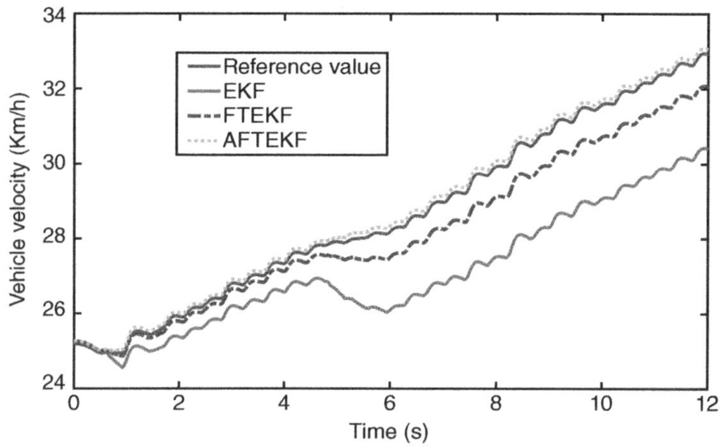

Figure 2.26 Vehicle velocity estimation on WAR.

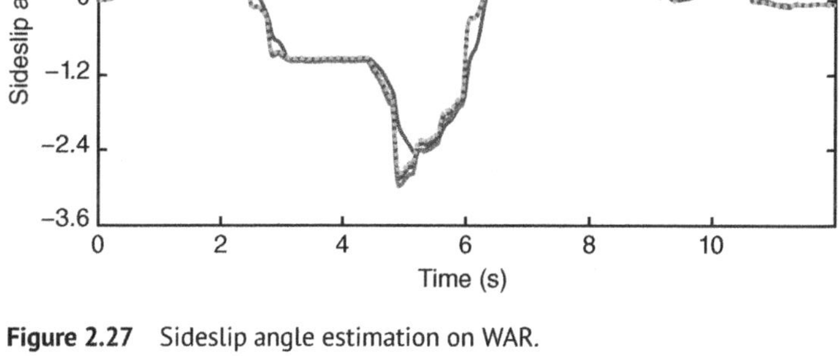

Figure 2.27 Sideslip angle estimation on WAR.

the three methods evaluated. The real-world steering inputs by the driver, though expert, introduce minor deviations that challenge the estimation algorithms to adapt to more erratic and less predictable inputs. This scenario highlights the robustness and adaptability of the AFTEKF. Its ability to maintain superior estimation accuracy amidst the real-world fluctuations is a testament to its advanced design and functionality. The fluctuations observed in the estimated curves are indicative of the algorithm's capacity to handle real-world driving dynamics, which often present more variability compared to simulations. Table 2.5 further solidifies this observation, presenting a quantitative comparison of the vehicle state estimation errors across the three methods. The data clearly shows that the AFTEKF has the smallest estimation error, reinforcing its superiority in accurately predicting vehicle states under varying conditions. The AFTEKF's minimal error margins underscore its effectiveness in processing the complex and sometimes noisy data that comes with real-world driving scenarios. This comprehensive assessment, depicted through the estimated curves and supported by the quantitative analysis in Table 2.5, underscores the AFTEKF's robust performance. It effectively manages the unpredictability of real-world driving, maintaining high accuracy and reliability. This robustness is crucial for real-world applications, where conditions can change rapidly and drivers' inputs may not always be as steady as those in simulation.

2.4.3.4 The Real Vehicle Test on the DAR

The evolving pattern of the steering wheel angle across the test scenarios is meticulously depicted in Fig. 2.28, offering a comprehensive visual insight into the variations in steering input throughout the tests. This figure not only illustrates the dynamic changes in steering angle but also provides a detailed timeline of how the driver interacts with the vehicle's steering system under different conditions. Figure 2.29 complements this depiction by presenting the lateral acceleration data, which includes partial missing measurements to simulate realistic conditions encountered in real-world driving scenarios. This visualization captures the fluctuations and responses of lateral acceleration during the tests, highlighting how the vehicle reacts to different driving maneuvers and environmental factors. Together, these figures contribute to a deeper understanding of the vehicle's

Table 2.5 RMSE on WAR.

Symbol	Sideslip angle	Vehicle velocity
EKF	0.2806	1.8230
FTEKF	0.2587	0.6110
AFTEKF	**0.2499**	**0.1217**

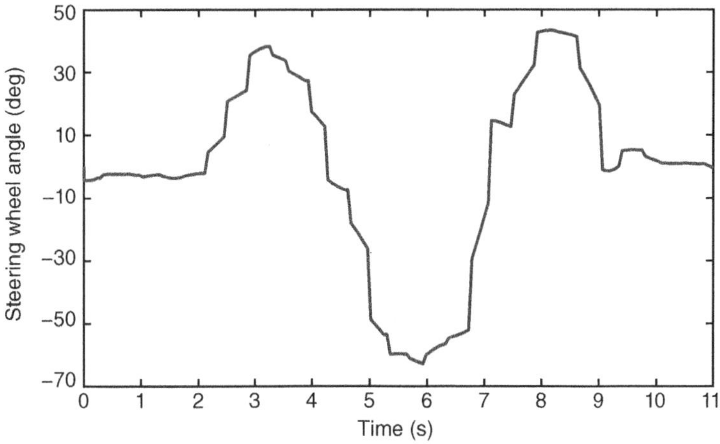

Figure 2.28 The steering wheel angle on the DAR.

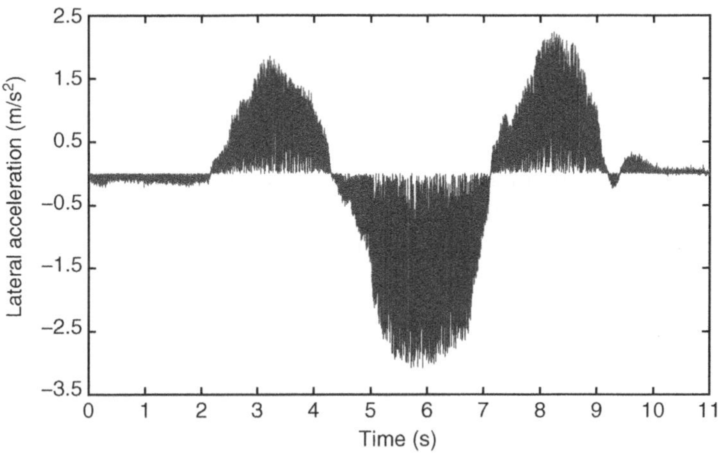

Figure 2.29 Lateral acceleration on DAR.

dynamic behavior, steering response dynamics, and lateral acceleration characteristics under varying operational conditions. They serve as essential tools for evaluating and refining vehicle control strategies and enhancing safety, stability, and overall performance in automotive engineering and research.

The probability distribution governing the absence of acceleration sensor data follows a uniform spread between [0,1], ensuring that data loss occurs randomly throughout the test. The sampling time is consistently set at 0.01 seconds to maintain high temporal resolution. Reflecting the conditions of a WAR, the

Table 2.6 RMSE on DAR.

Symbol	Sideslip angle	Vehicle velocity
EKF	0.1530	0.8024
FTEKF	0.1103	0.3596
AFTEKF	**0.0964**	**0.0435**

real vehicle mass parameter once again deviates from its nominal value in this scenario. Specifically, the actual total mass of the vehicle surpasses the nominal value due to the presence of three experimenters on board the test vehicle. This deviation from the expected mass parameter is a recurring theme in real-world testing, where the additional weight of occupants and equipment often impacts the vehicle's dynamics. The presence of additional mass highlights the importance of the AFTEKF algorithm's robustness and adaptability. The algorithm's ability to account for variations in the vehicle's mass parameter is crucial for maintaining accurate state estimations under varying real-world conditions. By introducing partial missing measurements in the lateral acceleration data, Fig. 2.29 emphasizes the AFTEKF algorithm's capability to handle incomplete or imperfect data inputs. This is particularly significant in real-world applications, where sensor data loss or corruption is a common issue. The uniform distribution model for missing data ensures that the algorithm is tested under conditions that mimic the random and unpredictable nature of real-world data loss.

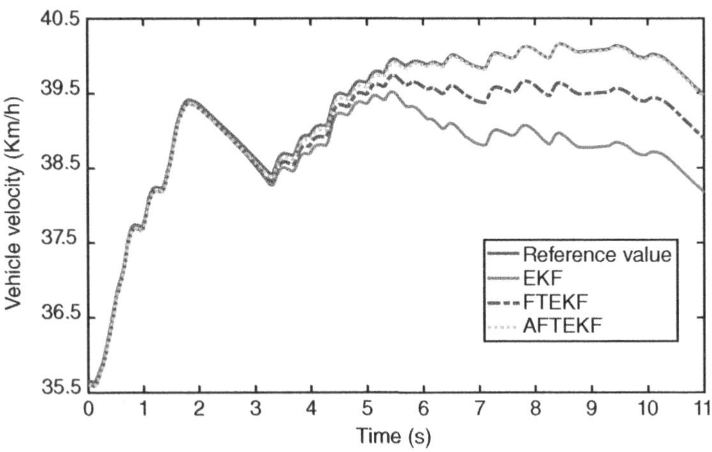

Figure 2.30 Vehicle velocity estimation on DAR.

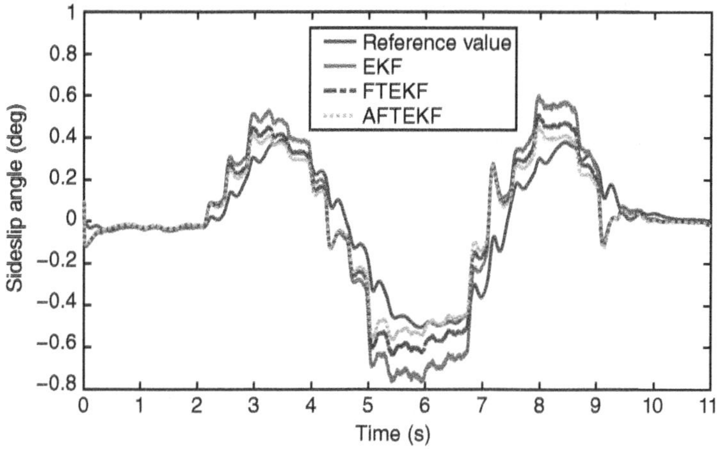

Figure 2.31 Sideslip angle estimation on DAR.

Figure 2.30 illustrates the estimation of vehicle velocity using the AFTEKF algorithm, showcasing its remarkable proximity to the reference value obtained from the DGPS. The graph demonstrates that the estimation accuracy of both AFTEKF and FTEKF surpasses that of the traditional EKF method. This significant improvement in accuracy underscores the advantages of the AFTEKF and FTEKF algorithms in handling real-world driving scenarios. A parallel observation is made in Fig. 2.31, which depicts the estimation of the sideslip angle by different methodologies. Once again, the AFTEKF algorithm stands out with the highest accuracy in estimating the sideslip angle, closely aligning with the reference values. This trend of superior performance is further validated by Table 2.6, where the quantitative assessment of estimation errors clearly indicates that the AFTEKF method achieves minimal error in vehicle state estimation. This consistent outperformance of the AFTEKF over the FTEKF and EKF methods highlights its robustness and precision in vehicle state estimation tasks. In real vehicle tests, the AFTEKF algorithm showcases its adaptability by dynamically fine-tuning the estimation error covariance of the prior state. This real-time adjustment capability ensures that the estimated values closely track the reference values, despite the inherent challenges posed by parameter perturbations, such as variations in vehicle mass. The adaptability of the AFTEKF method is particularly crucial when dealing with the dynamic and often unpredictable nature of real-world driving conditions. Moreover, the AFTEKF proves its effectiveness by consistently achieving optimal estimation results across various road friction conditions and initial states. Whether the vehicle is operating on wet asphalt, dry asphalt, or other surfaces, the AFTEKF algorithm maintains its accuracy and reliability. This robust performance underscores the efficacy and versatility of the AFTEKF method, making it a valuable tool for enhancing vehicle safety and performance.

2.5 Summary and Future Research

This section provides a systematic overview of the application and current research status of the KF in vehicle state estimation. We illustrate how the KF is employed for vehicle state estimation through a step-by-step derivation using a simple example. We apply the EKF to vehicle state estimation and validate its performance in a simulated environment. The results indicate that EKF can achieve high estimation accuracy in an ideal setting. Considering the likelihood of data loss in practical in-vehicle sensors and the potential mismatch of vehicle model parameters, a novel approach named AFTEKF has been developed for the estimation of vehicle states. This inventive method demonstrates its effectiveness in alleviating the impact of partial sensor data loss, parameter perturbations, and state mutations on estimation accuracy. Experimental findings substantiate that the AFTEKF outperforms the conventional EKF in terms of estimation accuracy. Additionally, the proposed methodology showcases robustness across a spectrum of tire-road friction coefficients and diverse initial states of the vehicle.

Nevertheless, in the conducted experiments, the assumption is made that the distribution of missing data is known. Identifying the distribution of missing data poses a challenge, and future research endeavors will prioritize the collection of sensor measurement data. The analysis of statistical characteristics will be conducted once the relevant sensors and testing equipment become available, enhancing the understanding and accuracy of the data distribution. Furthermore, we also assume that the process noise of the model is known. However, due to the complexity of actual driving conditions, various external disturbances and unpredictable driving road conditions can lead to changes in process noise. Therefore, we need to design an adaptive updating mechanism to accommodate such variations and enhance estimation accuracy. This targeted approach aims to further refine the estimation process and contribute to the advancement of state-of-the-art methodologies in the field.

References

1 Stephant, J., Charara, A., and Meizel, D. (2004). "Linear Observers for Vehicle Sideslip Angle: Experimental Validation," *in Proceedings of IEEE International Symposium on Industrial Electronics*, 341–346. Ajaccio.

2 Zhang, H., Zhang, G., and Wang, J. (2015). Sideslip angle estimation of an electric ground vehicle via finite-frequency H_∞ approach. *IEEE Transactions on Transportation Electrification* 2 (2): 200–209.
3 Ozkan, B., Margolis, D., and Pengov, M. (2008). The controller output observer: estimation of vehicle tire cornering and normal forces. *Journal of Dynamic Systems, Measurement, and Control-ASME* 130 (6): 061002.
4 Zhao, L.H., Liu, Z.Y., and Chen, H. (2010). Design of a nonlinear observer for vehicle velocity estimation and experiments. *IEEE Transactions on Control Systems Technology* 19 (3): 664–672.
5 Guo, H., Chen, H., Cao, D., and Jin, W. (2013). Design of a reduced-order non-linear observer for vehicle velocities estimation. *IET Control Theory Applications* 7 (8): 2056–2068.
6 Grip, H.F., Imsland, L., Johansen, T.A., and Fossen, T.I. (2008). Nonlinear vehicle side-slip estimation with friction adaptation. *Automatica* 44 (3): 611–622.
7 Gao, X., Yu, Z., Neubeck, J., and Wiedemann, J. (2010). Sideslip angle estimation based on input-output linearisation with tire-road friction adaptation. *Vehicle System Dynamics* 48 (2): 217–234.
8 Hashemi, E., Zarringhalam, R., Khajepour, A., and Melek, W. (2017). Real-time estimation of the road bank and grade angles with unknown input observers. *Vehicle System Dynamics* 55 (5): 648–667.
9 M'Sirdi, N.K., Rabhi, A., Fridman, L. et al. (2008). Second order sliding-mode observer for estimation of vehicle dynamic parameters. *International Journal of Vehicle Design* 48 (2–4): 190–207.
10 Chen, Y., Ji, Y., and Guo, K. (2014). A reduced-order nonlinear sliding mode observer for vehicle slip angle and tyre forces. *Vehicle System Dynamics* 52 (12): 1716–1728.
11 Imine, H., Benallegue, A., Madani, T., and Srairi, S. (2013). Rollover risk prediction of heavy vehicle using high-order sliding-mode observer: experimental results. *IEEE Transactions on Vehicular Technology* 63 (6): 2532–2543.
12 Boada, B.L., Boada, M.J.L., and Diaz, V. (2016). Vehicle sideslip angle measurement based on sensor data fusion using an integrated ANFIS and an unscented Kalman filter algorithm. *Mechanical Systems and Signal Processing* 72: 832–845.
13 Venhovens, P.J.T. and Naab, K. (1999). Vehicle dynamics estimation using Kalman filters. *Vehicle System Dynamics* 32 (2–3): 171–184.
14 Cho, W., Yoon, J., Yim, S. et al. (2009). Estimation of tire forces for application to vehicle stability control. *IEEE Transactions on Vehicular Technology* 59 (2): 638–649.
15 Zheng, H., Miao, Y., and Li, B. (2019). A heavy tractor semi-trailer stability control strategy based on electronic pneumatic braking system HIL test. *SAE International Journal of Vehicle Dynamics, Stability, and NVH* 3 (3): 237–249.

16 Anderson, R. and Bevly, D.M. (2010). Using GPS with a model-based estimator to estimate critical vehicle states. *Vehicle System Dynamics* 48 (11): 1412–1438.

17 Lee, S., Nakano, K., and Ohori, M. (2015). On-board identification of tyre cornering stiffness using dual Kalman filter and GPS. *Vehicle System Dynamics* 53 (4): 437–448.

18 Ryu, J. and Gerdes, J.C. (2004). Integrating inertial sensors with global positioning system (GPS) for vehicle dynamics control. *Journal of Dynamic Systems Measurement and Control-Transactions of the ASME* 126 (2): 242–254.

19 Wenzel, T.A., Burnham, K.J., Blundell, M.V., and Williams, R.A. (2006). Dual extended Kalman filter for vehicle state and parameter estimation. *Vehicle System Dynamics* 44 (2): 153–171.

20 Swami, A., Liu, C., Kubenz, J. et al. (2021). Experimental study on tire contact patch characteristics for vehicle handling with enhanced optical measuring system. *SAE International Journal of Vehicle Dynamics, Stability, and NVH* 5 (2): 332–350.

21 Baffet, G., Charara, A., and Dherbomez, G. (2007). An observer of tire-road forces and friction for active security vehicle systems. *IEEE/ASME Transactions on Mechatronics* 12 (6): 651–661.

22 Zong, C., Hu, D., and Zheng, H. (2013). Dual extended Kalman filter for combined estimation of vehicle state and road friction. *Chinese Journal of Mechanical Engineering* 26 (2): 312–324.

23 Tsunashima, H., Murakami, M., and Miyata, J. (2006). Vehicle and road state estimation using interacting multiple model approach. *Vehicle System Dynamics* 44 (10): 750–758.

24 Hang, P., Chen, X., Fang, S., and Luo, F. (2017). Robust control for four-wheel-independent-steering electric vehicle with steer-by-wire system. *International Journal of Automotive Technology* 18 (5): 785–797.

25 Naets, F., van Aalst, S., Boulkroune, B. et al. (2017). Design and experimental validation of a stable two-stage estimator for automotive sideslip angle and tire parameters. *IEEE Transactions on Vehicular Technology* 66 (11): 9727–9742.

26 Xia, X., Xiong, L., Lin, X., and Yu, Z. (2018). Vehicle sideslip angle estimation considering the tire pneumatic trail variation. *SAE Technical Paper* 2018-01-0571.

27 Hu, C., Wang, Z., Taghavifar, H., and Na, J. (2019). MME-EKF-based path-tracking control of autonomous vehicles considering input saturation. *IEEE Transactions on Vehicular Technology* 68 (6): 5246–5259.

28 Di Biase, F., Lenzo, B., and Timpone, F. (2020). Vehicle sideslip angle estimation for a heavy-duty vehicle via extended Kalman filter using a rational tyre model. *IEEE Access* 8: 142120–142130.

29 Li, X., Xu, N., Li, Q. et al. (2020). A fusion methodology for sideslip angle estimation on the basis of kinematics-based and model-based approaches.

Proceedings of the Institution of Mechanical Engineers, Part D: Journal of Automobile Engineering 234 (7): 1930–1943.

30 Sun, W., Wang, Z., Wang, J. et al. (2022). Research on a real-time estimation method of vehicle sideslip angle based on EKF. *Sensors* 22 (9): 3386.

31 Lu, H., Liu, Q., Shi, Y., and Yu, F. (2017). Estimation of vehicle sideslip angle and individual tyre-road forces based on tyre friction circle concept. *International Journal of Vehicle Autonomous Systems* 13 (4): 274–305.

32 Jiang, G., Liu, L., Guo, C. et al. (2017). A novel fusion algorithm for estimation of the side-slip angle and the roll angle of a vehicle with optimized key parameters. *Proceedings of the Institution of Mechanical Engineers, Part D: Journal of Automobile Engineering* 231 (2): 161–174.

33 Zhou, C., Liu, X.H., and Xu, F.X. (2021). Intervention criterion and control strategy of active front steering system for emergency rescue vehicle. *Mechanical Systems and Signal Processing* 148: 107160.

34 Wang, Y., Wei, M., Zhao, W. et al. (2017). Vehicle state estimation based on combined RLS and FAEKF. *China Mechanical Engineering* 28 (6): 750.

35 Zhang, F., Wang, Y., Hu, J. et al. (2021). A novel comprehensive scheme for vehicle state estimation using dual extended h-infinity Kalman filter. *Electronics* 10 (13): 1526.

36 Kalman, R.E. (1961). New results in linear filtering and prediction theory. *Journal of Basic Engineering, ASME Transactions Series D* 83 (1): 109.

37 Wang, X. and Yaz, E.E. (2014). Stochastically resilient extended Kalman filtering for discrete-time nonlinear systems with sensor failures. *International Journal of Systems Science* 1 (9): 1392–1401.

38 Sun, Y., Wang, Y., Wu, X.P., and Hu, Y.L. (2017). Robust extended fractional Kalman filter for nonlinear fractional systems with missing measurements. *Journal of the Franklin Institute – Engineering and Applied Mathematics* 355 (1): 361–380.

39 Wang, X. and Yaz, E.E. (2019). Second-order fault tolerant extended Kalman filter for discrete time nonlinear systems. *IEEE Transactions on Automatic Control* 64 (12): 5086–5093.

40 Dong, H., Wang, Z., Ho, D.W.C., and Gao, H. (2010). Variance-constrained H_∞ filtering for a class of nonlinear time-varying systems with multiple missing measurements: the finite-horizon case. *IEEE Transactions on Signal Processing* 58 (5): 2534–2543.

41 Hu, J., Wang, Z., Gao, H., and Stergioulas, L. (2012). Extended Kalman filtering with stochastic nonlinearities and multiple missing measurements. *Automatica* 48 (9): 2007–2015.

42 Yin, Z., Li, G., Zhang, Y., and Liu, J. (2019). Symmetric-strong-tracking-extended-Kalman-filter-based sensorless control of induction motor drives for modeling error reduction. *IEEE Transactions on Industrial Informatics* 15 (2): 650–662.

43 Zhou, D.H. and Frank, P.M. (1996). Strong tracking Kalman filtering of nonlinear time-varying stochastic systems with colored noise: application to parameter estimation and empirical adaptability analysis. *International Journal of Control* 65 (2): 295–307.

44 Jwo, D.J. and Wang, S.H. (2007). Adaptive fuzzy strong tracking extended Kalman filtering for GPS navigation. *IEEE Sensors Journal* 7 (5): 778–789.

45 Zhao, S. and Li, Y. (2009). Multi-sensor information fusion and strong tracking filter for vehicle nonlinear state estimation. *Proceedings of IEEE Intelligent Vehicles Symposium* 747–751.

46 Pacejka, H.B. (2012). *Tyre and Vehicle Dynamics*, 3e. Oxford: Butterworth-Heinemann.

47 Kiencke, U. and Nielsen, L. (2000). *Automotive Control Systems: For Engine, Driveline, and Vehicle*, 2e. Berlin: Springer-Verlag.

48 Dugoff, H., Fanches, P., and Segel, L. (1970). An analysis of tire properties and their influence on vehicle dynamic performance. *SAE Paper* 700377.

49 Mori, T., Fukuma, N., and Kuwahara, M. (1986). Explicit solution and eigenvalue bounds in the Lyapunov matrix equation. *IEEE Transactions on Automatic Control* 31 (7): 656–658.

50 Simon, D. (2006). *Optimal State Estimation: Kalman, H_∞, and Nonlinear Approaches*. Hoboken, NJ: Wiley.

51 Wang, Y., Yin, G., Li, Y. et al. (2021). Self-Learning control for coordinated collision avoidance of automated vehicles. *Proceedings of the Institution of Mechanical Engineers, Part D: Journal of Automobile Engineering* 235 (4): 1149–1163.

3

Ego-Vehicle State Estimation with Unknown Noise and Parameter Perturbations

3.1 Introduction

Sideslip angle, tire cornering stiffness, and vehicle velocity are crucial for both traditional and autonomous vehicles. They play essential roles in chassis stability control, as well as in tasks such as path planning and tracking control. However, these states cannot be directly measured by onboard sensors; therefore, various vehicle state estimation algorithms have been developed. Most of these algorithms assume that the noise characteristics are known, ignoring the impact of missing measurement data, and cannot simultaneously handle the effects of colored noise and white noise. In addition, the vehicle mass variation, which has a vital influence on the precision of the vehicle model, is rarely considered in these methods. Since the estimation accuracy of vehicle state highly depends on the accuracy of the vehicle model, the disturbance caused by vehicle mass change must be taken into account during the estimation process of vehicle state. In this chapter, a fuzzy adaptive robust cubature KF (FARCKF) and a fault-tolerant extended KF network are developed to estimate vehicle state. Experiments under different operating conditions will be implemented to verify the effectiveness of the algorithm.

3.2 Related Works

Advanced driver assistance systems, or autonomous driving systems, have become essential technologies in modern vehicles, significantly enhancing safety and reducing the likelihood of traffic accidents. These advanced systems not only help in preventing accidents but also play a crucial role in mitigating the substantial casualties and economic losses associated with traffic incidents [1–3]. Several advanced systems have already been integrated into mass-produced vehicles, exemplifying the current state of vehicle automation and safety technology. These include autonomous emergency braking [4], which automatically applies the

State Estimation of Multi-Agent Vehicle-Road Interaction Systems, First Edition.
Yan Wang, Guodong Yin, and Chao Huang.
© 2026 The Institute of Electrical and Electronics Engineers, Inc. Published 2026 by John Wiley & Sons, Inc.

brakes to prevent or mitigate collisions; platoon control [5, 6], which enables multiple vehicles to travel closely together in a coordinated manner to improve traffic flow and fuel efficiency; adaptive cruise control [7, 8], which adjusts the vehicle's speed to maintain a safe following distance from the vehicle ahead; and path tracking control [9, 10], which ensures the vehicle accurately follows a predetermined path, crucial for autonomous navigation and precise vehicle maneuvering. These systems typically set corresponding triggering conditions and decide whether to activate the system proactively through real-time monitoring of the vehicle state. Therefore, the accurate acquisition of the vehicle state is a prerequisite for the effective operation of these systems. Nevertheless, some critical states, such as sideslip angle and vehicle speed, are challenging to measure directly through onboard sensors. Therefore, the exclusive use of onboard sensors for vehicle state estimation has become an important area of research in autonomous driving. Traditionally, observer-based methods [11–13] have been employed to estimate vehicle states. These methods utilize mathematical models to estimate the state variables of a system from the input and output measurements. However, with the advent of more sophisticated sensors and increased computational power, model-based approaches have gained prominence. These approaches leverage detailed vehicle dynamics models to enhance the accuracy of state estimation. Model-based methods often involve the use of dynamic models of the vehicle to predict its behavior under various conditions. These models can be highly accurate but also require precise knowledge of vehicle parameters and environmental conditions. One of the main advantages of model-based methods is their ability to incorporate a wide range of physical phenomena, such as tire–road interactions and vehicle body dynamics, into the estimation process. In recent years, data-driven methods have emerged as a powerful alternative to traditional observer-based and model-based approaches. Data-driven methods utilize large datasets collected from vehicles and advanced machine learning algorithms to estimate vehicle states. These methods can learn complex patterns and relationships in the data, often achieving high accuracy without the need for detailed physical models. Machine learning techniques, such as neural networks and deep learning, have been particularly successful in this domain, providing robust state estimates even in challenging conditions. The evolution from traditional observer-based methods to model-based approaches and, more recently, data-driven methods represents a significant advancement in vehicle state estimation. Each of these approaches has its strengths and limitations. Observer-based methods are relatively simple and computationally efficient but may lack accuracy in highly nonlinear scenarios. Model-based methods offer high accuracy but require detailed knowledge of vehicle dynamics and parameters. Data-driven methods, while highly accurate and adaptive, often require large amounts of training data and can be computationally intensive.

Model-based methods for vehicle state estimation are primarily divided into two categories: kinematics-based and dynamics-based approaches. Kinematics-based estimation methods generally require the integration of a GPS and inertial measurement units (IMU) to provide accurate state information. This approach is evident in several studies, such as those conducted by [14–16]. However, the widespread adoption of kinematics-based methods in mass-produced vehicles faces significant challenges. Most commercially available vehicles do not come equipped with IMUs or similar high-precision devices, making it difficult to implement these methods on a large scale. Additionally, ensuring observability—a condition where the internal states of a system can be determined through its outputs—poses a significant challenge for kinematics-based methods, further hindering their adoption.

On the other hand, dynamics-based methods offer a more practical solution by relying on vehicle and tire models. These methods utilize the fundamental principles of vehicle dynamics to estimate states such as sideslip angle, yaw rate, and vehicle speed. Advanced filtering algorithms are often employed to enhance the accuracy of these estimations. Dynamics-based methods do not depend heavily on high-cost sensors like IMUs, making them more suitable for mass-produced vehicles. They achieve state estimation by processing readily available data from standard vehicle sensors, such as wheel speed sensors, steering angle sensors, and accelerometers, combined with mathematical models of vehicle behavior. Among these, the use of Kalman filtering for vehicle state estimation is a research hotspot. For instance, researchers have developed an EKF-based estimator specifically designed for predicting sideslip angles, as highlighted in [17, 18]. Building on this foundation, further advancements enabled the simultaneous estimation of both sideslip angle and lateral tire forces using EKF techniques, as demonstrated in [19]. Moreover, EKF has been successfully applied in real-world settings to achieve precise state estimation leveraging tire force sensor data [20]. Recognizing the need for enhanced adaptability, interactive multi-model EKF methodologies have been introduced to facilitate robust multi-target state estimation [21]. Additionally, the variable structure EKF approach [22] has emerged as an effective method, allowing for dynamic adjustments in model structure to accommodate varying environmental conditions. These innovations underscore the versatility and applicability of EKF in advancing the accuracy and reliability of vehicle state estimation across diverse operational scenarios. Moreover, in response to challenges posed by noise uncertainty in vehicle state estimation, researchers have introduced the extended H_∞ KF [23]. This method enhances prediction accuracy for vehicle speed and sideslip angle by incorporating additional factors to handle uncertainty more effectively. Additionally, various adaptations of the H_∞ filter have proven effective in improving estimation robustness [24]. Despite the fruitful results achieved through traditional EKF-based approaches in vehicle state estimation, there has been a growing awareness of the impact of data loss [25] on accuracy. Addressing this concern, the development of fault-tolerant EKF (FTEKF) estimators [26, 27] has emerged, specifically designed to maintain reliable vehicle state estimation even when encountering missing measurements. These advancements highlight ongoing efforts to enhance the reliability and resilience of estimation algorithms in real-world applications.

Additionally, advancements in the EKF methodology [28] have enabled its application in estimating vehicle states even when faced with partial loss of measurement data. Building on this progress, robust fault-tolerant estimation algorithms [29] have been developed, which not only handle data loss scenarios effectively but also account for parameter mismatches to improve sideslip angle estimation. Moreover, scholars have explored event-triggered mechanisms [30] to address challenges associated with data loss during vehicle state estimation. Experimental findings underscore the potential of these methods for practical applications, showcasing their ability to enhance estimation reliability in dynamic real-world conditions. Moreover, researchers have proposed the unscented KF (UKF) in recent studies [19, 31–34], demonstrating through experiments its superior estimation accuracy compared to the traditional EKF. In contrast to the UKF, the cubature KF (CKF) [35] has proven more effective and has found widespread application in accurately estimating sideslip angle and tire cornering stiffness [36–41]. These model-based estimation methods typically rely on a precise understanding of the underlying vehicle and tire models. Furthermore, methods employing Kalman filtering require precise characterization of measurement noise to achieve optimal estimation accuracy. These advancements underscore the critical role of accurate modeling and noise characterization in enhancing the reliability of vehicle state estimation techniques.

Recent advancements in machine learning have revolutionized the landscape of vehicle state estimation, captivating attention across industry and academia. At the forefront of these innovations are artificial neural networks (ANNs), celebrated for their unparalleled capability in universal function approximation. The integration of ANNs into vehicle state estimation methodologies represents a significant stride toward enhancing accuracy and robustness in real-world applications. Numerous scholars have advocated for ANN-based approaches to vehicle state estimation, leveraging onboard sensor data as inputs and actual vehicle states as reference outputs, as articulated in studies such as [42–45]. This methodology involves constructing extensive datasets from onboard sensor measurements and corresponding ground truth vehicle states, which are then used to train neural network models. Through iterative learning processes, these models effectively learn complex mappings between sensor inputs and desired state outputs, enabling accurate prediction of vehicle dynamics and state variables. Moreover, the application of recurrent neural networks (RNNs) has emerged as a promising avenue for vehicle motion prediction and continuous state updating [46]. Unlike traditional feedforward ANNs, RNNs possess the ability to retain the memory of past inputs, making them well-suited for sequential data processing tasks inherent in vehicle dynamics prediction. By capturing temporal dependencies in sensor data streams, RNNs can dynamically update state estimates based on evolving driving conditions, thereby enhancing prediction accuracy and responsiveness. In parallel, the fusion of Kalman filters with deep neural networks has garnered considerable interest in vehicle state estimation [47, 48]. Kalman filters excel in filtering noisy sensor data and modeling stochastic processes, while deep neural networks excel in learning intricate nonlinear mappings and patterns from data. By combining these strengths, hybrid Kalman filter-neural network architectures effectively mitigate the impact of sensor noise and model mismatches, thereby improving estimation accuracy under diverse operating conditions. Despite their promise, purely data-driven methods such as ANN-based approaches encounter challenges related to interpretability and generalization. The intricate architectures and vast number of parameters in neural networks can obscure the underlying relationships between inputs and outputs, making it challenging to interpret how decisions are made. Furthermore, ensuring the generalization of neural network models across diverse driving scenarios and environmental conditions remains an ongoing research challenge, necessitating robust validation and testing methodologies. Moreover, the integration of neural networks with Kalman filtering represents a straightforward yet powerful strategy to address the challenges posed by sensor noise and model discrepancies. This approach leverages the complementary strengths of both techniques: Kalman filters excel in handling noisy sensor measurements and modeling system dynamics, while neural networks are adept at learning complex nonlinear relationships from data. By combining these methodologies, the resulting hybrid approach enhances the robustness and accuracy of vehicle state estimation systems. Kalman filters provide a structured framework for state estimation by recursively updating predictions based on noisy sensor inputs and system dynamics models. However, their performance can be limited by inaccuracies in model assumptions or variations in real-world conditions. Neural networks, on the other hand, can adaptively learn from large datasets to capture intricate patterns and relationships that may not be explicitly modeled by traditional methods.

3.3 Fuzzy Adaptive Robust Cubature Kalman Filter for Vehicle State Estimation

In this section, we introduce a novel approach known as the FARCKF designed specifically for estimating sideslip angle and tire cornering stiffness. The method integrates several advanced techniques to enhance estimation accuracy and robustness across varying conditions. To begin with, the vehicle mass parameter undergoes dynamic updating using recursive least squares (RLS). This adaptive approach effectively mitigates the impact of vehicle mass variations, ensuring more reliable estimations throughout the operation. Next, a Takagi–Sugeno (T–S) fuzzy system is implemented to dynamically adjust the robust cubature KF (RCKF). This adaptation is crucial as it considers the dynamic nature of process noise, thereby improving the filter's resilience to unpredictable disturbances and environmental changes. The effectiveness of the FARCKF method is evaluated through a series of simulations and real-world experiments. These tests are designed to validate the algorithm's performance under diverse driving conditions and scenarios. By comparing the estimation results against ground truth data obtained from rigorous testing environments, the method's capability to accurately estimate sideslip angle and tire cornering stiffness is thoroughly assessed.

3.3.1 Vehicle Model and Problem Statement

As illustrated in Fig. 3.1, a bicycle model [49] and a linear tire model [50] are adopted for sideslip angle and tire cornering stiffness estimation. The bicycle model, also known as the single-track model, is a simplified representation used in vehicle dynamics to simulate the movement and behavior of vehicles, particularly in terms of lateral and yaw dynamics. It approximates a vehicle as having characteristics similar to a bicycle, with a single front wheel and a single rear wheel. This model is fundamental for analyzing steering control, stability, and maneuverability, making it valuable for applications in vehicle design, control system development, and autonomous driving simulations. The linear tire model simplifies tire behavior by assuming a linear relationship between tire forces and slip variables, such as slip angle and slip ratio. It is commonly used for vehicle design and applications where simplicity and quick evaluations of vehicle dynamics are sufficient.

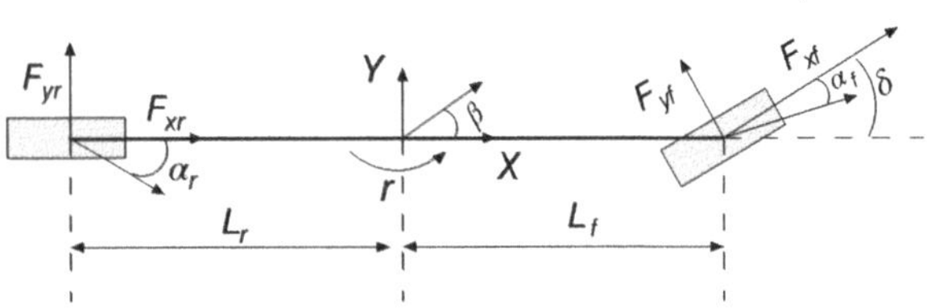

Figure 3.1 The bicycle model.

With a constant or slowly varying vehicle speed, the dynamic formulas of this vehicle model are as follows:

$$\dot{\beta} = \frac{F_{yf}\cos\delta + F_{yr}}{mv_x} - r \tag{3.1}$$

$$a_y = \frac{F_{yf}\cos\delta + F_{yr}}{m} \tag{3.2}$$

The description of variables in the bicycle model is shown in Table 3.1.

Under normal driving conditions, lateral tire forces are generally regarded to be linear with wheel sideslip angle. Thus, we employ a linear tire model to formulate the tire lateral forces.

$$F_{yf} = 2C_{yf}\alpha_f \tag{3.3}$$

$$F_{yr} = 2C_{yr}\alpha_r \tag{3.4}$$

where C_{yf} is the front tire cornering stiffness, C_{yr} is the rear tire cornering stiffness, α_f is the front wheel sideslip angle, and α_r is the rear wheel sideslip angle.

The front wheel sideslip angle and rear wheel sideslip angle are approximated as

$$\alpha_f = \delta - \beta - \frac{L_f r}{v_x} \tag{3.5}$$

$$\alpha_r = -\beta + \frac{L_r r}{v_x} \tag{3.6}$$

Table 3.1 The bicycle model variables.

Symbol	Description
β	Sideslip angle
v_x	Vehicle speed
r	Yaw rate
m	Vehicle mass
F_{yf}	Front axle lateral force
F_{yr}	Rear axle lateral force
δ	Front wheel steering angle
a_y	Lateral acceleration
L_f	Distance from the center of gravity to the front axle
L_r	Distance from the center of gravity to the rear axle

The vehicle model, i.e. formulations (Eqs. 3.1–3.6), can be converted to a continuous-time state-space model

$$\begin{cases} \dot{x} = f_c(x, u) + w \\ z = h_c(x, u) + v \end{cases} \tag{3.7}$$

$$x = [\beta, C_{yf}, C_{yr}]^T$$

$$z = a_y \quad u = [\delta, v_x, r]^T$$

The descriptions of variables in Eq. (3.7) are given in Table 3.2.

The vehicle model (Eqs. 3.1–3.6) can be employed to estimate sideslip angle and tire cornering stiffness, whereas it cannot estimate the vehicle mass parameter. Thus, we add Eq. (3.8) to estimate the vehicle mass parameter, as follows:

$$ma_y = \frac{(L_f C_{yf} - L_r C_{yr})r}{v_x} + \beta(C_{yf} + C_{yr}) - C_{yf}\delta \tag{3.8}$$

The regression model of RLS can be expressed as

$$\xi(t) = \omega^T(t)\alpha(t) + \eta(t) \tag{3.9}$$

where $\alpha(t)$ and $\omega(t)$ are the estimated parameters and regression vectors, respectively, $\xi(t)$ is the system output, and $\eta(t)$ is the error between $\xi(t)$ and $\omega^{\mathrm{T}}(t)\alpha(t)$. Therefore, the regression model for estimating the vehicle mass parameter can be given by Eq. (3.10).

$$\begin{cases} \omega^T(t) = a_y; \quad \alpha(t) = m \\ \xi(t) = \dfrac{(L_f C_{yf} - L_r C_{yr})r}{v_x} + \beta(C_{yf} + C_{yr}) - C_{yf}\delta \end{cases} \tag{3.10}$$

Table 3.2 Variables in space model.

Symbol	Description
z	Measurement vector
x	State vector
u	Input vector
$f_c(\cdot)$	State transition function
$h_c(\cdot)$	Output function
w	Process noise
v	Measurement noise

3.3.2 Methodology

The framework of the proposed FARCKF method, depicted in Fig. 3.2, integrates five essential modules: signal measurement, RCKF, vehicle mass estimation, T-S fuzzy system, and real-time calculation of wheel sideslip angle. Each module plays a crucial role in enhancing the accuracy and robustness of vehicle state estimation. The signal measurement module serves as the foundational component, capturing real-time data such as front wheel angle, yaw rate, vehicle speed, and lateral acceleration. These measurements are essential inputs for subsequent estimation processes within the FARCKF framework. Central to the FARCKF method, the RCKF module operates through a structured sequence of initialization, time update, and measurement update steps. This module dynamically updates the sideslip angle and tire cornering stiffness, leveraging the latest vehicle model parameters retrieved from the vehicle mass estimation module. This iterative process ensures that the estimation remains accurate and responsive to dynamic changes in vehicle conditions. Real-time calculation of the wheel sideslip angle is facilitated by integrating data from both the sideslip angle module and the signal measurement module. This computation provides critical feedback for ongoing vehicle state estimation and control strategies. The T-S fuzzy system within the FARCKF framework acts as an adaptive mechanism, continuously updating parameters to adjust the process noise within the RCKF algorithm dynamically. This adaptive capability enhances the algorithm's robustness, enabling it to maintain accurate estimation across varying operational conditions and environmental factors. Furthermore, the estimation procedure for the vehicle mass parameter is meticulously outlined, emphasizing its RLS approach for continuous adjustment based on real-time data inputs. This adaptive strategy ensures that the vehicle mass estimation remains responsive to changes, further enhancing the overall performance of the FARCKF method.

The simplified lateral dynamic models can be represented by a parameter identification form, as seen in Eq. (3.9). RLS is a powerful algorithm used in signal processing and estimation to sequentially update estimates of unknown parameters based on incoming data. At its core, RLS operates by recursively minimizing the sum of squared errors between predicted and observed data points, continually refining its estimates as new information becomes available. Unlike batch least squares methods that require processing all data at once, RLS updates estimates incrementally, making it well-suited for real-time applications where data arrives sequentially and processing time is critical. The principle behind RLS involves maintaining an estimate of the parameter vector and an associated covariance matrix, which encapsulates uncertainty in parameter estimates. With each new data point, RLS updates these estimates using the Kalman filtering framework, adapting the estimates to changes in the underlying

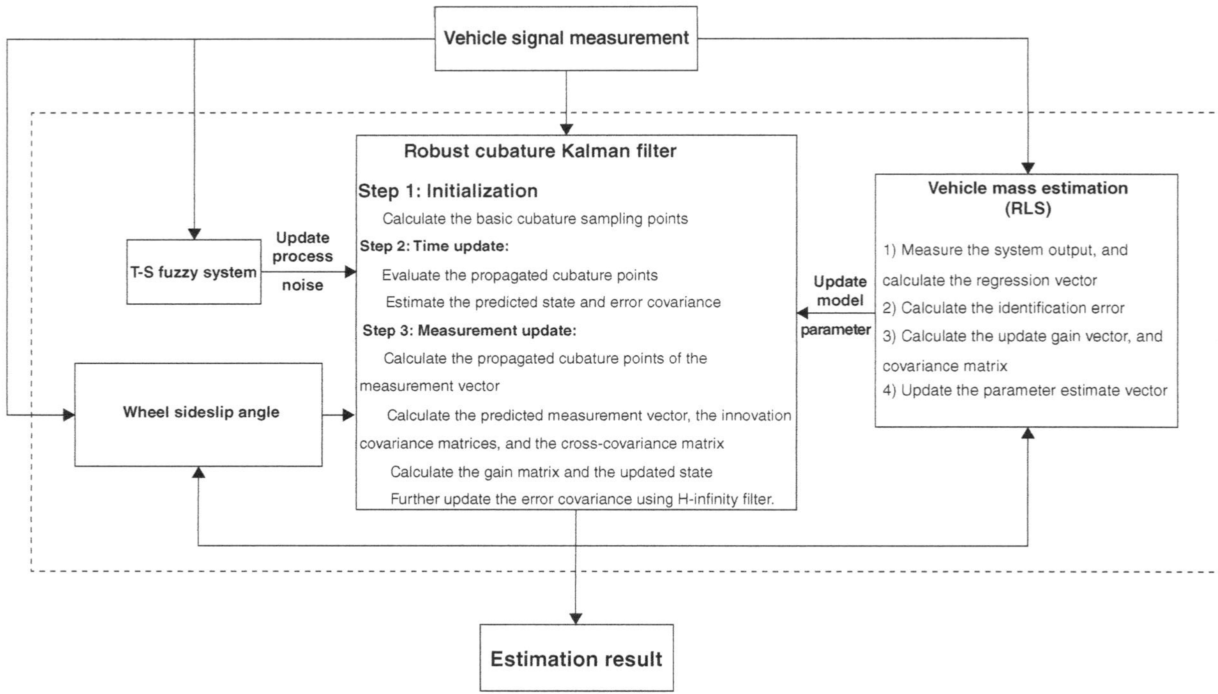

Figure 3.2 The framework of the proposed estimation algorithm.

system dynamics. This adaptability makes RLS particularly effective in tracking time-varying parameters and handling nonstationary data, characteristics often encountered in fields such as adaptive signal processing, control systems, and financial modeling. One of the key advantages of RLS lies in its ability to provide optimal estimates with minimal computational resources compared to batch methods. By leveraging the recursive nature of the algorithm and updating estimates based only on recent data, RLS achieves efficient parameter tracking and prediction. Moreover, RLS offers superior numerical stability compared to other online estimation techniques, such as stochastic gradient descent, by incorporating the covariance matrix to regulate parameter updates and mitigate the effects of noise and outliers in the data. Applications of RLS span various domains, including adaptive filtering for noise reduction in signals, system identification for model parameter estimation in control systems, and adaptive equalization in telecommunications. Its versatility and efficiency have made RLS a cornerstone in real-time processing tasks where accurate and timely parameter estimation is crucial. Ongoing research continues to enhance RLS by integrating it with advanced machine learning techniques, exploring adaptive strategies for covariance matrix updates, and extending its applicability to large-scale and distributed systems. As such, RLS remains a foundational tool in the arsenal of techniques for dynamic parameter estimation and adaptive signal processing in modern technological applications. The main steps of the RLS algorithm [51] are presented as follows:

The output information of the system $\xi(t)$ and the information of the regression vector $\omega(t)$ are first obtained.

Calculate the error $\eta(t)$ using Eq. (3.11).

$$\eta(t) = \xi(t) - \omega^T(t)\alpha(t) \tag{3.11}$$

The gain vector $\delta(t)$ and covariance matrix $\Psi(t)$ can be updated using Eqs. (3.12) and (3.13), respectively.

$$\delta(t) = \frac{\Psi(t-1)\omega(t)}{\lambda + \omega^T(t)\Psi(t-1)\omega(t)} \tag{3.12}$$

$$\Psi(t) = \frac{1}{\lambda}\left[\Psi(t-1) - \frac{\Psi(t-1)\omega^T(t)\omega(t)\Psi(t-1)}{\lambda + \omega^T(t)\Psi(t-1)\omega(t)}\right] \tag{3.13}$$

$\alpha(t)$ is estimated using Eq. (3.14).

$$\alpha(t) = \alpha(t-1) + \delta(t)\eta(t) \tag{3.14}$$

In Eqs (3.12) and (3.13), λ is referred to as the forgetting factor. Note that λ is commonly taken in the interval (0.9, 1) [52], and we set it to be 0.99 in this study.

The vehicle mass parameter can be updated dynamically to improve the accuracy of the vehicle model. Based on this more accurate vehicle model, we estimate sideslip angle and tire cornering stiffness using the RCKF algorithm with a T-S fuzzy system.

To estimate the sideslip angle and tire cornering stiffness using discrete measurements, the continuous-time vehicle model is discretized.

$$\begin{cases} x_{\theta+1} = f(x_\theta, u_\theta) + w_\theta \\ z_{\theta+1} = h(x_{\theta+1}, u_{\theta+1}) + v_{\theta+1} \end{cases} \tag{3.15}$$

where $f(\cdot)$ is the state transition function, θ represents the index of intervals, i.e. the θth Δt, Δt is a sampling interval, $h(\cdot)$ is the output function, w_θ is the system process noise, the covariance matrix of the process noise is Q_θ, $v_{\theta+1}$ is the system measurement noises, and the covariance matrix of the measurement noises is $R_{\theta+1}$. In this study, we assume that these noises follow a Gaussian distribution. $z_{\theta+1} = [a_{y,\theta+1}]^T$ is the measurement vector, and the system state vector is denoted by $x_{\theta+1} = [\beta_{\theta+1}, C_{yf,\theta+1}, C_{yr,\theta+1}]^T$. The nonlinear system can be calculated as

$$\begin{bmatrix} \beta_{\theta+1} \\ C_{yf,\theta+1} \\ C_{yr,\theta+1} \end{bmatrix} = \begin{bmatrix} \beta_\theta + \left[2C_{yf,\theta} \left(\delta_\theta - \beta_\theta - \dfrac{L_f r_{x,\theta}}{v_{x,\theta}} \right) \cos\delta_\theta + \right. \\ \left. 2C_{yr,\theta} \left(\dfrac{L_r r_{x,\theta}}{v_{x,\theta}} - \beta_\theta \right) - r_{x,\theta} m v_{x,\theta} \right] \Delta t / m v_{x,\theta} \\ C_{yf,\theta} \\ C_{yr,\theta} \end{bmatrix} \tag{3.16}$$

$$a_{y,\theta+1} = a_{y,\theta} + \left[\frac{F_{yf,\theta} \cos\theta + F_{yr,\theta}}{m} \right] \Delta t \tag{3.17}$$

The CKF stands as a pivotal advancement in nonlinear state estimation, overcoming the limitations of the traditional EKF. Developed in response to the growing need for accurate estimation in systems exhibiting significant nonlinearities, CKF utilizes sigma-point filters to approximate the posterior distribution of state variables without resorting to linearization. Unlike EKF, which hinges on linear approximations prone to inaccuracies in nonlinear regimes, CKF integrates deterministic cubature rules or other sigma-point methods to propagate and update state estimates robustly through nonlinear process and measurement models. This approach not only enhances estimation accuracy by more faithfully approximating the true posterior distribution but also improves computational efficiency by circumventing the need for Jacobian matrices and numerical differentiation, thereby catering to real-time applications in domains such as robotics, aerospace, automotive, and finance. The theoretical underpinnings of CKF involve selecting sigma points based on the current state estimate and covariance, followed by predicting the state using nonlinear process models and propagating these sigma points through the state function. Subsequently, the algorithm estimates the covariance of the predicted state and incorporates measurement information to update the state estimate, thus refining the estimation accuracy iteratively. This methodological framework, rooted in the principles of sigma-point filters, has revolutionized state estimation by offering a more robust alternative to traditional methods, particularly in scenarios where nonlinear dynamics prevail.

Advantages of CKF include its ability to handle highly nonlinear systems more effectively than EKF, ensuring better accuracy without sacrificing computational efficiency. This capability has fueled its widespread adoption across diverse fields, including robotics for localization and navigation, aerospace for satellite trajectory prediction and attitude determination, automotive for autonomous vehicle control in dynamic environments, and finance for forecasting market trends based on nonlinear dynamics. Despite its strengths, CKF faces challenges such as complex implementation requiring careful sigma-point selection and assumptions of Gaussian distributions, which may not always hold true in practice, potentially affecting estimation accuracy. Recent developments in CKF research focus on enhancing adaptability through adaptive sigma-point selection strategies, integrating with deep learning techniques to improve estimation performance in complex scenarios, and exploring parallelization methods to optimize computational efficiency for real-time applications. As these advancements continue to evolve, CKF remains at the forefront of nonlinear state estimation, pushing boundaries and setting new standards in accuracy, reliability, and applicability across a wide range of dynamic systems and environments. Based on the RCKF framework [53], the iterative procedure of the RCKF with a T-S fuzzy system is depicted as follows.

3.3.2.1 Initialization

$$\hat{x}_0 = E(x_0) \tag{3.18}$$

$$P_0 = E\left[(x_0 - \hat{x}_0)(x_0 - \hat{x}_0)^T\right] \tag{3.19}$$

where E is a mathematical expectation and P is the covariance matrix of x.

The cubature sampling points τ_i can be updated using Eq. (3.20), and weight ϕ_i is equal to $\frac{1}{c}$.

$$\phi_i = \frac{1}{c}, \tau_i = \sqrt{\frac{c}{2}} \left[\begin{pmatrix} 1 \\ 0 \\ \vdots \\ 0 \end{pmatrix} \cdots \begin{pmatrix} 0 \\ 0 \\ \vdots \\ 1 \end{pmatrix} \begin{pmatrix} -1 \\ 0 \\ \vdots \\ 0 \end{pmatrix} \cdots \begin{pmatrix} 0 \\ 0 \\ \vdots \\ -1 \end{pmatrix} \right] \quad i = 1, 2, \ldots c, \quad c = 2n \tag{3.20}$$

where n is the dimension of x and c is the total number of cubature points.

3.3.2.2 Time Update

Singular value decomposition of $P_{\theta-1/\theta-1}$ is computed as

$$P_{\theta-1/\theta-1} = U \begin{bmatrix} S & 0 \\ 0 & 0 \end{bmatrix} V^T \tag{3.21}$$

$$P_{\theta-1/\theta-1} = U_{\theta-1/\theta-1} S_{\theta-1/\theta-1} V^T_{\theta-1/\theta-1} \tag{3.22}$$

where S is a diagonal matrix and $P_{\theta-1/\theta-1}$ is the symmetric covariance matrix.

The cubature points can be obtained from Eq. (3.23)

$$\chi^{(i)}_{\theta-1/\theta-1} = S_{\theta-1/\theta-1}\tau_i + \hat{x}_{\theta-1/\theta-1} \tag{3.23}$$

where $\hat{x}_{\theta-1/\theta-1}$ is the estimated state at a time step $\theta-1$. $\chi^{(i)}_{\theta-1/\theta-1}$ is the cubature point of $\hat{x}_{\theta-1/\theta-1}$.

The propagated cubature points can be calculated via Eq. (3.24)

$$\chi^{*(i)}_{\theta/\theta-1} = f\left(\chi^{(i)}_{\theta-1/\theta-1}, u_{\theta-1}\right) \tag{3.24}$$

$\hat{x}_{\theta/\theta-1}$ and $P_{\theta/\theta-1}$ are updated using Eqs. (3.25) and (3.26), respectively.

$$\hat{x}_{\theta/\theta-1} = \sum_{i=1}^{c} \phi_i \chi^{*(i)}_{\theta-1/\theta-1} \tag{3.25}$$

$$P_{\theta/\theta-1} = \sum_{i=1}^{c} \phi_i \chi^{*(i)}_{\theta/\theta-1} \chi^{*(i)T}_{\theta/\theta-1} - \hat{x}_{\theta/\theta-1}\hat{x}^T_{\theta/\theta-1} + Q_{\theta-1} \tag{3.26}$$

3.3.2.3 Measurement Update

Singular value decomposition of $P_{\theta/\theta-1}$ is calculated as

$$P_{\theta/\theta-1} = U_{\theta/\theta-1} S_{\theta/\theta-1} V^T_{\theta/\theta-1} \tag{3.27}$$

The cubature points can be calculated using Eq. (3.28)

$$\chi^{(i)}_{\theta/\theta-1} = S_{\theta/\theta-1}\tau_i + \hat{x}_{\theta/\theta-1} \tag{3.28}$$

The propagated cubature points can be given by Eq. (3.29)

$$Z^{(i)}_{\theta/\theta-1} = h\left(\chi^{(i)}_{\theta/\theta-1}, u_\theta\right) \tag{3.29}$$

The predicted measurement vector $\hat{z}_{\theta/\theta-1}$ can be computed via Eq. (3.30). The innovation covariance matrix $P_{zz,\theta/\theta-1}$ and the cross-covariance matrix $P_{xz,\theta/\theta-1}$ are updated by Eqs. (3.31) and (3.32), respectively.

$$\hat{z}_{\theta/\theta-1} = \sum_{i=1}^{c} \phi_i Z^{(i)}_{\theta-1/\theta-1} \tag{3.30}$$

$$P_{zz,\theta/\theta-1} = \sum_{i=1}^{c} \phi_i Z^{(i)}_{\theta/\theta-1} Z^{(i)T}_{\theta/\theta-1} - \hat{z}_{\theta/\theta-1}\hat{z}^T_{\theta/\theta-1} + R_\theta \tag{3.31}$$

$$P_{xz,\theta/\theta-1} = \sum_{i=1}^{c} \phi_i \chi_{\theta/\theta-1}^{(i)} Z_{\theta/\theta-1}^{(i)T} - \hat{x}_{\theta/\theta-1} \hat{z}_{\theta/\theta-1}^{T} \tag{3.32}$$

The gain matrix W_θ can be calculated using Eq. (3.33) and the updated state $\hat{x}_{\theta/\theta}$ can be calculated using Eq. (3.34).

$$W_\theta = P_{xz,\theta/\theta-1} P_{zz,\theta/\theta-1}^{-1} \tag{3.33}$$

$$\hat{x}_{\theta/\theta} = \hat{x}_{\theta/\theta-1} + W_\theta (z_\theta - \hat{z}_{\theta/\theta-1}) \tag{3.34}$$

In the measurement update step, error covariance $P_{\theta/\theta}$ needs to be further updated using H_∞ filter (see Eq. (3.35)).

$$P_{\theta/\theta} = P_{\theta/\theta-1} - \left\{ \begin{bmatrix} P_{xz,\theta/\theta-1} & P_{\theta/\theta-1} \end{bmatrix} * \begin{bmatrix} P_{zz,\theta/\theta-1} - R_\theta + I & P_{xz,\theta/\theta-1}^T \\ P_{xz,\theta/\theta-1} & P_{\theta/\theta-1} - \gamma^2 I \end{bmatrix}^{-1} * \begin{bmatrix} P_{xz,\theta/\theta-1}^T \\ P_{\theta/\theta-1}^T \end{bmatrix} \right\} \tag{3.35}$$

where I represents an identity matrix.

To ensure the existence of H_∞ filter, $P_{\theta/\theta}$ must be positive definite, which is determined by γ (see Eq. (3.36)). As described in reference [54], γ should meet the following inequality.

$$P_{\theta/\theta}^{-1} = P_{\theta/\theta-1}^{-1} + H_\theta^T R_\theta^{-1} H_\theta - \gamma^2 I > 0 \tag{3.36}$$

$$\gamma^2 > \max \left\{ eig\left(P_{\theta/\theta-1}^{-1} + H_\theta^T R_\theta^{-1} H_\theta \right)^{-1} \right\} \tag{3.37}$$

where $\max\{eig(\cdot)^{-1}\}$ is referred to as the maximum eigenvalue of the matrix. Thus, the value γ can be obtained from Eq. (3.38)

$$\gamma^2 = \Omega \max \left\{ eig\left(P_{\theta/\theta-1}^{-1} + H_\theta^T R_\theta^{-1} H_\theta \right)^{-1} \right\} \tag{3.38}$$

where Ω is as scalar larger than one.

According to the reference [55], we can approximate the linearized measurement equation $H_\theta^T = P_{\theta/\theta-1}^{-1} P_{xz,\theta/\theta-1}$ and substitute it into Eq. (3.38) to derive

$$\gamma^2 = \Omega \max \left\{ eig\left(P_{\theta/\theta-1}^{-1} + P_{\theta/\theta-1}^{-1} P_{xz,\theta/\theta-1} R_\theta^{-1} \left[P_{\theta/\theta-1}^{-1} P_{xz,\theta/\theta-1} \right]^T \right)^{-1} \right\} \tag{3.39}$$

Also, the process noise variance matrix Q_θ should be adaptively adjusted in the RCKF algorithm. Referring to the relevant conclusions in [56], we propose the adjustment method of Q_θ as follows:

$$Q_{\theta,new} = \rho(e)Q_{\theta,old} \tag{3.40}$$

where ρ is a positive number. ρ takes a large value if the difference between the actual measured value and their estimated values is large; otherwise, it takes a small value. As shown in Eq. (3.40), it is challenging to select ρ, which is associated with e. e is computed via Eq. (3.41).

$$e = (z_\theta - \hat{z}_{\theta/\theta-1})^T(z_\theta - \hat{z}_{\theta/\theta-1}) \tag{3.41}$$

A fuzzy system is used to adaptively adjust the parameter ρ, which helps improve the estimation accuracy. To implement the fuzzy inference system, three fuzzy sets are considered as S = small, M = medium, and B = big. The membership function is shown in Fig. 3.3.

A T–S fuzzy system is used to adjust the process noise variance matrix Q_θ with the general rules as

R_j: If x_1 is A_1^j and x_2 is A_2^j and … and x_n is A_n^j, then $y_j = f_j(X)$ $j = 1, 2 \ldots M$, $X = (x_1, x_2, x_3, \ldots x_n)^T$

where x_i is the input variable, $i = 1, 2 \cdots n$, A_i^j is the fuzzy subset, M is the total number of fuzzy rules, $f_j(\cdot)$ is a consequent function of rule R_j, and y_j is the output. The degree of activation of the jth rule is given by

$$\mu_j(X) = \prod_{i=1}^{n} \mu_{A_i^j}(x_i) \tag{3.42}$$

where $\mu_{A_i^j}(x_i)$ is the membership function of x_i in A_i^j.

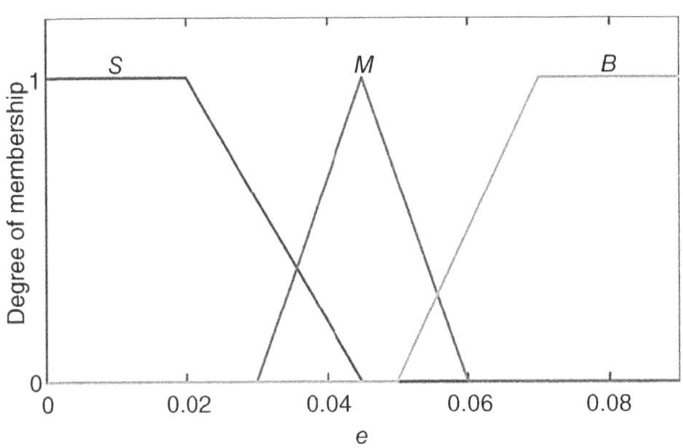

Figure 3.3 Membership function for fuzzy rules.

The output Y is computed by aggregating the individual rule's contributions.

$$Y(X) = \frac{\sum_{j=1}^{M} \mu_j(X) f_j(X)}{\sum_{j=1}^{M} \mu_j(X)} \tag{3.43}$$

In this work, we define three rules as follows:

R_1 : If e is small then $y_1(e) = 0.3$

R_2 : If e is medium then $y_2(e) = 1$

R_3 : If e is big then $y_3(e) = 3$

The updated formula for final ρ is given by

$$\rho(e) = \frac{\sum_{t=1}^{3} \mu_t(e) y_t(e)}{\sum_{t=1}^{3} \mu_t(e)} \tag{3.44}$$

Based on the fuzzy rules and membership functions, Q_θ can be adaptively updated. The pseudocode of the FARCKF algorithm is provided by Algorithm 3.1.

Algorithm 3.1 FARCKF Algorithm

Initialize: Set $\hat{x}_0 = E(x_0)$ $P_0 = E\left[(x_0 - \hat{x}_0)(x_0 - \hat{x}_0)^T\right]$

The cubature sampling points τ_i and weights ϕ_i can be updated using Eq. (3.20).

```
for θ ← 0 to T do
    Estimate vehicle mass parameter using Eqs. (3.11)~(3.14)
    for each time update process do
        Singular value decomposition P_{θ-1/θ-1} using Eqs. (3.21) and (3.22)
        Calculate the cubature points using Eqs. (3.23) and (3.24)
        Estimate x̂_{θ/θ-1} and P_{θ/θ-1} using Eqs. (3.25) and (3.26)
    end for
    for each measurement update process do
        Singular value decomposition P_{θ/θ-1} using Eq. (3.27)
        Calculate the cubature points using Eqs. (3.28) and (3.29)
        Calculate ẑ_{θ/θ-1}, P_{zz,θ/θ-1}, and P_{xz,θ/θ-1} using Eqs. (3.30)~(3.32)
        Calculate W_θ and x̂_{θ/θ} using Eqs. (3.33) and (3.34)
        Further update the error covariance P_{θ/θ} using Eq. (3.35)
    end for
    Update process noise variance matrix Q_θ using Eqs. (3.40) and (3.41) and
  (3.44)
end for
```

3.3.3 Simulation and Experiment Tests

To validate the effectiveness of the FARCKF method, comprehensive CarSim-MATLAB simulations and real-world vehicle experiments were conducted. The study compared FARCKF against other methods like CKF, RCKF, and ARCKF (RCKF with vehicle mass estimation) in estimating sideslip angle and tire cornering stiffness. In both simulation and real vehicle tests conducted on dry and wet asphalt roads, FARCKF consistently demonstrated superior performance. The DGPS provided accurate reference values for sideslip angle measurements, while tire cornering stiffness reference values were sourced from suppliers. The simulations and experiments underscored FARCKF's capability to dynamically adjust vehicle mass parameters using RLS, enhance accuracy through the RCKF, and adapt process noise with the T-S fuzzy system. These features collectively improved estimation accuracy under varying road conditions, highlighting FARCKF's potential for enhancing vehicle control and safety systems in practical applications.

3.3.3.1 Double Lane Change Test on High Friction Coefficient Road

The tire–road friction coefficient remains constant at 0.85, while the initial vehicle speed is set at 40 km/h for this test scenario. Figs. 3.4 and 3.5 illustrate the dynamic behaviors of the vehicle speed, front wheel angle, and lateral acceleration. Analysis of these figures reveals minimal fluctuation in vehicle speed throughout the test period. Notably, the evolutions of the front wheel angle and lateral acceleration exhibit striking similarities, showcasing nearly identical patterns over time.

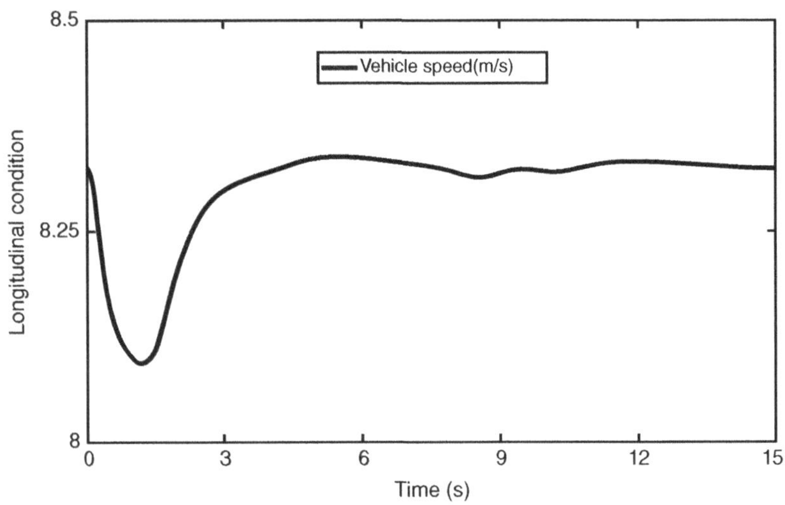

Figure 3.4 Vehicle speed on high friction coefficient road.

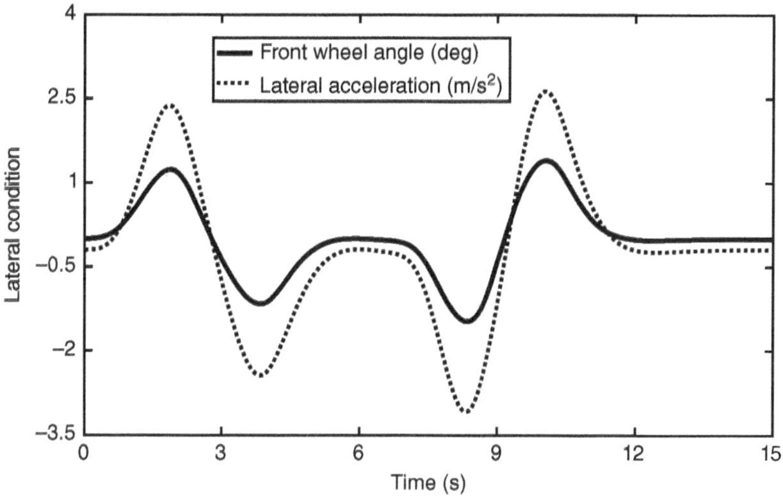

Figure 3.5 Lateral driving condition on high friction coefficient road.

This synchronous behavior underscores the coordinated response of the vehicle's steering and lateral dynamics under the specified testing conditions.

To assess the accuracy of the vehicle mass estimation algorithm, we initialized the vehicle with a nominal mass of 1195 kg, while the actual mass of the vehicle is 995 kg. Figures 3.6 and 3.7 present detailed dynamics of the vehicle mass estimation and associated parameters throughout the experimental period. In Fig. 3.6, the vehicle mass estimation curve exhibits a rapid decline during

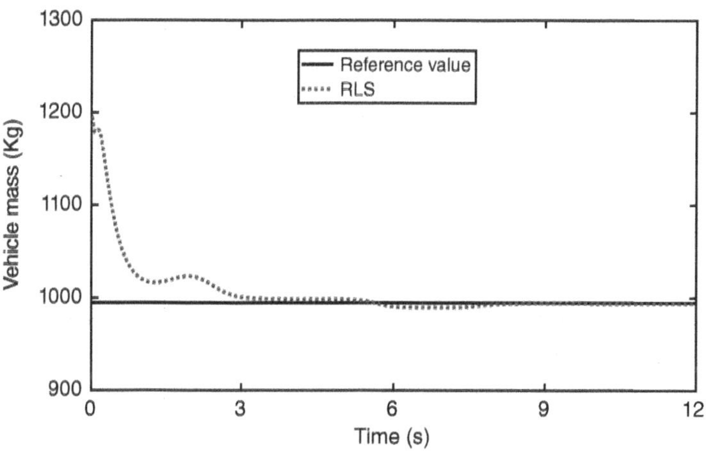

Figure 3.6 Vehicle mass estimation on high friction coefficient road.

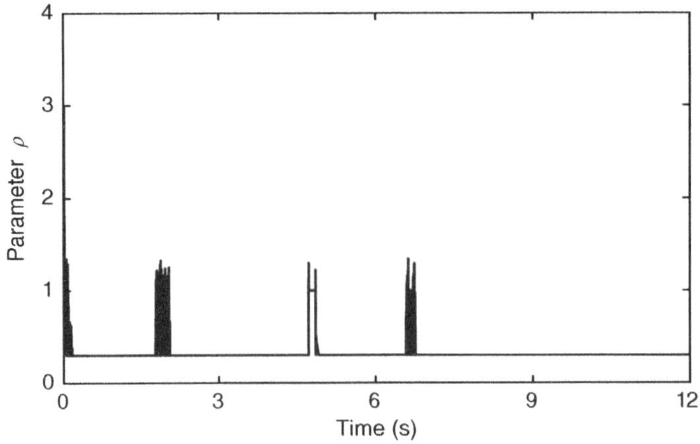

Figure 3.7 The parameter on high friction coefficient road.

the initial 0–3.1 seconds, reflecting the algorithm's adjustment phase to match the actual vehicle mass. Subsequently, the curve stabilizes with slight fluctuations around the reference value of 995 kg, ultimately converging to this value. This behavior illustrates the algorithm's capability to adapt and correct its estimate over time, showcasing its effectiveness in accurately estimating the true vehicle mass despite initial discrepancies. Figure 3.7 focuses on the initial conditions of the estimated sideslip angle and tire cornering stiffness, which are initialized with random values. Consequently, there is a noticeable deviation between the predicted and measured values of lateral acceleration at the beginning of the simulation. This discrepancy underscores the algorithm's transient response as it initializes and calibrates these parameters based on real-time data inputs and model predictions. These figures collectively demonstrate the algorithm's ability to converge toward accurate estimations of vehicle mass and dynamic parameters over time, highlighting its reliability in practical applications for vehicle dynamics and control. Therefore, the value of ρ is relatively large at the beginning. As the estimation error decreases, ρ is rapidly reduced to a small value. In the subsequent estimation process, if there is an error between the estimated value and the true value of the lateral acceleration, ρ can dynamically adjust its value such that the estimation accuracy can be guaranteed.

The estimation results of front tire cornering stiffness and rear tire cornering stiffness obtained from the RLS, RCKF, ARCKF, and FARCKF methods are compared in Figs. 3.8 and 3.9. RLS serves as a benchmark method for estimating tire cornering stiffness, making it a suitable comparison reference instead of the CKF. In the experiments, we consistently use RLS alongside our methods to evaluate the effectiveness of tire cornering stiffness estimation across varying

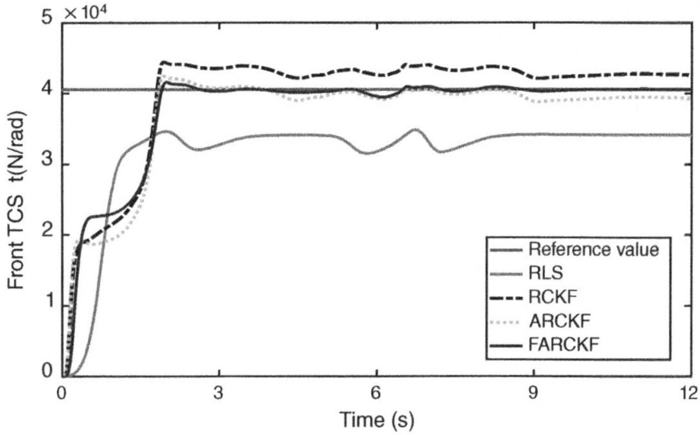

Figure 3.8 Front tire cornering stiffness estimation on high friction coefficient road.

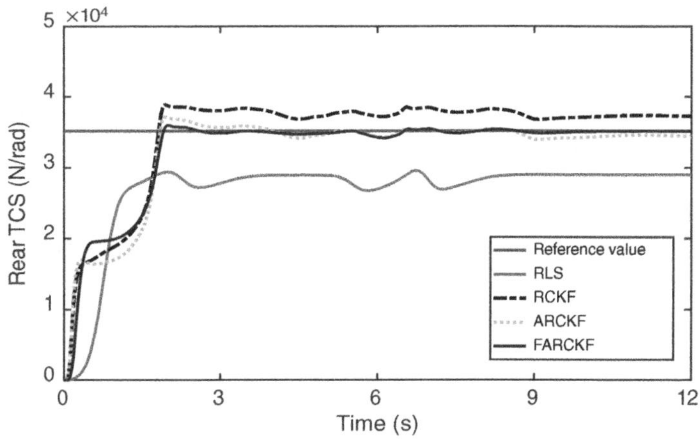

Figure 3.9 Rear tire cornering stiffness estimation on high friction coefficient road.

operational conditions. Figures 3.8 and 3.9 illustrate that the estimated front tire cornering stiffness achieved by the FARCKF algorithm converges to approximately 40,500 N/rad, while the estimated rear tire cornering stiffness approaches about 35,200 N/rad. These values closely align with the reference measurements. Notably, the FARCKF method demonstrates the closest proximity to the reference values compared to other methods. Moreover, the ARCKF method also exhibits superior estimation accuracy compared to RCKF. This enhancement can be attributed to ARCKF's capability to dynamically update the vehicle mass parameter during the estimation process. In contrast, RCKF utilizes a static mass parameter, which can lead to less accurate estimates, especially under varying vehicle conditions. The effectiveness of ARCKF and FARCKF lies in their adaptive mechanisms, particularly in adjusting to real-time changes in vehicle dynamics and environmental conditions. This adaptability ensures that the estimation algorithms can maintain accuracy even when faced with uncertainties or variations in the operational environment. By continuously updating the vehicle mass parameter and employing robust filtering techniques, ARCKF and FARCKF not only enhance the precision of tire cornering stiffness estimation but also improve the overall reliability of vehicle state estimation in dynamic scenarios. These advancements underscore the practical utility of adaptive and robust estimation methods in modern vehicle dynamics and control applications.

The sideslip angle estimation results obtained from the CKF, RCKF, ARCKF, and FARCKF methods are presented in Fig. 3.10. This figure provides a visual comparison of how each method performs in estimating the sideslip angle under controlled conditions. To quantitatively assess the performance, Table 3.3 displays the Mean Absolute Error (MAE) and RMSE metrics for each estimation method. These metrics are crucial for evaluating the accuracy and reliability of the estimation algorithms. It is observed that the ARCKF method consistently outperforms both RCKF and CKF in terms of estimation accuracy, as indicated by lower MAE and RMSE values. The FARCKF method, incorporating a fuzzy adaptive approach to adjust the process noise dynamically, further enhances the estimation accuracy compared to ARCKF. Specifically, Table 3.3 highlights that the FARCKF achieves MAE and RMSE values of 0.0160 and 0.0166, respectively. These metrics indicate a significant improvement in sideslip angle estimation accuracy, demonstrating at least a 54% enhancement compared to CKF and

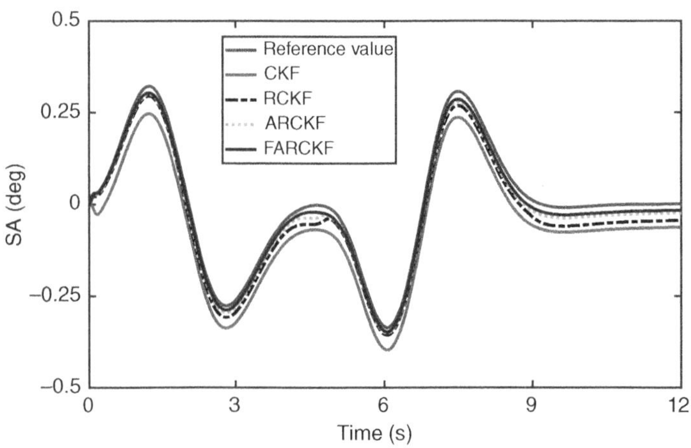

Figure 3.10 Sideslip angle estimation on high friction coefficient road.

Table 3.3 Estimation error on high friction coefficient road.

Symbol	MAE	RMSE
CKF	0.0660	0.0662
RCKF	0.0349	0.0366
ARCKF	0.0221	0.0230
FARCKF	**0.0160**	**0.0166**

RCKF. The superiority of FARCKF can be attributed to its ability to adaptively adjust the filtering parameters based on real-time sensor data and environmental conditions. By dynamically tuning the process noise parameters through a fuzzy logic framework, FARCKF mitigates the impact of uncertainties and disturbances, thereby improving the robustness and accuracy of sideslip angle estimation.

3.3.3.2 Double Lane Change Test on Low Friction Coefficient Road

The tire–road friction coefficient is set to 0.2, and the initial vehicle speed is 30 km/h. Figure 3.11 visually illustrates the dynamic evolution of the vehicle speed throughout the test period. It is evident from the figure that the vehicle speed undergoes significant fluctuations, characteristic of low-friction road conditions where traction is limited. In Fig. 3.12, the variations in the front wheel angle and lateral acceleration are presented. These curves differ notably from those depicted in Fig. 3.5, which showcased data from a higher friction coefficient road. On low-friction surfaces, vehicles tend to operate more frequently within

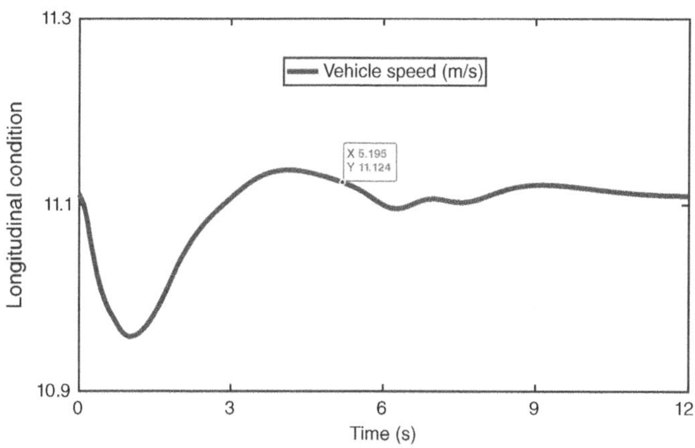

Figure 3.11 Vehicle speed on low friction coefficient road.

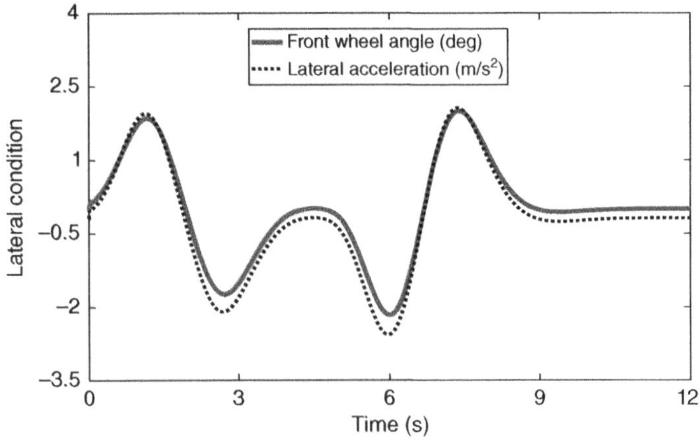

Figure 3.12 Lateral driving condition on low friction coefficient road.

nonlinear regimes, where small changes in conditions can lead to pronounced effects on vehicle dynamics. The front wheel angle and lateral acceleration curves in Fig. 3.12 exhibit non-consistent patterns. Concurrently, the lateral acceleration fluctuates more prominently, underscoring the dynamic challenges imposed by the low friction coefficient environment.

The estimated vehicle mass curve drops rapidly from 0 second to about 3.9 seconds and then gradually stabilizes around 995 kg, as shown in Fig. 3.13. The dynamic change of parameter ρ in the FARCKF algorithm is shown in Fig. 3.14. At the beginning of the simulation, since the initial values of the estimated sideslip angle and tire cornering stiffness are three random numbers, there is a

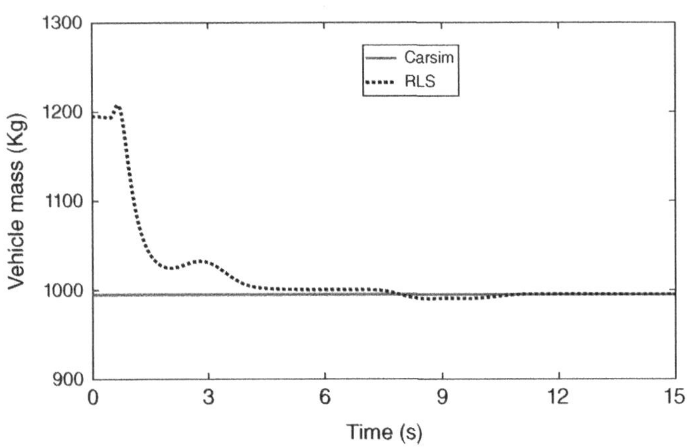

Figure 3.13 Vehicle mass estimation on low friction coefficient road.

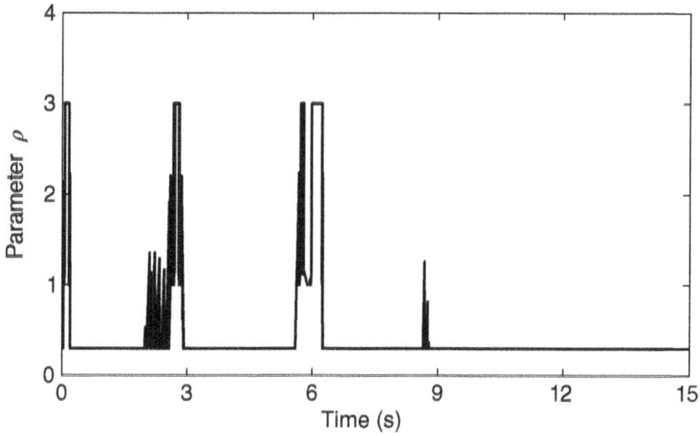

Figure 3.14 The parameter on low friction coefficient road.

large deviation between the measured value and the predicted value of the lateral acceleration. Therefore, the value of the parameter ρ is relatively large at the beginning of the simulation. Different from Fig. 3.7, ρ has multiple larger values in Fig. 3.14, which indicates that the larger estimation error occurs more than once in the subsequent estimation process.

The estimation results of front and rear tire cornering stiffness from various methods are analyzed in Figs. 3.15 and 3.16. The figures illustrate that the FACKF algorithm achieves a convergence of approximately 37,670 N/rad for front tire cornering stiffness and about 32,740 N/rad for rear tire cornering stiffness. Notably, the estimated values obtained by FARCKF closely approach the reference

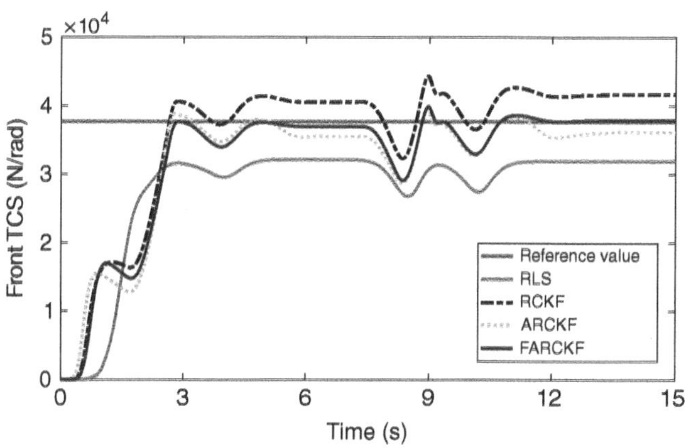

Figure 3.15 Front tire cornering stiffness estimation on low friction coefficient road.

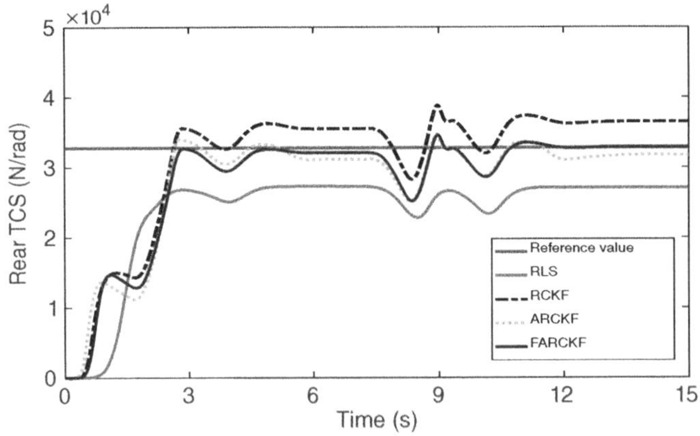

Figure 3.16 Rear tire cornering stiffness estimation on low friction coefficient road.

values, demonstrating its superior accuracy in comparison to CKF, ARCKF, and ARKF methods. On low friction coefficient roads, as depicted in these figures, the curves depicting tire cornering stiffness estimation exhibit more pronounced fluctuations compared to those observed on higher friction coefficient roads. This variability is attributed to the vehicle operating within nonlinear regions more frequently on low-friction surfaces. The dynamic nature of tire behavior under these conditions necessitates robust estimation techniques to accurately capture and predict tire cornering stiffness variations.

Figure 3.17 further validates the effectiveness of the FARCKF method by illustrating its estimation curve in close proximity to the reference values, outperforming CKF, ARCKF, and ARKF. Table 3.4 provides MAE and RMSE

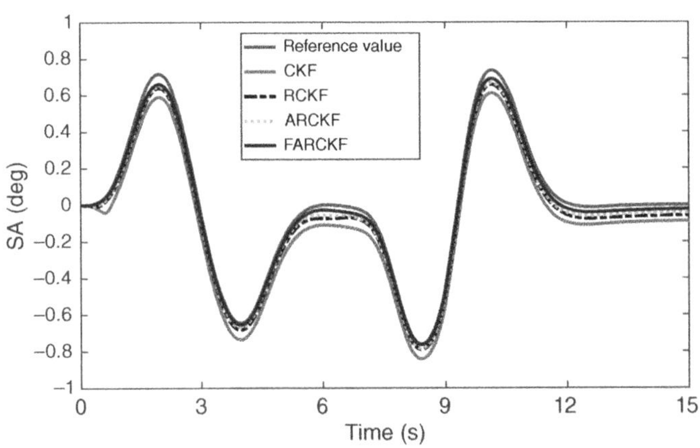

Figure 3.17 Sideslip angle estimation on low friction coefficient road.

Table 3.4 Estimation error on low friction coefficient road.

Symbol	MAE	RMSE
CKF	0.1001	0.1022
RCKF	0.0537	0.0560
ARCKF	0.0431	0.0455
FARCKF	**0.0255**	**0.0286**

metrics for comparison. The proposed FARCKF algorithm achieves MAE and RMSE values of 0.0255 and 0.0286, respectively, indicating significant improvements in estimation accuracy for tire cornering stiffness compared to traditional CKF and RCKF methods. Specifically, the FARCKF method enhances sideslip angle estimation accuracy by over 48%, highlighting its robustness and superior performance across different road friction conditions. Furthermore, the consistency in estimation accuracy between sideslip angle and tire cornering stiffness under varying road friction coefficients underscores the robustness of the FARCKF method. This resilience suggests that FARCKF effectively adapts to changes in tire-road friction coefficients, crucial for maintaining reliable vehicle dynamics estimation in diverse driving environments.

3.3.3.3 The Real Vehicle Test on the Dry Asphalt Road

In the double lane change test, the steering control of the vehicle is completed by an experienced driver. Figure 3.18 shows the dynamic change of vehicle speed

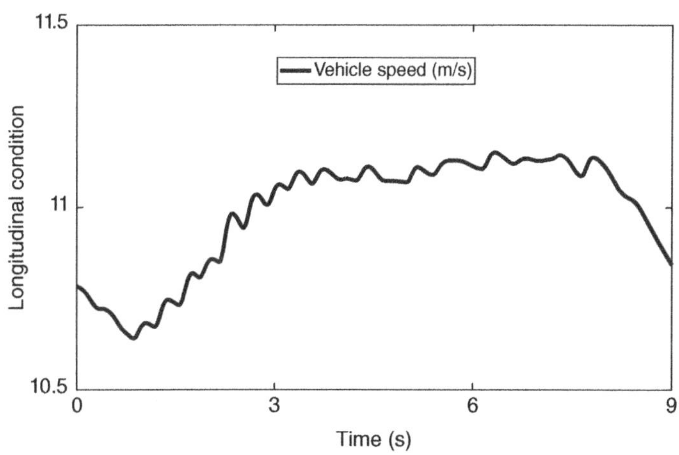

Figure 3.18 Vehicle speed on dry asphalt road.

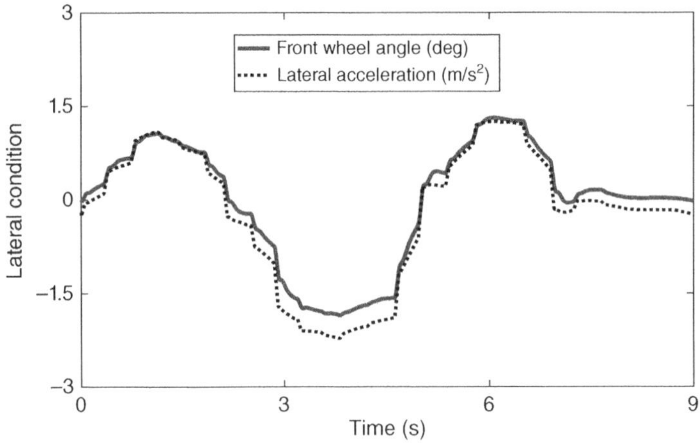

Figure 3.19 Lateral driving condition on dry asphalt road.

during the test. The fluctuation of the vehicle speed is larger than that of the simulation. This is because it is difficult for the driver to ensure the stability of the vehicle speed when steering. The front wheel angle and lateral acceleration are shown in Fig. 3.19. Similar to Fig. 3.5, the trends of the two curves are almost identical.

Figures 3.20 and 3.21 show the dynamic variations of the vehicle mass estimation and the parameter ρ. The real vehicle mass is 995 kg, so the initial vehicle mass is set to be 995 kg in the estimation process. Three experimenters are loaded on the electric vehicle, whose total mass is 195 kg. Thus, the total mass of the vehicle and the experimenters is 1190 kg. From Fig. 3.20, we can observe that the

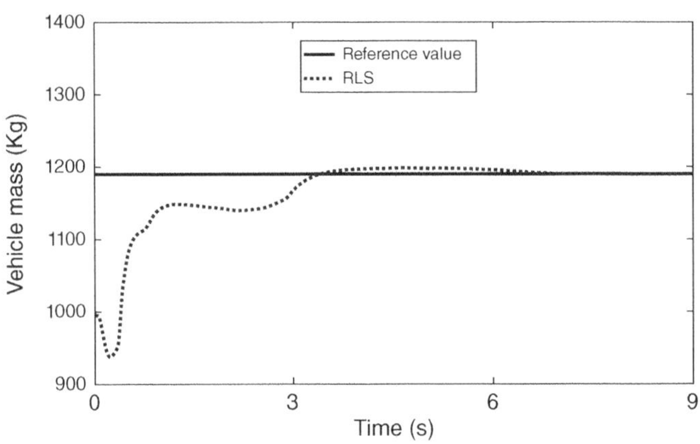

Figure 3.20 Vehicle mass estimation on dry asphalt road.

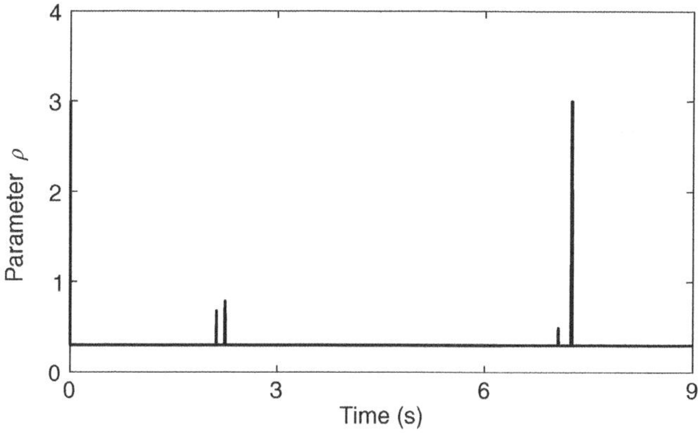

Figure 3.21 The parameter on dry asphalt road.

estimated vehicle mass curve rises rapidly from 0 second to 3.6 seconds and then gradually stabilizes around 1190 kg. In Fig. 3.21, at the beginning of the experiment, since the initial values of the estimated sideslip angle and tire cornering stiffness are three random numbers, there is a large difference between the measured value and the predicted value of the lateral acceleration. Therefore, the value of the parameter ρ is relatively large at the beginning.

The estimated tire cornering stiffness curves reveal slight fluctuations, underscoring the challenges associated with achieving smooth and consistent steering using manual control methods. As illustrated in Figs. 3.22 and 3.23, the application of the FARCKF algorithm demonstrates its capability to effectively converge toward precise estimates of tire properties. Specifically, the algorithm successfully

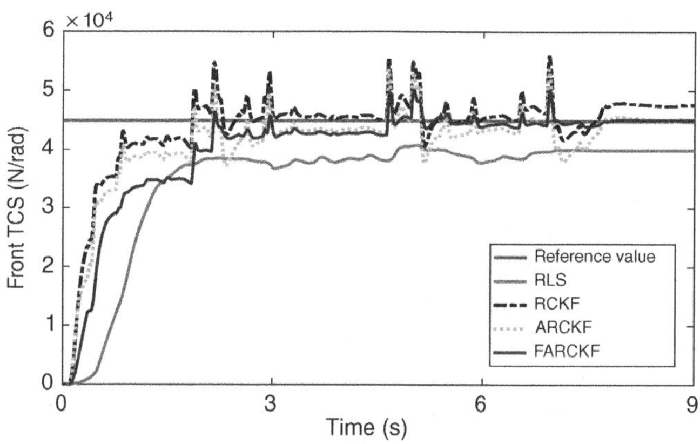

Figure 3.22 Front tire cornering stiffness estimation on dry asphalt road.

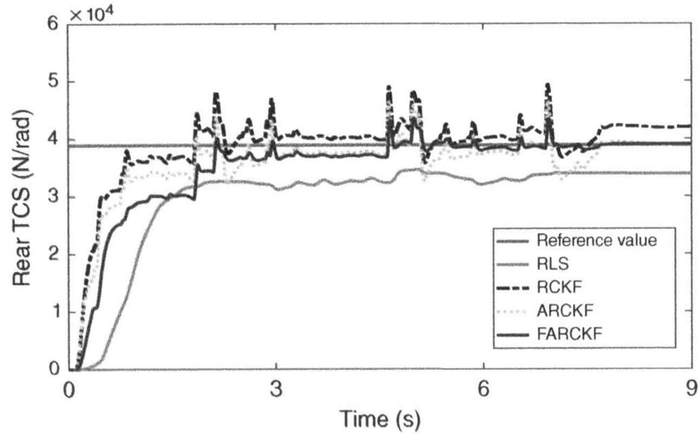

Figure 3.23 Rear tire cornering stiffness estimation on dry asphalt road.

converges to an estimated front tire cornering stiffness of 44,800 N/rad and a rear tire cornering stiffness approaching 38,800 N/rad. These values closely approximate the reference values derived from extensive simulation tests, validating the algorithm's robustness and accuracy in estimating critical parameters crucial for vehicle dynamics and control. Moreover, the comparison between estimated and reference values highlights the FARCKF algorithm's superior performance in minimizing estimation errors, thereby enhancing the reliability of vehicle state prediction and control strategies. By mitigating fluctuations in estimated tire characteristics, the algorithm contributes significantly to improving the overall stability and responsiveness of the vehicle under varying driving conditions. Furthermore, the alignment between estimated and simulated results underscores the algorithm's ability to adapt to real-world complexities and variations, such as dynamic changes in tire behavior due to road conditions and vehicle load.

In Fig. 3.24, the estimated sideslip angle curves are not smooth. This is due to the fact that manual steering cannot achieve smooth and steady steering. The MAE and RMSE of different estimation methods are shown in Table 3.5. The MAE and RMSE indexes of the proposed algorithm are 0.0166 and 0.0178, respectively. The estimation accuracy of the sideslip angle is at least improved by more than 62% compared with the existing CKF and RCKF. In summary, the errors of other estimation methods are larger than that of our proposed algorithm since they do not take into account the dynamics of the vehicle mass parameter or the process noise parameter. During the real-world vehicle experiments, an adept driver executes the lane-changing maneuver, as showcased in Fig. 3.21. The dynamic alteration of the steering wheel angle is visibly more pronounced compared to the simulation test. This is because the driver cannot perform smooth and stable steering maneuvers like a virtual driver in a simulation test.

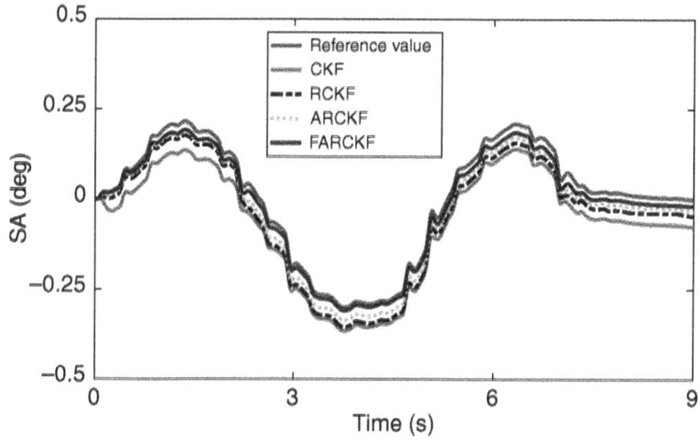

Figure 3.24 Sideslip angle estimation on dry asphalt road.

Table 3.5 Estimation error of sideslip angle on dry asphalt road.

Symbol	MAE	RMSE
CKF	0.0699	0.0706
RCKF	0.0464	0.0478
ARCKF	0.0322	0.0332
FARCKF	**0.0166**	**0.0178**

3.3.3.4 The Real Vehicle Test on the Wet Asphalt Road

Figures 3.25 provides a visual representation of the dynamic variations in vehicle speed observed during the testing phase. It is noteworthy that the fluctuations in vehicle speed are notably greater than those observed in simulations. This discrepancy underscores the complexities and unpredictable nature of real-world driving conditions compared to controlled simulation environments. In Fig. 3.26, the plots of front wheel angle and lateral acceleration further highlight the challenges encountered during testing. Unlike the trends depicted in Fig. 3.19, these curves exhibit inconsistencies and deviations. The irregularities stem from the vehicle's operation on a wet asphalt road, where conditions often push the vehicle into nonlinear operating regimes. Such conditions introduce variable friction coefficients and unpredictable surface interactions, significantly influencing the vehicle's handling dynamics.

The estimated vehicle mass curve rises rapidly from 0 second to 5.2 seconds and then gradually stabilizes around 1190 kg, as shown in Fig. 3.27. The dynamic

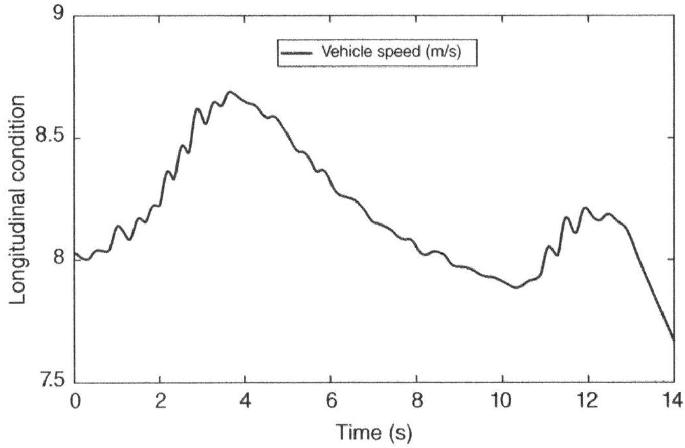

Figure 3.25 Vehicle speed on wet asphalt road.

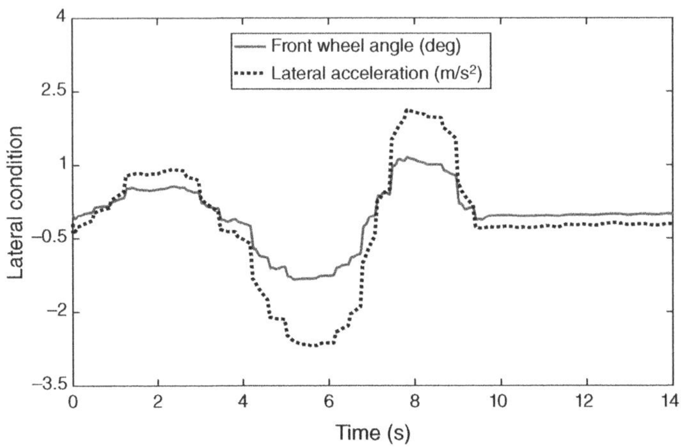

Figure 3.26 Lateral driving condition on wet asphalt road.

change of parameter ρ in the FARCKF algorithm is shown in Fig. 3.28. Similar to Fig. 3.21, ρ is very large at the beginning of the test. However, ρ also reaches larger values more than once in the subsequent estimation process. It shows the larger estimation error occurs more than once when a double lane change test is performed on a wet asphalt road. The estimation results of tire cornering stiffness obtained from different methods are compared in Figs. 3.29 and 3.30. The estimated value of front tire cornering stiffness using the FARCKF algorithm can converge to 43,900 N/rad and the estimated value of rear tire cornering stiffness

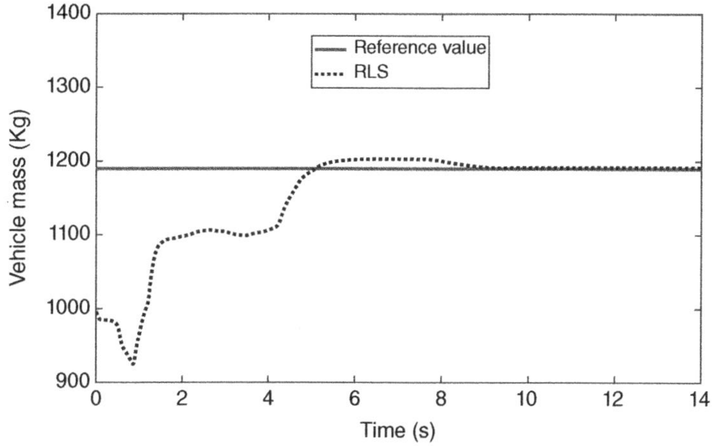

Figure 3.27 Vehicle mass estimation on wet asphalt road.

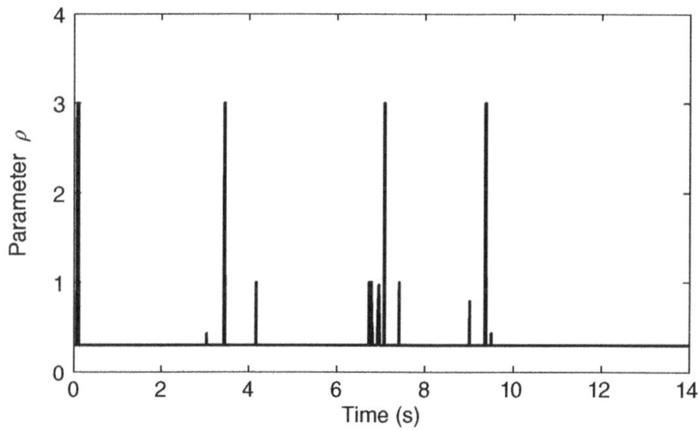

Figure 3.28 The parameter on wet asphalt road.

converges to 38,000 N/rad. The estimated value of FARCKF is also closest to the reference value.

In Fig. 3.31, the plots depicting estimated sideslip angles reveal noticeable irregularities, characterized by non-smooth curves. This phenomenon arises due to the inherent limitations of manual steering, which struggles to maintain consistent smoothness and stability during maneuvers. Table 3.6 provides a comparative analysis of MAE and RMSE metrics across various estimation methods. For our proposed algorithm, the MAE and RMSE values are recorded as 0.0142 and 0.0208, respectively. These results signify a significant enhancement in sideslip angle estimation accuracy—improving by more than 54% compared

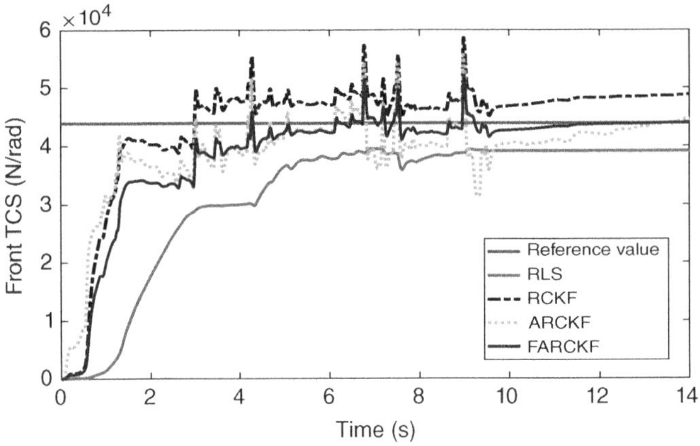

Figure 3.29 Front tire cornering stiffness estimation on wet asphalt road.

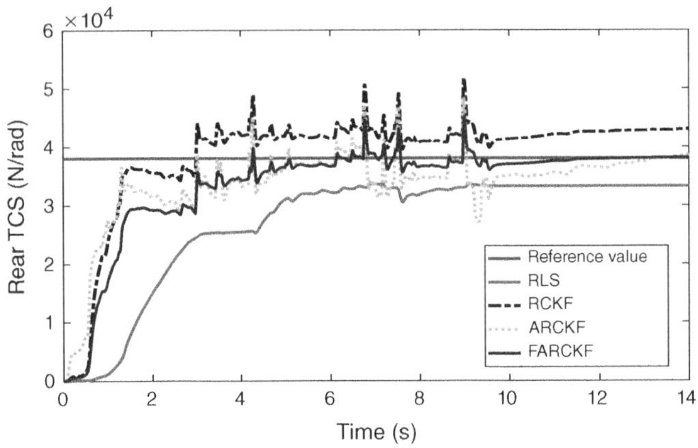

Figure 3.30 Rear tire cornering stiffness estimation on wet asphalt road.

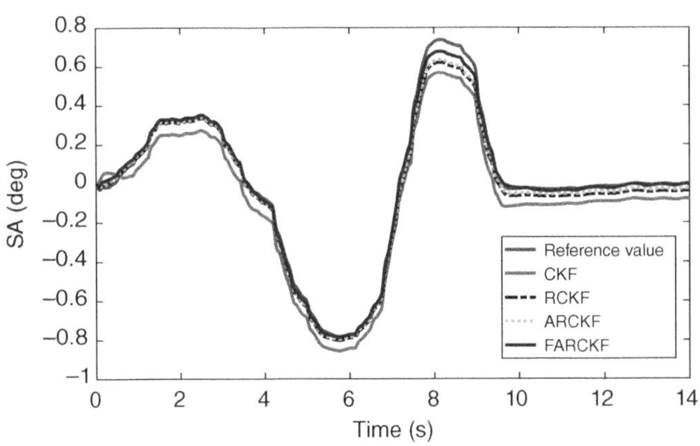

Figure 3.31 Sideslip angle estimation on wet asphalt road.

Table 3.6 Estimation error of sideslip angle on wet Asphalt road.

Symbol	MAE	RMSE
CKF	0.0869	0.0925
RCKF	0.0349	0.0456
ARCKF	0.0269	0.0373
FARCKF	**0.0142**	**0.0208**

to conventional CKF and RCKF methods. This improvement underscores the efficacy of our approach in accurately predicting vehicle dynamics, crucial for advanced driver assistance systems and vehicle stability control. Moreover, the consistency in estimation accuracy between sideslip angle and tire cornering stiffness across different frictional road conditions is notable. This suggests that the FARCKF method exhibits robustness against variations in the tire-road friction coefficient.

3.4 Hybridizing Physical and Data-Driven Methods for Vehicle State

This section introduces an innovative approach known as the FTEKFNet, designed to enhance vehicle state estimation by integrating both physics-based principles and data-driven methodologies. At its core, FTEKFNet leverages the iterative framework of the FTEKF, augmenting it with a pre-trained ANN. This neural network plays a pivotal role by directly predicting the Kalman gain, which is then seamlessly integrated into the FTEKF structure to form FTEKFNet. The key innovation of FTEKFNet lies in its ability to address challenges such as unknown noise characteristics and data loss, which are prevalent in real-world scenarios. By employing the predictive capabilities of the neural network, FTEKFNet offers robust solutions that adapt to varying levels of noise, including colored noise typically encountered in practical applications. Experimental validations conducted under diverse conditions underscore the efficacy of FTEKFNet. These tests demonstrate its capacity to handle uncertainties in real-time data streams, ensuring reliable vehicle state estimation even in challenging operational environments. By simultaneously mitigating the impact of unknown noise sources and accommodating data loss scenarios, FTEKFNet enhances the overall reliability and adaptability of vehicle state estimation systems.

3.4.1 Vehicle Model and Problem Statement

In prior approaches to vehicle state estimation, the yaw rate of the vehicle has often been the primary parameter of interest. However, with advancements in automotive electronics technology, obtaining the yaw rate has become straightforward due to the availability of highly integrated on-board gyroscopes [57]. Consequently, our focus shifts to estimating the longitudinal velocity and sideslip angle, which are crucial for comprehensive vehicle dynamics analysis. To achieve this, we utilize the bicycle model, a widely accepted representation for capturing the dynamic response of a vehicle. The bicycle model, a simplified representation of vehicle dynamics, relies on several key assumptions to characterize the behavior of a vehicle. It represents the vehicle with two wheels, one at the front and one at the rear, assuming lateral symmetry, which means the left and right sides of the vehicle behave identically. The vehicle body is considered rigid, with no deformation under load, and small angle approximations are used, allowing for linear approximations of trigonometric functions. The forces generated by the tires are assumed to be linear with respect to slip angle and slip ratio within a certain range, and parameters such as tire stiffness, vehicle mass, and moments of inertia are considered constant during the analysis. Vertical load transfer between wheels during acceleration, braking, or cornering is typically neglected, as are aerodynamic effects such as drag and downforce. The model also assumes that changes in the steering angle are instantaneous, with no lag or delay. By relying on these assumptions, the bicycle model provides a tractable and effective means of analyzing vehicle dynamics, particularly for control design and state estimation purposes. Figure 3.32 provides a visual representation of the bicycle model, illustrating its key components and layout. The mathematical foundations of the bicycle model are detailed in Eqs. (3.45)–(3.47). These equations describe the relationships between various dynamic parameters, including forces, velocities, and angles, providing a basic model for estimating the longitudinal velocity and sideslip angle. By employing this model, we can accurately characterize the vehicle's behavior under different driving conditions, enhancing the precision of our state estimation.

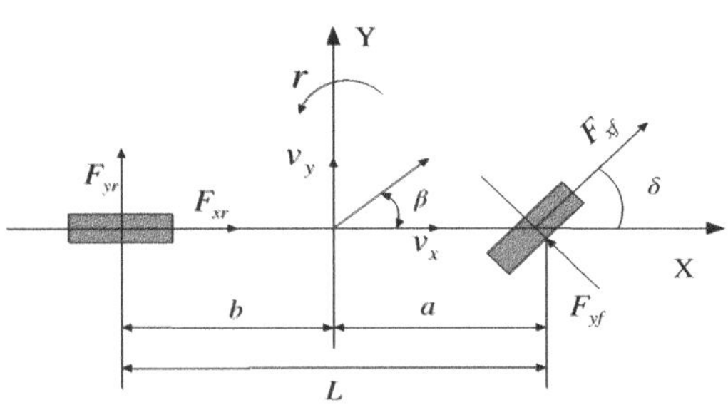

Figure 3.32 The bicycle model.

$$\dot{\beta} = \left(\frac{aC_f - bC_r}{mv_x^2} - 1\right) r + \frac{C_f + C_r}{mv_x}\beta - \frac{C_f}{mv_x}\delta \tag{3.45}$$

$$\dot{v}_x = r\beta v_x + a_x \tag{3.46}$$

$$a_y = \left(\frac{aC_f - bC_r}{mv_x}\right) r + \frac{C_f + C_r}{m}\beta - \frac{C_f}{m}\delta \tag{3.47}$$

where β is sideslip angle, v_x represents longitudinal vehicle velocity, C_f and C_r are the front and rear tire cornering stiffnesses, a and b are distances from the center of gravity to front axle and rear axle, a_x and a_y are longitudinal acceleration and lateral acceleration, m is vehicle mass, δ is front wheel steering angle, r is yaw rate, L is wheelbases, v_y is lateral vehicle velocity, F_{xf}, F_{yf} are longitudinal and lateral forces of the front axle, F_{xr}, F_{yr} are longitudinal and lateral forces of the rear axle.

In this approach, the front wheel steering angle, longitudinal acceleration, and yaw rate are utilized as input variables, while lateral acceleration is employed as the measured output variable. The system state is characterized by two primary variables: longitudinal velocity and sideslip angle. These variables are crucial for understanding and predicting the vehicle's dynamic behavior under various driving conditions. However, it is important to acknowledge that onboard sensors, which provide these measurements, are subject to external influences such as electromagnetic interference and mechanical vibrations. These factors can lead to occasional data loss or inaccuracies in the sensor readings, posing a challenge for the controller to maintain accurate state estimation. Considering this influence, we establish the state-space equations with data loss as follows.

$$\begin{cases} X_\tau = f(X_{\tau-1}, u_{\tau-1}) + W_{\tau-1} \\ Z_\tau = \eta_\tau h(X_\tau, u_\tau) + V_\tau \end{cases} \tag{3.48}$$

$$X_\tau = [\beta, v_x]^{\mathrm{T}}, Z_\tau = [a_y]^{\mathrm{T}}, u_\tau = [r, \delta, a_x]^{\mathrm{T}}$$

where W_τ represents a process noise with a covariance of Q_τ, V_τ signifies a measurement noise with a covariance of R_τ, f denotes the state transition function, τ corresponds to the sampling instant, h represents the measurement output function, X_τ stands for the state vector, and both the process noise and measurement noise are assumed to be uncorrelated. The measurement vector with data loss is represented by Z_τ, u_τ is the input vector, $\eta_\tau = diag\left\{\varepsilon_\tau^1, \varepsilon_\tau^2, \ldots \varepsilon_\tau^g\right\}$, $\varepsilon_\tau^i (i = 1, 2 \ldots g)$ are g-independent random variables. Furthermore, $diag\{\cdot\}$ is a diagonal matrix, and ε_τ^i represents the probability density function. The initial state is assumed to be independent of all noise signals.

In the context of nonlinear system state estimation based on the aforementioned model, it is typically assumed that comprehensive knowledge is available regarding the model described by Eq. (3.48). However, in practical applications, both the process noise and the measurement noise are often time-varying, which introduces significant complexities. Moreover, the common assumption that these noises follow a Gaussian white noise distribution does not always hold true in real-world scenarios, where colored noise is also prevalent. This indicates that our understanding of the model described by Eq. (3.48) is actually partial and imperfect. In addition to the challenges posed by time-varying and colored noise, the loss of measurement data is a frequent occurrence that directly impacts the accuracy of state estimation. These factors underscore the need for a robust and adaptable filtering algorithm capable of handling such uncertainties and data imperfections. To address these challenges and achieve high-precision vehicle state estimation, we propose a unified filtering algorithm that combines the FTEKF with an ANN. This hybrid approach, termed FTEKFNet, leverages the strengths of both methods. The FTEKF provides a robust framework for dealing with model uncertainties and noise, while the ANN enhances the system's adaptability and predictive capabilities by learning from data. By integrating these two methodologies, FTEKFNet is capable of accurately estimating vehicle states even under challenging conditions characterized by time-varying noise, colored noise, and data loss. This innovative approach ensures that the estimation process remains reliable and precise, thereby improving the overall performance and safety of vehicle control systems in real-world applications.

3.4.2 Methodology

In this section, the architecture of FTEKFNet will be comprehensively outlined. The overall structure of the FTEKF and its multiple functional modules will be presented first. In addition, detailed explanations of model training and data collection will then be provided. Compared to the traditional EKF, it is crucial for the FTEKF to be able to deal with the state estimation problem under data loss by first developing a state-space model that takes into account data loss, i.e. Eq. (3.48). Subsequently, the formulas for the Kalman filter gain, the posterior state update, and the covariance matrix computation are rederived based on the new state-space model. Therefore, the initialization and time update steps of the FTEKF are the same as those of the conventional EKF, while there are significant differences in the measurement update step. The FTEKF unfolds through the following iterative steps.

3.4.2.1 Initialization

$$\hat{X}_0 = E(X_0) \tag{3.49}$$

$$P_0 = E\left[(X_0 - \hat{X}_0)(X_0 - \hat{X}_0)^T\right] \tag{3.50}$$

where E is a mathematical expectation and P is the covariance matrix.

3.4.2.2 Time Update

Prediction of states

$$\hat{X}_{\tau|\tau-1} = f(X_{\tau-1|\tau-1}, u_{\tau-1|\tau-1}) \tag{3.51}$$

Prediction of the covariance matrix of states

$$\begin{cases} P_{\tau|\tau-1} = A_\tau P_{\tau-1|\tau-1} A_\tau^T + Q \\ A_\tau = \dfrac{\partial f(X_{\tau-1|\tau-1}, u_{\tau-1|\tau-1})}{\partial X_{\tau-1|\tau-1}} \end{cases} \tag{3.52}$$

3.4.2.3 Measurement Update

Prediction of measurements

$$\hat{Z}_{\tau|\tau-1} = \eta_\tau h(X_{\tau|\tau-1}, u_{\tau|\tau-1}) \tag{3.53}$$

Calculate the Kalman gain

$$\begin{cases} K_\tau = \left(P_{\tau|\tau-1} H_\tau^T \eta_\tau^T\right) \times \left[\eta_\tau H_\tau P_{\tau|\tau-1} H_\tau^T \eta_\tau^T \right. \\ \qquad \left. + E\left(\tilde{\eta}_\tau h(X_{\tau|\tau-1}, u_{\tau|\tau-1})\, h^T\,(X_{\tau|\tau-1}, u_{\tau|\tau-1}) \tilde{\eta}_\tau^T\right) + R_\tau\right]^{-1} \\ H_\tau = \dfrac{\partial h(X_{\tau|\tau-1}, u_{\tau|\tau-1})}{\partial X_{\tau|\tau-1}} \end{cases} \tag{3.54}$$

where $\tilde{\eta}_\tau = \eta_\tau - \overline{\eta}_\tau$ and $\overline{\eta}_\tau = E(\eta_\tau) = diag\left\{\overline{\varepsilon}_\tau^1, \overline{\varepsilon}_\tau^2, \cdots \overline{\varepsilon}_\tau^g\right\}$.

Update the system state

$$\begin{aligned} \hat{X}_{\tau|\tau} &= \hat{X}_{\tau|\tau-1} + K_k \Delta\tilde{Z} \\ &= \hat{X}_{\tau|\tau-1} + K_\tau (Z_\tau - \hat{Z}_{\tau/\tau-1}) \end{aligned} \tag{3.55}$$

Update the covariance matrix

$$\begin{aligned} P_{\tau|\tau} = &(I - K_\tau \eta_\tau H_\tau) P_{\tau-1|\tau} \times (I - K_\tau \eta_\tau H_\tau)^T \\ &+ K_\tau R_\tau K_\tau^T + K_\tau E[\tilde{\eta}_\tau h(X_{\tau|\tau-1}, u_{\tau|\tau-1}) \\ &\times h^T (X_{\tau|\tau-1}, u_{\tau|\tau-1}) \tilde{\eta}_\tau^T] K_\tau^T \end{aligned} \tag{3.56}$$

We have meticulously structured the entire filtering recursion process of the FTEKF into distinct modules, as depicted in Fig. 3.33. This modular approach highlights the critical importance of precise knowledge of the process noise matrix during the computation of the error covariance. Accurate estimation of this matrix is essential for maintaining the integrity of the filter's performance. The calculation of the Kalman gain, which is a pivotal step in the filtering process, relies on Eq. (3.54). This equation, in turn, necessitates accurate information about the measurement noise matrix. The Kalman gain determines how much the predictions should be adjusted based on new measurements, making it a cornerstone of the estimation process. Any inaccuracies in the noise matrices

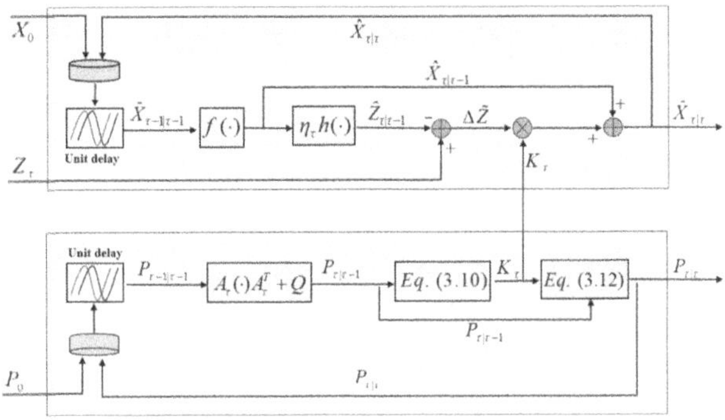

Figure 3.33 The block flowchart of FTEKF.

directly affect the computation of the Kalman gain, thereby impacting the overall estimation accuracy. Furthermore, the challenge of data loss introduces additional complexity to the estimation process. Unlike traditional EKF, the FTEKF incorporates more intricate update formulas for the Kalman gain and error covariance to mitigate the adverse effects of data loss. These formulas are designed to handle the dynamic and unpredictable nature of real-world environments, where sensor data can be intermittently lost or corrupted. Given the unknown statistical characteristics of the noise and the uncertainty regarding the type of noise, computations based on the traditional FTEKF recursive formulas can lead to decreased estimation accuracy. In severe cases, this may even result in filter divergence. To address these challenges, we propose leveraging deep learning techniques to enhance the FTEKF. By training a deep neural network to learn and directly predict the Kalman gains from available data in real time, we can significantly improve the filter's adaptability and accuracy. This approach circumvents the limitations of traditional methods, which rely heavily on precise noise statistics that are often unavailable or difficult to obtain in real-world scenarios. To this end, we have designed the FTEKFNet, a hybrid filtering framework that integrates the robust principles of FTEKF with the predictive power of ANN. The ANN component of FTEKFNet is pre-trained to understand the complex relationships between the input data and the optimal Kalman gains, enabling it to make real-time predictions that enhance the filter's performance. This integration allows FTEKFNet to dynamically adjust to varying noise conditions and data loss, maintaining high estimation accuracy even under challenging circumstances.

FTEKFNet integrates the FTEKF with an ANN to achieve precise vehicle state estimation in situations where the statistical characteristics of noise are unknown and measurement data may be lost. Additionally, it effectively addresses the challenges associated with vehicle state estimation under colored noise conditions. By examining the architecture of FTEKF, it becomes evident that the computation of

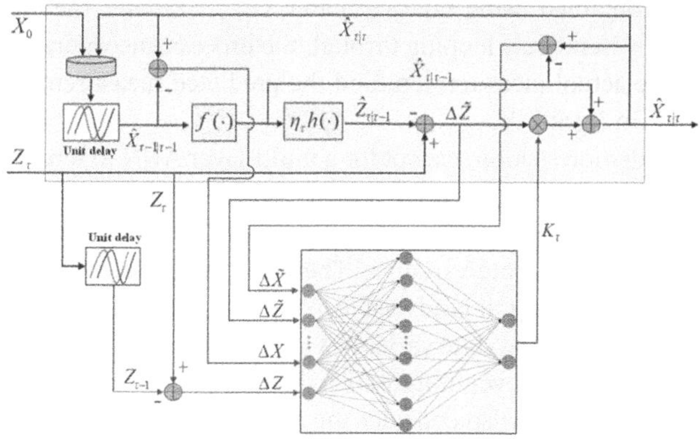

Figure 3.34 The block flowchart of FTEKFNet.

the Kalman gain is crucial for system state updates. Therefore, we have removed the module in FTEKF responsible for computing the error covariance. Instead, by providing appropriate inputs to the ANN model, it directly predicts the optimal Kalman gain for iterative updates of the system state.

The specific structure is illustrated in Fig. 3.34. The entire FTEKFNet consists of two main modules: the system state update and the Kalman gain prediction. The system state update section corresponds to the light orange box in the diagram, while the Kalman gain prediction section is primarily implemented by the ANN model. In the detailed design, the system state update module still follows the traditional FTEKF framework, but when calculating the Kalman gain, we no longer use the traditional error covariance matrix computation method. Instead, we directly utilize the Kalman gain predicted by the ANN. The advantage of this approach is that the ANN can learn complex noise patterns and data loss patterns, providing more accurate gain values and thereby enhancing the robustness and accuracy of the system. During the training process, we use a large amount of simulation data and real driving data to fully train and validate the ANN model, ensuring its reliability under various noise conditions and data loss scenarios. In summary, FTEKFNet, by integrating FTEKF with ANN, overcomes the limitations of traditional Kalman filters in situations with unknown noise statistical characteristics and measurement data loss. It significantly improves the estimation performance under colored noise conditions. Its modular design ensures good scalability and maintainability, providing an innovative and effective solution for achieving high-precision vehicle state estimation.

In the system state update module, the process begins by using the state values from the previous time step, denoted as $X_{\tau-1|\tau-1}$, to obtain a prior estimation, $X_{\tau|\tau-1}$, through the use of Eq. (3.51). Subsequently, the prediction of the measurement variable is obtained using Eq. (3.52) and $X_{\tau|\tau-1}$. Finally, the ultimate estimate is derived by iteratively looping through the process, incorporating the deviation between the actual measurement and the predicted measurement, along with the Kalman gain K_τ and $X_{\tau|\tau-1}$.

In the Kalman gain prediction section, we opt for a multi-layer ANN to forecast real-time Kalman gains. Based on the traditional FTEKF gain calculation formula, it is known that accurate gain computation relies on the statistical characteristics of the observed variables and state variables. Therefore, when selecting the input features for the neural network, it is advisable to include information about both the observed variable Z_τ and the state variable $X_{\tau-1|\tau-1}$. Further, taking the discrete difference of these variables over unit time can effectively capture the variations in noise. Simultaneously, we also consider the impact of data loss, where the deviation between adjacent moments of observed variables can well reflect the occurrence of data loss. Therefore, the network's inputs primarily consist of four variables.

(a) The forward evolution error: $\Delta X = X_{\tau|\tau} - X_{\tau-1|\tau-1}$
(b) The forward update error: $\Delta\tilde{X} = X_{\tau|\tau} - X_{\tau|\tau-1}$
(c) The observation error: $\Delta Z = Z_\tau - Z_{\tau-1}$
(d) The innovation error: $\Delta\tilde{Z} = Z_\tau - \hat{Z}_{\tau|\tau-1}$

The structure of the ANN model is depicted in Fig. 3.35. It can be observed that the ANN takes ΔX, $\Delta\tilde{X}$, ΔZ, and $\Delta\tilde{Z}$ as inputs. Subsequently, it generates corresponding filtering gains through an activation function. Since the estimated state is two-dimensional, there are two gain values here. The backpropagation algorithm is employed to optimize the performance function, which is a nonlinear function of the deviation between the predicted value of the ANN and the ground truth of the Kalman gains. It should be noted that there is no special requirement for the architecture of neural networks; in addition to the basic neural network model, some RNNs can also be used for Kalman gain prediction. Furthermore, the

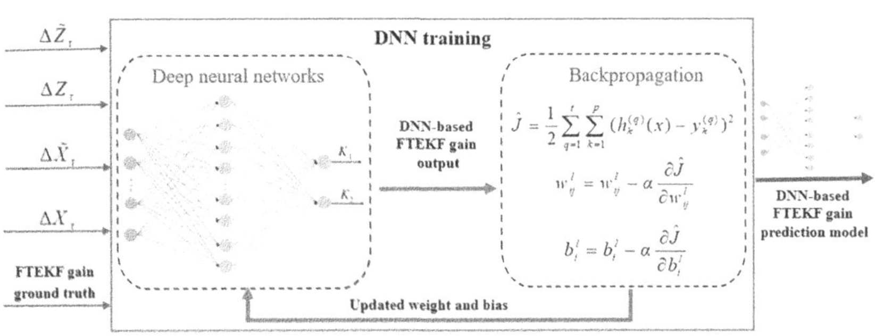

Figure 3.35 The block flowchart of FTEKFNet.

entire neural network training utilizes the backpropagation learning algorithm. The backpropagation learning algorithm, a cornerstone of modern neural network training, is a method used to minimize the error in output predictions by adjusting the weights of the network through a process known as gradient descent. This algorithm operates in two main phases: the forward pass and the backward pass. During the forward pass, input data is fed into the network, and computations flow through the layers to produce an output. This output is then compared to the actual target values using a loss function to calculate the error. In the backward pass, this error is propagated backward through the network. This involves computing the gradient of the loss function with respect to each weight by applying the chain rule of calculus. The gradient essentially indicates how much a change in each weight will affect the overall error. By updating the weights in the opposite direction of the gradient, typically scaled by a learning rate, the network iteratively reduces the error. This process continues until the error converges to a minimum value, ideally leading to improved accuracy in the network's predictions. One of the key advantages of backpropagation is its efficiency in handling large datasets and complex networks. It is computationally feasible, particularly when combined with optimization techniques such as stochastic gradient descent, which updates weights using small batches of data. This not only speeds up convergence but also makes the computation more manageable. Backpropagation has enabled the training of deep neural networks, which consist of many layers, thereby allowing for the high-level abstraction necessary for tasks like image and speech recognition. Moreover, it supports online learning, where the model can continuously improve as new data becomes available, making it suitable for real-time applications. Despite challenges like vanishing or exploding gradients, which can occur in very deep networks, advancements such as the introduction of normalization techniques, advanced activation functions, and adaptive learning rate methods have mitigated these issues, enhancing the stability and performance of the algorithm. Backpropagation's systematic approach to error minimization, adaptability to various types of neural architectures, and its role in enabling the deep learning revolution underscore its fundamental importance in the field of artificial intelligence and machine learning, driving advancements across a multitude of applications.

After iterative training with a substantial amount of data, the model is eventually developed to directly predict Kalman gains. Given that a single hidden layer in a neural network is generally adept at managing complex functions, this study employs an ANN structured with one hidden layer. The configuration of an ANN with a solitary hidden layer is illustrated in Fig. 3.35. Typically, a neuron is characterized by having multiple inputs. It receives n inputs $x_1, x_2 \ldots x_n$ from the preceding neuron, each associated with weights denoted as ω_i. The neuron also includes a bias term, μ, which, upon summation with the weighted inputs, yields the net input, represented as s. This relationship can be formulated as:

$$s = \sum_{i=1}^{n} \omega_i x_i + \mu \tag{3.57}$$

Therefore, Eq. (3.57) can describe the output of each node in the hidden layer of the ANN, from which we can easily derive the following formula for the Kalman gains.

$$K_1 = \ell_{12}\left(\sum_{j=1}^{q} w_{j1}^{23} \Gamma_{12}\left(\sum_{i=1}^{n} w_{ij}^{12} x_i + \mu_i^{12}\right) + \mu_1^{23}\right) \tag{3.58}$$

$$K_2 = \ell_{23}\left(\sum_{j=1}^{q} w_{j2}^{23} \Gamma_{12}\left(\sum_{i=1}^{n} w_{ij}^{12} x_i + \mu_i^{12}\right) + \mu_2^{23}\right) \tag{3.59}$$

where K_1, K_2 represent the Kalman gains, the subscripts, and superscripts 12 and 23 denote the connections from the input layer to the hidden layer neurons and from the hidden layer to the output layer, respectively. q, n are the number of neurons in the output and hidden layers, μ_1^{23} and μ_2^{23} are the biases of the K_1 and K_2 in the output layer, and ℓ, Γ are the active functions in the output and hidden layers.

The backpropagation algorithm is employed for training the ANN. The cost function for the rth sample is denoted as J and shown in Eq. (3.60).

$$J(\omega, \mu, x, y) = \frac{1}{2}\sum_{i=1}^{m}\left(\hat{y}_i^{(r)}(x) - y_i^{(r)}\right)^2 \tag{3.60}$$

$x = [\Delta X, \Delta\tilde{X}, \Delta Z, \Delta\tilde{Z}]$ is the input data for the ANN, y_i is the ground truth of Kalman gains, and $\hat{y}_i$ is the prediction result of the ANN.

Datasets consisting of p samples, the overarching cost function is derived as follows

$$\tilde{J}(\omega, \mu, x, y) = \frac{1}{2}\sum_{m=1}^{p}\sum_{i=1}^{m}\left(\hat{y}_i^{(r)}(x) - y_i^{(r)}\right)^2 \tag{3.61}$$

For the weights ω_{ij}^l and biases μ_i^l of the lth layer, they can be obtained using the gradient descent method.

$$w_{ij}^l = w_{ij}^l - \alpha\frac{\partial}{\partial w_{ij}^l}\tilde{J}(w, b, x, y) \tag{3.62}$$

$$b_i^l = b_i^l - \alpha\frac{\partial}{\partial b_i^l}\tilde{J}(w, b, x, y) \tag{3.63}$$

where α is the learning rate.

The overall backpropagation learning algorithm can be summarized by the following steps: 1) first, the input is propagated forward through the network; 2) second, the partial derivative of each neuron is calculated; 3) the weights and biases are updated using the gradient descent method from the last layer to the first layer.

To ensure a high level of predictive accuracy in the ANN, ample training data is imperative. We established a comprehensive simulation platform integrating CarSim and Simulink. In Simulink, the FTEKF algorithm is implemented, and noise interference is artificially introduced with known statistical characteristics. The Kalman gain data collected in this controlled environment serves as the true value, or ground truth, for the training data. We began by categorizing vehicle operating conditions into regular and extreme scenarios, followed by systematic data collection. Vehicle speed was sampled at intervals of 3 km/h within the range of 0–120 km/h. White noise was added with a mean of zero and variance within the range of [0.00001, 0.01]. This setup allowed us to simulate and collect input data for the neural network and corresponding ground truth values for Kalman gains under various conditions. Additionally, we collected data under conditions of abrupt changes in noise to ensure the robustness of our model. Furthermore, similar data collection was conducted under the influence of colored noise interference to cover a broader spectrum of real-world scenarios. In addition to simulation data, we also conducted real vehicle experiments to expand the dataset and enhance its representativeness. The combination of simulated and real-world data ensures that the ANN is exposed to a wide variety of conditions and noise characteristics, improving its generalization capabilities. By the end of the data collection phase, we had gathered a total of 35,600 samples. To facilitate effective training and evaluation, we allocated 80% of the data for training, 10% for testing, and the remaining 10% for validation. This comprehensive data collection approach ensures that the ANN is well-trained and capable of accurately predicting Kalman gains even under varied and challenging conditions. The joint simulation platform and extensive real-world data collection provide a robust foundation for developing a high-precision vehicle state estimation system.

3.4.3 Simulation and Experiment Tests

To assess the efficacy of FTEKFNet thoroughly, we embarked on a series of rigorous simulation experiments. During these simulations, output values generated by the CarSim software served as a baseline reference against which the estimates produced by FTEKFNet were meticulously compared with those derived from the recursive EKF (REKF) and the traditional EKF. These comparisons were conducted across a variety of maneuvers, including DLC and J-turn tests, each performed under controlled conditions to evaluate the algorithm's performance across different dynamic scenarios. The simulation phase provided a controlled environment where the maneuver-specific outputs from CarSim facilitated a direct comparison of the estimation accuracy among FTEKFNet, REKF, and EKF.

This allowed us to analyze how well FTEKFNet adapted to the complexities of vehicle dynamics compared to more conventional filtering methods. In addition to simulation experiments, we expanded our evaluation by integrating offline data collected from real-world vehicle tests conducted on dry asphalt roads. This real-world data collection phase was crucial as it provided empirical insights into how well FTEKFNet could generalize its performance beyond idealized simulation scenarios. By capturing the nuances and variations inherent in real-world driving conditions, such as road surface irregularities and environmental factors, this dataset enabled a more comprehensive assessment of the algorithm's robustness and accuracy. This hybrid approach of combining simulated and real-world data aimed to provide a holistic evaluation of FTEKFNet's capabilities, highlighting its potential advantages over traditional methods in terms of accuracy, adaptability, and robustness in varying operational conditions. In summary, the combination of simulation experiments with real vehicle data collection allowed us to comprehensively demonstrate the effectiveness of FTEKFNet. By leveraging both controlled simulations and real-world driving data, we could validate its performance across different test scenarios, affirming its potential as an advanced vehicle state estimation technique capable of addressing the complexities of modern driving environments.

3.4.3.1 The Double Lane Change Test

In the DLC test, we implemented a PID steering controller to autonomously steer the vehicle through predefined maneuvers. The experiment commenced with an initial velocity of 42 km/h, chosen to simulate real-world driving conditions. Key metrics such as the front wheel steering angle and lateral acceleration were recorded and visualized in Figs. 3.36 and 3.37, respectively, to evaluate the performance of our proposed algorithm, FTEKFNet, in estimating vehicle state under dynamic operational scenarios. Upon close inspection of Fig. 3.36, it becomes evident that the steering angle curves exhibit deviations influenced by the injected noise interference. These deviations illustrate the challenges posed by noise in accurate state estimation, highlighting the need for advanced filtering techniques like FTEKFNet to mitigate such effects effectively. Initially, the noise matrices used in the experiment were predefined based on assumed covariance values, which differed from the true covariance matrix of the noise observed during testing. This discrepancy underscores the importance of adaptive algorithms capable of adjusting to varying noise characteristics encountered in real-world scenarios.

In Fig. 3.37, the red solid line represents the lateral acceleration curve with introduced noise, while the blue line illustrates partial data loss. This data loss is simulated by multiplying the lateral acceleration values by a uniformly distributed random variable. As a result, at certain moments, the data abruptly drops to zero

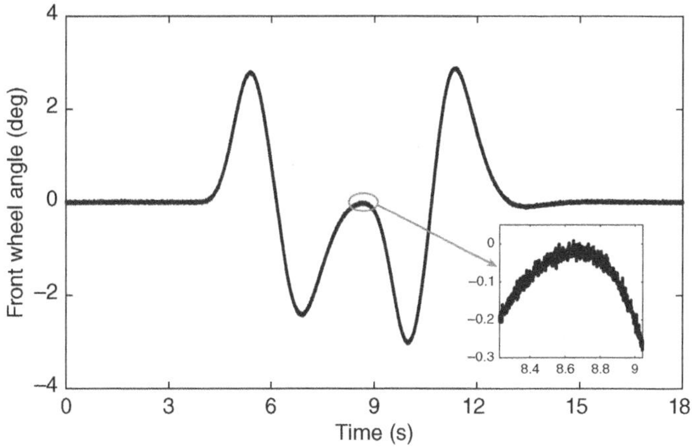

Figure 3.36 The front wheel angle in the DLC test.

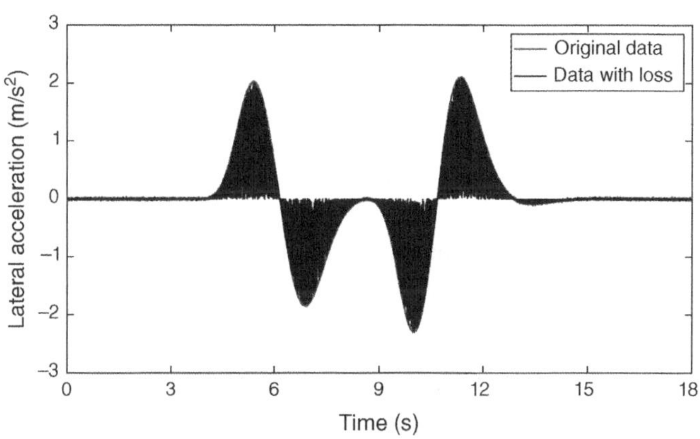

Figure 3.37 Lateral acceleration with data loss in the DLC test.

or becomes less than its original value, clearly indicating varying degrees of data loss. This behavior is crucial to study as it mimics real-world scenarios where sensor data might be intermittent or corrupted. Moving on to Fig. 3.38, the yaw rate curve fed into the estimator is depicted. Similar to the lateral acceleration data, this curve also exhibits variations under the influence of noise. The magnified view of the yaw rate curve highlights these variations more clearly, showing irregularities that disrupt the expected smoothness of the curve. These disruptions are indicative of the noise's impact on the yaw rate measurements, which in turn affects the accuracy of the vehicle state estimation.

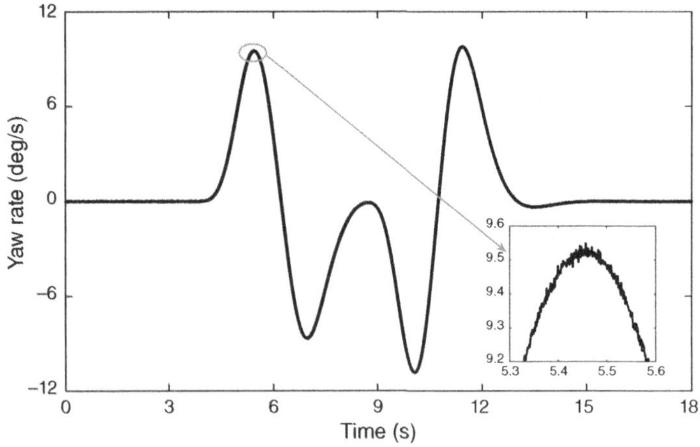

Figure 3.38 The yaw rate in the DLC test.

Figure 3.39 illustrates the estimation results for the sideslip angle using different methods, providing a comprehensive comparison of their effectiveness. The red solid line represents the vehicle state output from the CarSim software, which is considered the reference value or ground truth. At the specific moment analyzed, the vehicle is operating at a low speed with a small front wheel steering angle, placing it within a linear region. Consequently, the trend of the sideslip angle curve closely follows the pattern of the front wheel steering angle, making deviations easier to detect and analyze. The estimated curve produced by the EKF shows noticeable deviations from the reference value. This deviation arises because

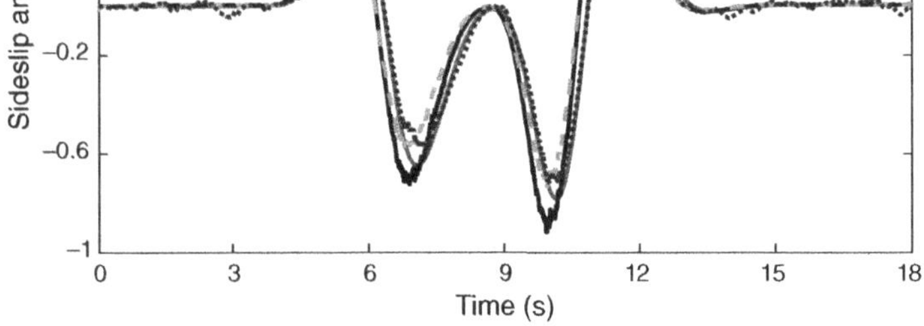

Figure 3.39 The estimated sideslip angle in the DLC test.

the EKF performs optimally only when the measured data and the statistical properties of the noise are accurately obtained in real time. In practical scenarios, obtaining real-time, precise noise characteristics can be challenging, leading to inaccuracies in the EKF's estimations. This limitation is particularly evident under conditions of data loss or noise variability, where the EKF struggles to maintain accuracy. In contrast, the FTEKF demonstrates superior estimation accuracy compared to the EKF. FTEKF accounts for the impact of missing measurements, a crucial factor that the EKF does not adequately address. By considering missing data, FTEKF provides more reliable estimations. However, since the noise parameters in FTEKF are often arbitrarily assigned, it does not always achieve optimal estimation performance. This arbitrary assignment can lead to suboptimal results, particularly when the noise characteristics deviate significantly from the assumed values. FTEKFNet, on the other hand, represents a significant advancement by directly training to predict Kalman gains based on diverse data sets encompassing different noise conditions and test scenarios. This approach eliminates the need for error covariance computation during the estimation process, thereby avoiding iterative loops affected by unknown noise biases. By bypassing these biases, FTEKFNet enhances both the efficiency and accuracy of state estimation. Moreover, FTEKFNet effectively addresses the issue of state estimation in the presence of data loss. During the model training phase, the effect of observation errors is incorporated, enabling the network to learn and adapt to the uncertainties caused by partial or complete data loss. This capability ensures that FTEKFNet maintains high estimation accuracy even under challenging conditions where traditional methods like EKF and REKF falter. The overall estimation performance of FTEKFNet surpasses that of both REKF and EKF for several reasons. First, its ability to adapt to varying noise conditions and data loss scenarios through advanced training techniques ensures robust performance. Second, by eliminating the need for real-time noise parameter computation, FTEKFNet reduces the computational load and potential for errors associated with traditional recursive methods. Finally, the direct prediction of Kalman gains enhances the precision of state updates, leading to more accurate and reliable vehicle state estimations.

Figure 3.40 illustrates the estimation results for vehicle speed, showcasing the performance of different estimation methods. Among the methods tested, FTEKFNet stands out as exhibiting the highest estimation performance. This can be attributed to its advanced architecture, which effectively combines the strengths of fault-tolerant filtering and deep learning techniques. To quantitatively express the estimation errors and highlight the performance differences among the methods, we use the root mean square error (RMSE) as a key metric. The RMSE is a widely accepted measure of the differences between values predicted by a model and the actual observed values, providing a clear indication of estimation accuracy. Table 3.7 presents the RMSE values for the estimated vehicle speed

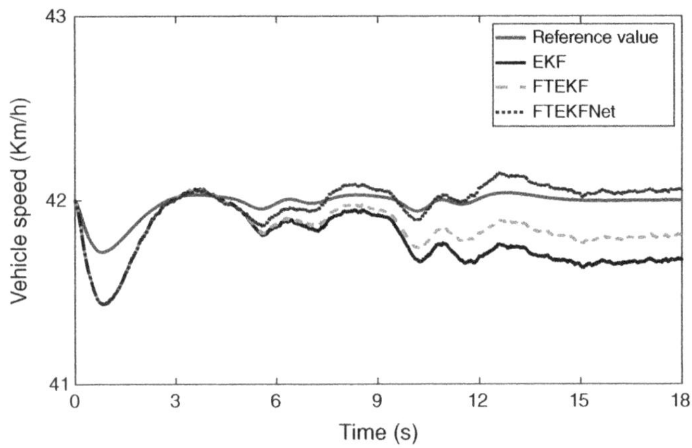

Figure 3.40 The estimated vehicle velocity in the DLC test.

Table 3.7 RMSE of different methods in the DLC test.

Symbol	v_x	β
EKF	0.2311	0.1076
FTEKF	0.1567	0.0415
FTEKFNet	0.0854	0.0276

and sideslip angle across the different algorithms. Specifically, the RMSE for the estimated vehicle speed using FTEKFNet is 0.0854, and the RMSE for the sideslip angle is 0.0276. These figures are significantly lower compared to those obtained using other methods, underscoring the superior performance of FTEKFNet. The lower RMSE values indicate that FTEKFNet has a better capability to accurately track the true vehicle state, minimizing deviations and errors in its estimations.

3.4.3.2 The J-Turn Test

The initial velocity for this experiment is set at 53 km/h. In Figs. 3.41 and 3.42, the variations in the front wheel angle and lateral acceleration are presented. Unlike the scenario depicted in Fig. 3.36, the front wheel steering angle in this case reaches a certain peak value before gradually decreasing to zero. This behavior is indicative of the vehicle's dynamic response under the test conditions. Furthermore, to assess the algorithm's adaptability to noise changes during the estimation process, we deliberately doubled the amplitude of the noise from the third second onward. This increase in noise amplitude is clearly evident in

Table 3.8 RMSE of different methods in the J-turn test.

Symbol	v_x	β
EKF	0.3821	0.1684
FTEKF	0.1308	0.0826
FTEKFNet	0.0572	0.0251

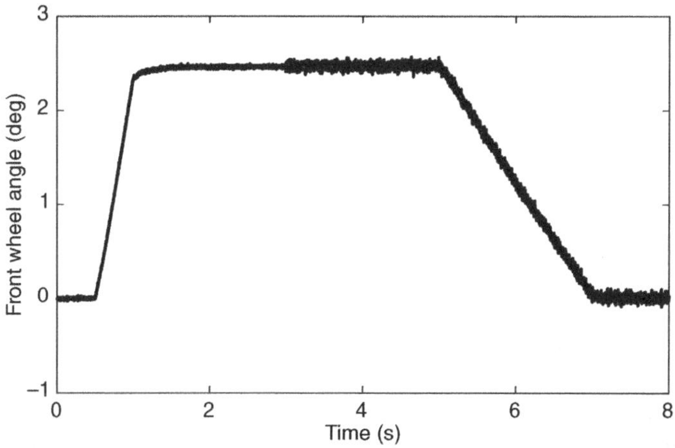

Figure 3.41 The front wheel angle in the J-turn test.

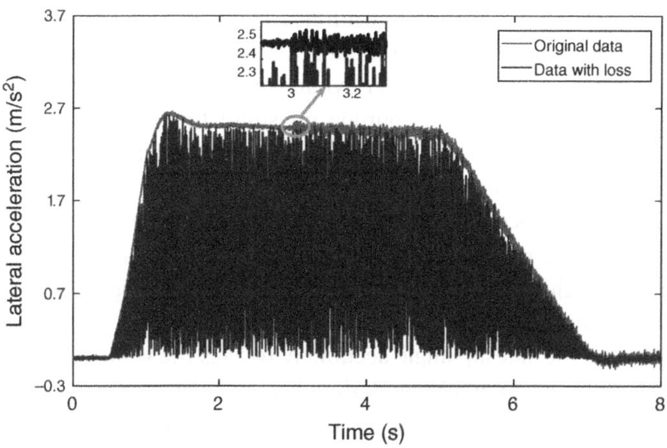

Figure 3.42 Lateral acceleration with data loss in the J-turn test.

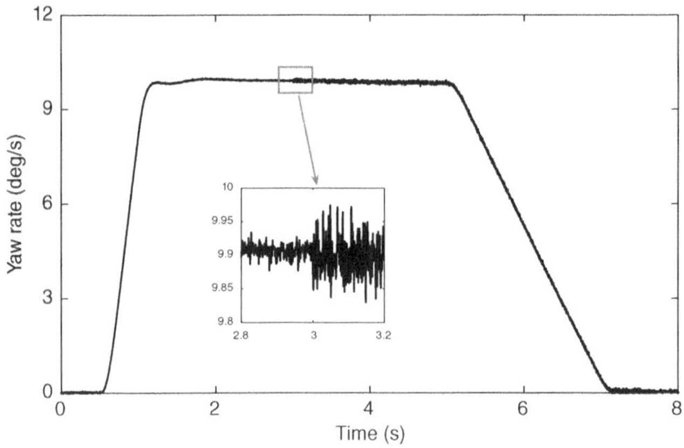

Figure 3.43 The yaw rate in the J-turn test.

Fig. 3.42. To simulate data partial loss, similar to the approach in Fig. 3.37, we multiplied the uniformly distributed noise by the original data. From Fig. 3.42, it is apparent that the lateral acceleration suddenly drops to zero at certain points, signifying the moments of measurement data loss. A closer inspection of the magnified sections reveals discrepancies between some data points and the reference values, further indicating instances of partial data loss. In Fig. 3.43, the yaw rate curve input to the estimator is depicted, with a zoomed-in view illustrating the significant changes in noise occurring at the 3-second mark. This plot highlights how the noise amplitude variation impacts the yaw rate measurements. The initial noise matrices were set based on assumed conditions but turned out to deviate from the actual noise characteristics during the test. This deviation underscores the importance of adaptive algorithms capable of handling varying noise conditions. The purpose of this experiment is to validate the robustness and adaptability of the proposed algorithm under dynamic noise conditions and data loss scenarios. By increasing the noise amplitude and simulating partial data loss, we aim to challenge the algorithm's ability to maintain accurate state estimation.

Figures 3.44 and 3.45 illustrate the variations in the sideslip angle and the estimated curve for vehicle speed, respectively. The coherence between the variation of the sideslip angle curve and the front wheel steering angle can be attributed to the factors elucidated in the relevant description in Fig. 3.39. This relationship indicates that the vehicle's behavior in response to steering inputs remains consistent, and the sideslip angle accurately reflects the vehicle's lateral dynamics. From the estimated curves of sideslip angle and vehicle speed, it can be observed that despite changes in noise, FTEKFNet still exhibits the highest estimation accuracy and closely matches the reference curves provided by the CarSim software. The robustness of FTEKFNet is quantitatively confirmed by

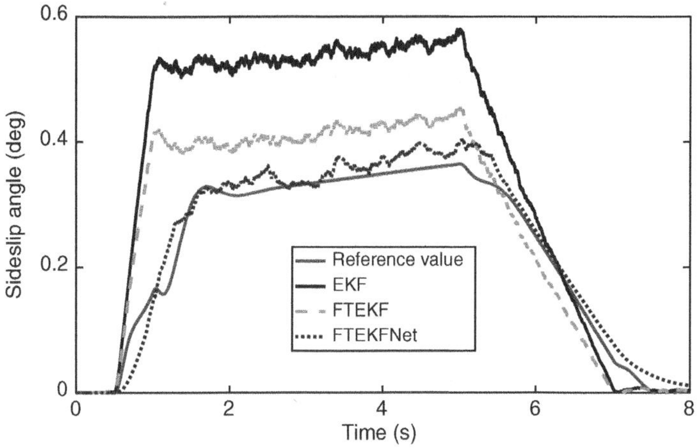

Figure 3.44 The estimated sideslip angle in the J-turn test.

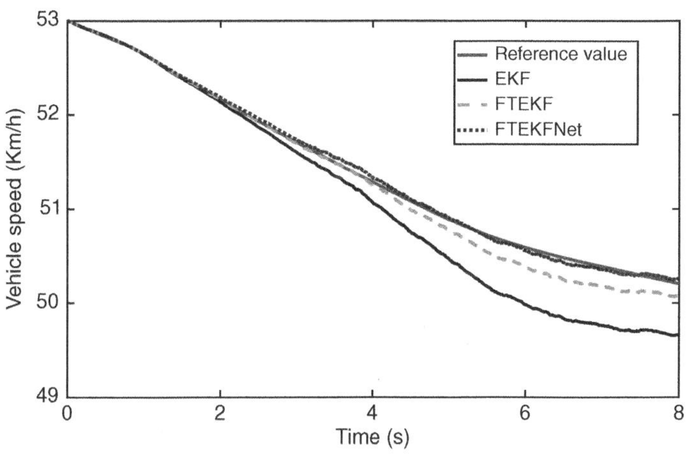

Figure 3.45 The estimated vehicle velocity in the J-turn test.

the RMSE values presented in Table 3.8. Specifically, the RMSE for the estimated vehicle speed using FTEKFNet is 0.0572, while the RMSE for the sideslip angle is 0.0251. These RMSE values are the lowest among the methods compared, indicating that FTEKFNet consistently achieves the most accurate estimations. This low RMSE highlights the algorithm's capability to minimize estimation errors, ensuring that the estimated values remain close to the actual vehicle states. The robustness and accuracy of FTEKFNet can be further appreciated by examining the performance under different noise conditions. The neural network model within FTEKFNet was trained using a variety of noise scenarios, including white noise with different variances and colored noise, to ensure that it could handle real-world uncertainties effectively. This extensive training allows FTEKFNet to predict Kalman gains that adapt dynamically to the noise characteristics present in the data, leading to more reliable state estimations. The superior performance of FTEKFNet is primarily due to its neural network model, which was trained to account for the influence of different noise levels and types during the training phase. This comprehensive training enables the Kalman gain predicted by the neural network model to exhibit strong robustness to changes in noise. Consequently, FTEKFNet can maintain high estimation accuracy even under varying noise conditions, which is a significant advantage over traditional methods. Moreover, the performance of the algorithm under these challenging conditions is critical for real-world applications where sensor noise and data loss are common. The ability to double the noise amplitude and still achieve accurate estimations demonstrates the algorithm's robustness and practical applicability. This experiment also emphasizes the need for advanced filtering techniques that can dynamically adjust to changing conditions, a capability that is essential for autonomous systems operating in unpredictable environments.

3.4.3.3 The Real Vehicle Test on the Dry Asphalt Road

The test scenario is depicted in Fig. 3.46. For safety reasons, all steering operations during the actual vehicle tests were conducted using a steering robot. This ensures precise and consistent steering inputs, which are crucial for obtaining reliable data. Measurements derived from a differential GPS served as benchmark values, providing high-accuracy reference points for comparing the estimation outcomes produced by various algorithms, including FTEKFNet, REKF, and EKF. In real-vehicle experiments, our primary focus was on evaluating the estimation performance of FTEKFNet, particularly under the influence of colored noise.

Figure 3.46 The test vehicle on a dry asphalt road.

Colored noise, which has a non-uniform frequency spectrum, is prevalent in real-world conditions and poses a significant challenge for state estimation algorithms. FTEKFNet is designed to address this challenge effectively by leveraging its neural network model, which has been trained on a wide range of noise scenarios. One of the significant advantages of our model is its flexibility regarding the vehicle's power source. The algorithm does not impose any specific requirements on whether the vehicle is fuel-powered or electric. This versatility allows for the collection of validation data from a diverse range of vehicles, thereby enhancing the robustness and generalizability of the FTEKFNet algorithm. To ensure a fair and accurate comparison of the estimation performance, it is essential to obtain certain vehicle parameters in advance. These parameters include the mass of the vehicle and the cornering stiffness of the tires. Accurate knowledge of these parameters is crucial as they significantly influence the vehicle's dynamic behavior and, consequently, the performance of the state estimation algorithms. By standardizing these parameters across different tests, we can ensure that the comparisons between FTEKFNet and other algorithms like REKF and EKF are valid and meaningful.

Figure 3.47 represents the spectral analysis of colored noise, with a particular focus on purple noise. Unlike white noise, which maintains consistent power across all frequencies, purple noise exhibits a power spectrum that is proportional to the square of the frequency. This characteristic implies that the power of purple noise increases as the frequency increases. This behavior is significantly different from that of white noise and presents unique challenges and considerations for state estimation algorithms. In this analysis, we specifically selected purple noise due to its distinct frequency-dependent power characteristics, which can

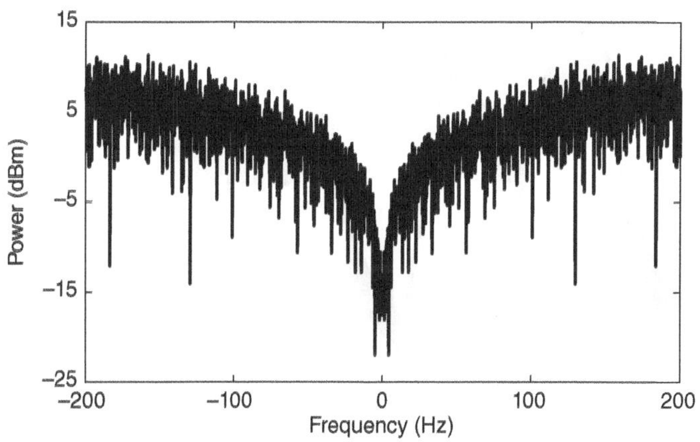

Figure 3.47 The spectral analysis of colored noise.

better simulate real-world scenarios where noise intensity varies with frequency. The spectral analysis shown in Fig. 3.47 highlights how the energy distribution of purple noise is skewed toward higher frequencies. This type of noise is relevant in various practical applications, particularly in automotive environments where different components and external factors can introduce frequency-specific disturbances. Understanding the impact of purple noise on vehicle state estimation is crucial for developing robust algorithms that can handle such conditions. The initial velocity was set at 64.7 km/h, which is representative of typical urban driving speeds. This speed setting provides a realistic context for evaluating the performance of the state estimation algorithms under the influence of purple noise. The choice of this specific speed ensures that the results are applicable to everyday driving scenarios, making the findings more relevant for practical implementation.

The curves of the front wheel steering angle, lateral acceleration, and yaw rate, overlaid with colored noise interference, are illustrated in Figs. 3.48–3.50, respectively. These figures present the same types of data as in the simulation tests, providing a direct comparison of how different noise types affect the measurements. The primary distinction here is the introduction of purple noise, which has a significantly greater impact on the true values compared to Gaussian white noise. In Fig. 3.48, the front wheel steering angle curve shows noticeable deviations from the original angle due to the influence of purple noise. This type of noise, with its frequency-dependent power increase, causes more pronounced fluctuations, making it evident that the vehicle's steering response is more susceptible to high-frequency disturbances. The steering angle oscillations are more erratic, indicating the challenges posed by colored noise in maintaining precise control.

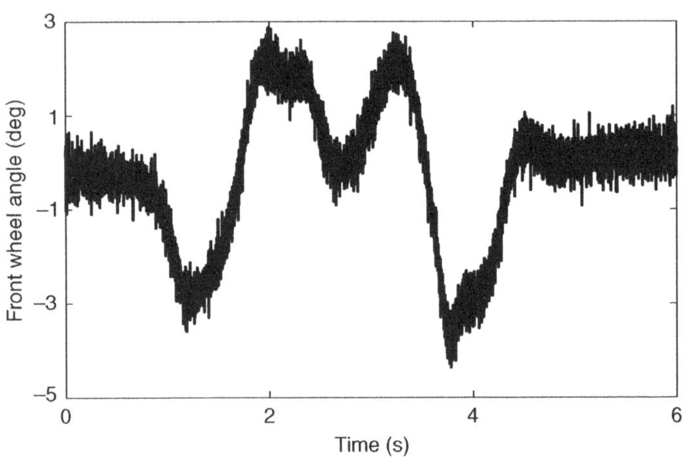

Figure 3.48 The front steering wheel angle in the real vehicle test.

Figure 3.49 highlights the lateral acceleration curve. Here, the effect of purple noise is even more apparent. The curve exhibits abrupt changes, with the acceleration suddenly dropping to zero or falling below the expected values at certain points. This phenomenon indicates the occurrence of data loss, where the measurements become unreliable or completely absent due to the interference. Such interruptions in the data stream can severely impact the performance of state estimation algorithms, highlighting the need for robust methods like FTEKFNet that can handle these anomalies effectively. The yaw rate curve in Fig. 3.50 also demonstrates the disruptive influence of purple noise. Similar to the other parameters, the yaw rate experiences significant deviations from the

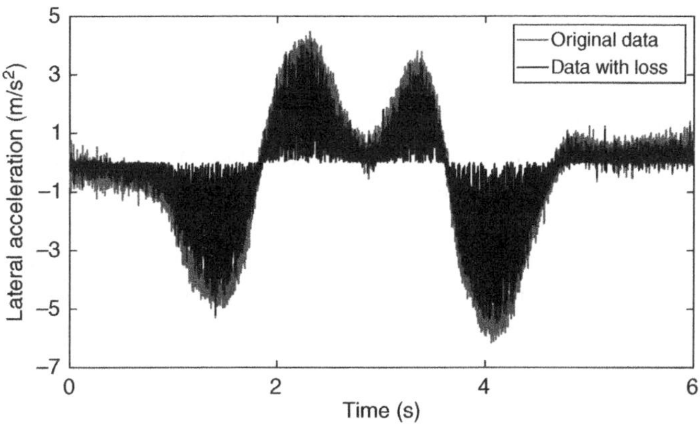

Figure 3.49 Lateral acceleration with data loss in the real vehicle test.

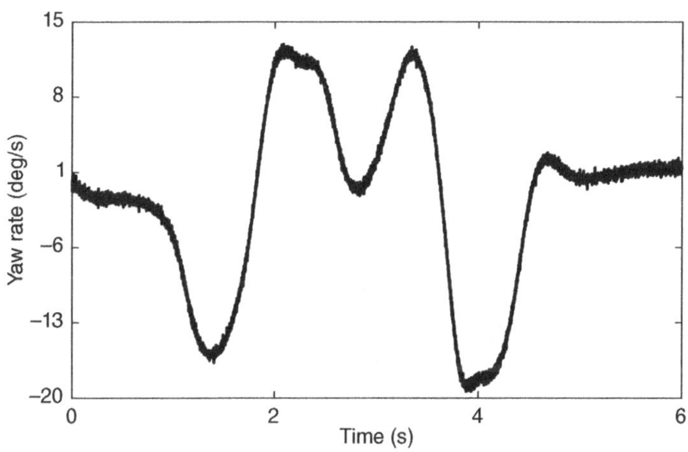

Figure 3.50 The yaw rate in the real vehicle test.

original values. The curve's erratic nature underlines the difficulty in maintaining accurate state estimates when subjected to high-frequency noise. The magnified sections of the graph clearly show how the noise affects the stability and accuracy of the yaw rate measurements, further emphasizing the challenges in real-world driving conditions. The impact of purple noise on these curves underscores the limitations of traditional noise assumptions in vehicle state estimation. While Gaussian white noise assumes a consistent power distribution across frequencies, purple noise's increasing power with frequency presents a more complex scenario. This complexity necessitates advanced algorithms capable of adapting to such conditions, ensuring reliable performance even under adverse noise environments.

Figure 3.51 illustrates the estimated results of the sideslip angle using different methods. In this scenario, the vehicle speed is relatively higher, and continuous steering is performed, increasing the vehicle's nonlinearity. As a result, there is a corresponding decrease in the consistency between the sideslip angle curve and the front wheel steering angle curve. The EKF, with its limited capability to handle colored noise and inability to address data loss issues, shows significant deviations from the reference values in its estimation curve. This limitation is evident in the discrepancies observed, particularly during high-speed maneuvers where the vehicle's dynamics become more complex. The FTEKF, which considers the impact of data loss, surpasses EKF in terms of estimation accuracy. However, when facing colored noise interference, FTEKF's accuracy diminishes. Despite its improvements over EKF, FTEKF cannot fully mitigate the effects of colored noise, leading to less accurate estimations compared to the FTEKFNet. The highest estimation

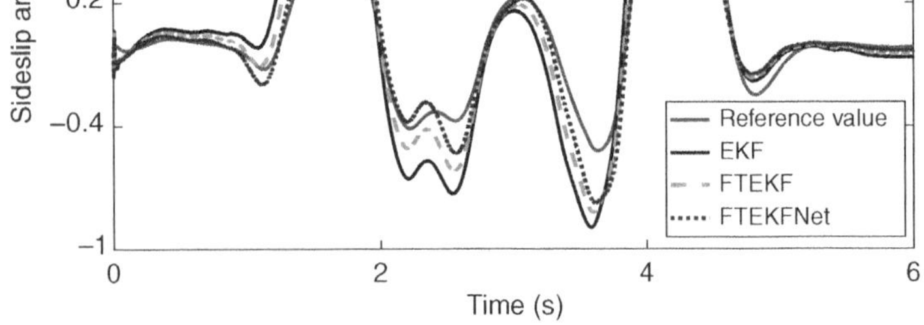

Figure 3.51 The estimated sideslip angle in the real vehicle test.

accuracy is achieved by FTEKFNet, as it can handle the effects of various factors on estimation accuracy simultaneously. This robustness is due to the incorporation of a neural network trained to predict Kalman gains, effectively addressing both noise variability and data loss.

Figure 3.52 illustrates the predicted results of vehicle speed using different methods. Consistent with the estimation results for the sideslip angle, FTEKFNet consistently achieves the most accurate estimation compared to the other two methods. The vehicle speed estimation curve produced by FTEKFNet closely matches the reference values, demonstrating its superior performance. This consistency in high accuracy across different parameters highlights the effectiveness of FTEKFNet in providing reliable vehicle state estimations. The results presented in Table 3.9 further underscore the efficacy of the FTEKFNet algorithm. The table shows significantly lower RMSE values for FTEKFNet compared to EKF and FTEKF. These lower RMSE values indicate that FTEKFNet provides more precise estimations of both sideslip angle and vehicle speed. The consistently better performance of FTEKFNet under various noise interference conditions, including Gaussian white noise, purple noise, and scenarios with data loss, emphasizes its robustness and adaptability. FTEKFNet's superior performance can be attributed to its hybrid approach, which combines the strengths of both physics-based and data-driven methods. By leveraging deep learning to predict Kalman gains, FTEKFNet eliminates the need for error covariance computation, thereby avoiding iterative loops with unknown noise bias. This approach allows FTEKFNet to maintain high estimation accuracy even when faced with unpredictable noise and incomplete data. The incorporation of a data-driven

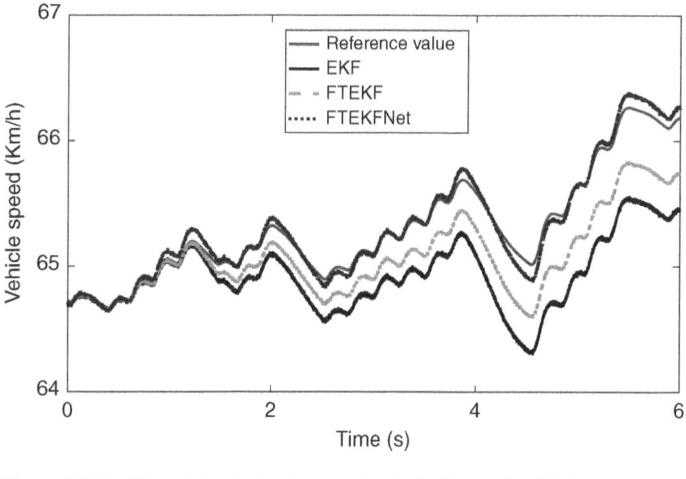

Figure 3.52 The estimated vehicle velocity in the real vehicle test.

Table 3.9 RMSE of different methods in the real vehicle test.

Symbol	v_x	β
EKF	0.4563	0.2259
FTEKF	0.2724	0.1366
FTEKFNet	0.0531	0.0929

approach significantly enhances the adaptability of the FTEKFNet algorithm. The neural network model within FTEKFNet is trained on a diverse set of data, including various noise conditions and test scenarios. This training enables FTEKFNet to generalize well across different real-world conditions, ensuring reliable performance. The ability to accurately predict Kalman gains in real-time, based on the current noise characteristics and data availability, sets FTEKFNet apart from traditional methods. In conclusion, the analysis of Figs. 3.51 and 3.52, along with the results in Table 3.9, clearly demonstrates that FTEKFNet offers the highest estimation accuracy among the tested methods. Its ability to handle colored noise and data loss, combined with the robust performance across different vehicle states, makes FTEKFNet a highly effective solution for vehicle state estimation. The integration of data-driven techniques within the FTEKF framework significantly enhances the algorithm's adaptability and overall performance, ensuring accurate and reliable estimations in various operating conditions.

3.5 Summary and Future Research

In this chapter, a novel approach, FARCKF, is introduced for the estimation of sideslip angle and tire cornering stiffness. This method incorporates dynamic updates of the vehicle mass parameter and process noise, aiming to enhance estimation performance throughout the estimation process. Through comprehensive simulation and experimental tests conducted under varied road conditions, the estimation accuracy and robustness of the proposed FARCKF approach are thoroughly evaluated and analyzed. Comparative assessments against existing methods such as the CKF and RCKF demonstrate superior estimation accuracy achieved by the FARCKF. Additionally, the FARCKF exhibits enhanced robustness to variations in vehicle mass and tire-road friction coefficient, making it a promising advancement in vehicle state estimation technology. Looking ahead, there are several intriguing avenues for future research. First, in this study, we assume sensor noise to follow Gaussian white noise characteristics, whereas practical scenarios may involve colored noise, which can significantly impact estimation accuracy. Addressing this discrepancy will be crucial for enhancing the applicability and reliability of the FARCKF in real-world settings. Furthermore, the influence of road-bank angles on estimation accuracy has been omitted in this current research. Exploring the effects of road-bank angles and integrating them into the estimation framework could further refine the accuracy of sideslip angle and tire cornering stiffness estimations. These future research directions aim to extend the capabilities of the FARCKF method, ensuring its effectiveness across diverse environmental and operational conditions in automotive applications.

Furthermore, a hybrid estimation algorithm, FTEKFNet, is introduced for the simultaneous estimation of vehicle velocity and sideslip angle, leveraging a fusion of physical modeling and data-driven approaches. This innovative approach incorporates a neural network to directly predict Kalman gains, thereby enhancing robustness against uncertainties associated with Gaussian noise and improving the algorithm's capability to handle colored noise. By integrating features related to data loss in the input data of the ANN, FTEKFNet effectively addresses challenges in vehicle state estimation under conditions of incomplete or unreliable data. Comparative evaluations against traditional EKF and FTEKF methods reveal that FTEKFNet achieves significantly enhanced estimation accuracy and adaptability across diverse operational conditions. Experimental validations, including simulations and preliminary real-vehicle tests under controlled conditions, substantiate the efficacy of FTEKFNet. However, further extensive testing across a broader spectrum of vehicle models and real-world environments is necessary to validate its universal applicability. Moreover, the current approach assumes precise knowledge of vehicle model parameters, which may not always be available in practical scenarios. Future enhancements will focus on integrating real-time parameter identification methods into the FTEKFNet framework, aiming to enhance algorithmic accuracy and adaptability by dynamically adjusting model parameters based on real-time data inputs. In summary, FTEKFNet represents a promising advancement in vehicle state estimation technology, offering robust performance in the face of noise uncertainties and data irregularities. Continued research and development efforts will refine its capabilities and broaden its applicability in various automotive engineering applications.

References

1 Wang, Y., Zhang, Z., Wei, H. et al. (2024). A novel fault-tolerant scheme for multi-model ensemble estimation of tire road friction coefficient with missing measurements. *IEEE Transactions on Intelligent Vehicles* 9 (1): 1066–1078.

2 Lu, J., Han, L., Wei, Q. et al. (2023). Event-triggered deep reinforcement learning using parallel control: A case study in autonomous driving. *IEEE Transactions on Intelligent Vehicles* 8 (4): 2821–2831.

3 Wei, H., Lou, B., Zhang, Z. et al. (2024). Autonomous navigation for eVTOL: Review and future perspectives. *IEEE Transactions on Intelligent Vehicles* https://doi.org/10.1109/TIV.2024.33526-13.

4 Zhang, Z., Wang, J., Guo, L. et al. (2023). Research on minimum non-collision distance and protection strategy for normal train to avoid rear-end accidents with braking-fault train. *IEEE Transactions on Intelligent Transportation Systems* 24 (4): 4306–4319.

5 Zhan, J., Ma, Z., and Zhang, L. (2023). Data-driven modeling and distributed predictive control of mixed vehicle platoons. *IEEE Transactions on Intelligent Vehicles* 8 (1): 573–582.

6 Wei, H., Zhang, H., AI-Haddad, K., and Shi, Y. (2023). Ensuring secure platooning of constrained intelligent and connected vehicles against byzantine attacks: A distributed MPC framework. *Engineering*. https://doi.org/10.1016/j.eng.2023.10.007.

7 Ampountolas, K. (2023). The unscented Kalman filter for nonlinear parameter identification of adaptive cruise control systems. *IEEE Transactions on Intelligent Vehicles* 8 (8): 4094–4104.

8 Xu, H., Liu, S., Wang, B., and Wang, J. (2023). Distributed-observer-based distributed control law for affine nonlinear systems and its application on interconnected cruise control of intelligent vehicles. *IEEE Transactions on Intelligent Vehicles* 8 (2): 1874–1888.

9 Zhou, X., Wang, Z., Shen, H., and Wang, J. (2022). Robust adaptive path-tracking control of autonomous ground vehicles with considerations of steering system backlash. *IEEE Transactions on Intelligent Vehicles* 7 (2): 315–325.

10 Chen, G., Zhao, X., Gao, Z., and Hua, M. (2023). Dynamic drifting control for general path tracking of autonomous vehicles. *IEEE Transactions on Intelligent Vehicles* 8 (3): 2527–2537.

11 Zhao, L., Liu, Z., and Chen, H. (2011). Design of a nonlinear observer for vehicle velocity estimation and experiments. *IEEE Transactions on Control Systems Technology* 19 (3): 664–672.

12 Guo, H., Chen, H., Cao, D., and Jin, W. (2013). Design of a reduced-order non-linear observer for vehicle velocities estimation. *IET Control Theory & Applications* 7 (17): 2056–2068.

13 Chen, Y., Ji, Y., and Guo, K. (2014). A reduced-order nonlinear sliding mode observer for vehicle slip angle and tyre forces. *Vehicle System Dynamics* 52 (12): 1716–1728.

14 Ding, X., Wang, Z., and Zhang, L. (2022). Event-triggered vehicle sideslip angle estimation based on low-cost sensors. *IEEE Transactions on Industrial Informatics* 18 (7): 4466–4476.

15 Ping, X., Cheng, S., Yue, W. et al. (2020). Adaptive estimations of tyre–road friction coefficient and body's sideslip angle based on strong tracking and interactive multiple model theories. *Proceedings of the Institution of Mechanical Engineers, Part D: Journal of Automobile Engineering* 234 (14): 3224–3238.

16 Xia, X., Hang, P., Xu, N. et al. (2021). Advancing estimation accuracy of sideslip angle by fusing vehicle kinematics and dynamics information with fuzzy logic. *IEEE Transactions on Vehicular Technology* 70 (7): 6577–6590.

17 Naets, F., van Aalst, S., Boulkroune, B. et al. (2017). Design and experimental validation of a stable two-stage estimator for automotive sideslip angle and tire parameters. *IEEE Transactions on Vehicular Technology* 66 (11): 9727–9742.

18 Li, X., Chan, C.-Y., and Wang, Y. (2016). A reliable fusion methodology for simultaneous estimation of vehicle sideslip and yaw angles. *IEEE Transactions on Vehicular Technology* 65 (6): 4440–4458.

19 Doumiati, M., Victorino, A.C., Charara, A., and Lechner, D. (2011). Onboard real-time estimation of vehicle lateral tire-road forces and sideslip angle. *IEEE/ASME Transactions on Mechatronics* 16 (4): 601–614.

20 Nam, K., Fujimoto, H., and Hori, Y. (2012). Lateral stability control of in-wheel-motor-driven electric vehicles based on sideslip angle estimation using lateral tire force sensors. *IEEE Transactions on Vehicular Technology* 61 (5): 1973–1985.

21 Kim, B., Yi, K., Yoo, H.-J. et al. (2015). An IMM/EKF approach for enhanced multitarget state estimation for application to integrated risk management system. *IEEE Transactions on Vehicular Technology* 64 (3): 876–889.

22 Li, L., Jia, G., Ran, X. et al. (2014). A variable structure extended Kalman filter for vehicle sideslip angle estimation on a low friction road. *Vehicle System Dynamics* 52 (2): 280–308.

23 Zhang, F., Wang, Y., Hu, J. et al. (2021). A novel comprehensive scheme for vehicle state estimation using dual extended H-infinity Kalman filter. *Electronics* 10 (1526): 1–17.

24 Zhang, H., Zhang, G., and Wang, J. (2016). Sideslip angle estimation of an electric ground vehicle via finite-frequency H_∞ approach. *IEEE Transactions on Transportation Electrification* 2 (2): 200–209.

25 Wang, X. and Yaz, E.E. (2014). Stochastically resilient extended Kalman filtering for discrete-time nonlinear systems with sensor failures. *International Journal of Systems Science* 1 (9): 1393–1401.

26 Wang, Y., Xu, L., Zhang, F. et al. (2021). An adaptive fault-tolerant EKF for vehicle state estimation with partial missing measurements. *IEEE/ASME Transactions on Mechatronics* 26 (3): 1318–1327.

27 Wang, Y., Wei, H., Hu, B., and Lv, C. (2023). Robust estimation of vehicle dynamic state using a novel second-order fault-tolerant extended Kalman filter. *SAE International Journal of Vehicle Dynamics, Stability, and NVH* 7 (3): 301–311.

28 Yuan, H. and Song, X. (2022). A modified EKF for vehicle state estimation with partial missing measurements. *IEEE Signal Processing Letters* 29: 1594–1598.

29 Shangguan, J., Yue, M., Xu, C., and Zhao, J. (2023). Robust fault-tolerant estimation of sideslip and roll angles for distributed drive electric buses with stochastic passenger mass. *IEEE Transactions on Intelligent Transportation Systems* 24 (12): 14480–14489.

30 Wang, Y., Chen, H., Yin, G. et al. (2023). Motion state estimation of preceding vehicles with packet loss and unknown model parameters. *IEEE/ASME Transactions on Mechatronics* https://doi.org/10.1109/TMECH.2023.3345956.

31 Boada, B.L., Boada, M.J.L., and Diaz, V. (2016). Vehicle sideslip angle measurement based on sensor data fusion using an integrated ANFIS and an Unscented Kalman Filter algorithm. *Mechanical Systems and Signal Processing* 72 (73): 833–845.

32 Li, Y., Yin, G., Zhuang, W. et al. (2018). Compensating delays and noises in motion control of autonomous electric vehicles by using deep learning and unscented Kalman predictor. *IEEE Transactions on Systems, Man, and Cybernetics: Systems* doi: 10.1109/TSMC.2018.2850367.

33 Novi, T., Capitani, R., and Annicchiarico, C. (2019). An integrated artificial neural network–unscented Kalman filter vehicle sideslip angle estimation based on inertial measurement unit measurements. *Proceedings of the Institution of Mechanical Engineers, Part D: Journal of Automobile Engineering* 233 (7): 1864–1878.

34 Antonov, S., Fehn, A., and Kugi, A. (2011). Unscented Kalman filter for vehicle state estimation. *Vehicle System Dynamics* 49 (9): 1497–1520.

35 Arasaratnam, I. and Haykin, S. (2009). Cubature Kalman filters. *IEEE Transactions on Automatic Control* 54 (6): 1254–1269.

36 Cheng, S., Li, L., and Chen, J. (2018). Fusion algorithm design based on adaptive SCKF and integral correction for sideslip angle observation. *IEEE Transactions on Industrial Electronics* 65 (7): 5754–5763.

37 Li, J. and Zhang, J. (2016). Vehicle sideslip angle estimation based on hybrid Kalman filter. *Mathematical Problems in Engineering* 2016 (1): 3269142.

38 Jin, X., Yin, G., and Hanif, A. (2016). Cubature kalman filter-based state estimation for distributed drive electric vehicles. *In: Proceedings of the Chinese Control Conference (CCC)*, Chengdu, China, pp. 9038–9042.

39 Wei, W., Bei, S., Zhu, K., Zhang, L., and Wang, Y. (2016). Vehicle state and parameter estimation based on adaptive cubature kalman filter. *In: Proceedings of the International Conference on Innovative Computing, Information and Control (ICIC)*, pp. 1871–1877.

40 Sun, Y. and Chen, Q. (2016). Joint estimation of states and parameters of vehicle model using cubature kalman filter. *In: Proceedings of the IEEE International Conference on Systems, Man, and Cybernetics (SMC)*, Budapest, Hungary, pp. 000977–000982.

41 Xin, X., Chen, J., and Zou, J. (2014). Vehicle state estimation using cubature kalman filter. *In: Proceedings of the IEEE 17th International Conference on Computational Science and Engineering (CSE)*, pp. 44–48.

42 Xing, Y. and Lv, C. (2020). Dynamic state estimation for the advanced brake system of electric vehicles by using deep recurrent neural networks. *IEEE Transactions on Industrial Electronics* 67 (11): 9536–9547.

43 Boada, B., Boada, M.-J., Vargas-Melendez, L., and Diaz, V. (2018). A robust observer based on H_∞ filtering with parameter uncertainties combined with neural networks for estimation of vehicle roll angle. *Mechanical Systems and Signal Processing* 99: 611–623.

44 Boada, B., Boada, M.-J., Gauchia, A. et al. (2015). Sideslip angle estimator based on ANFIS for vehicle handling and stability. *Journal of Mechanical Science and Technology* 29 (4): 1473–1481.

45 Melzi, S. and Sabbioni, E. (2011). On the vehicle sideslip angle estimation through neural networks: Numerical and experimental results. *Mechanical Systems and Signal Processing* 25 (6): 2005–2019.

46 Dahal, P., Mentasti, S., Paparusso, L. et al. (2024). RobustStateNet: robust ego vehicle state estimation for autonomous driving. *Robotics and Autonomous Systems* 172.

47 Kim, D., Min, K., Kim, H., and Huh, K. (2020). Vehicle sideslip angle estimation using deep ensemble-based adaptive Kalman filter. *Mechanical Systems and Signal Processing* 144: 106862.

48 Kim, D., Kim, G., Choi, S., and Huh, K. (2021). An integrated deep ensemble-unscented Kalman filter for sideslip angle estimation with sensor filtering network. *IEEE Access* 9: 149681–149689.

49 Badji, B., Fenaux, E., Bagdouri, M.E., and Miraoui, A. (2009). Nonlinear single track model analysis using Volterra series approach. *Vehicle System Dynamics* 47 (1): 81–98.

50 Berntorp, K. and Di Cairano, S. (2019). Tire-stiffness and vehicle-state estimation based on noise-adaptive particle filtering. *IEEE Transactions on Control Systems Technology* 27 (3): 1100–1114.

51 Han, K., Choi, M., and Choi, S. (2018). Estimation of tire cornering stiffness as a road surface classification indicator using understeering characteristics. *IEEE Transactions on Vehicular Technology* 67 (8): 6851–6860.

52 Rajamani, R., Phanomchoeng, G., Piyabongkarn, D. et al. (2012). Algorithms for real-time estimation of individual wheel tire-road friction coefficients. *IEEE/ASME Transactions on Mechatronics* 17 (6): 1183–1195.

53 Zhang, Q., Meng, X., Zhang, S., and Wang, Y. (2015). Singular value decomposition-based robust Cubature Kalman filtering for an integrated GPS/SINS navigation system. *Journal of Navigation* 68: 549–562.

54 Li, W. and Jia, Y. (2010). H-infinity filtering for a class of nonlinear discrete-time systems based on unscented transform. *Signal Processing* 90 (12): 3301–3307.

55 Sibley, G., Sukhatme, G.S., and Matthies, L. (2006). The iterated sigma point kalman filter with applications to long range stereo. *Presented at the 2nd Robotics Science and Systems Conference*, Philadelphia, USA, (16–19 August 2006).

56 Zarei, J. and Shokri, E. (2015). Convergence analysis of nonlinear filtering based on cubature Kalman filter. *IET Science, Measurement and Technology* 9 (3): 294–305.

57 Bonci, A., Di Biase, A., Giannini, M.C., and Longhi, S. (2023). Yaw rate-based PID control for lateral dynamics of autonomous vehicles, design, and implementation. *In 2023 IEEE 28th International Conference on Emerging Technologies and Factory Automation (ETFA)*, Sinaia, Romania, pp. 1–8.

4

State Estimation of the Preceding Vehicle with Data Loss and Parameter Perturbations

4.1 Introduction

To optimize planning and decision-making in autonomous vehicles and thereby enhance road safety, accurate knowledge of the motion state of PVs is crucial. Current methodologies typically rely on specialized sensors mounted on the ego vehicle for estimating PVs' states. However, with advancements in information technology, there is a growing interest in using V2V communication for PVs' state estimation. This approach offers potential benefits but raises significant challenges that need addressing. One primary challenge is reducing the communication rate while ensuring accurate PVs' state estimation within the constraints of limited communication bandwidth. V2V communication must efficiently transmit and receive data without compromising the fidelity of state estimates. Moreover, wireless communication is susceptible to data packet loss, which can stem from various factors, including cyberattacks or network congestion. Packet loss disrupts the transmission of key sensor data, directly impacting the accuracy of vehicle state estimation algorithms.

Additionally, uncertainties in system dynamics and imprecise vehicle model parameters pose substantial obstacles to accurate state estimation. Factors such as unknown variations in road conditions, unpredictable weather impacts, or vehicle wear and tear can all affect system dynamics, complicating state estimation and control processes. Inaccurate vehicle model parameters, such as erroneous mass estimates or uncertain tire-road friction coefficients, further degrade the performance of state estimation algorithms. While considerable research has focused on state estimation for connected vehicles, few studies comprehensively address the simultaneous challenges posed by model parameter uncertainties and packet loss. This gap highlights a critical need for robust methodologies capable of mitigating these combined effects to ensure reliable vehicle state estimation under practical driving conditions. Furthermore, simulation studies and field experiments are essential to validate the effectiveness of proposed methodologies under diverse

State Estimation of Multi-Agent Vehicle-Road Interaction Systems, First Edition.
Yan Wang, Guodong Yin, and Chao Huang.
© 2026 The Institute of Electrical and Electronics Engineers, Inc. Published 2026 by John Wiley & Sons, Inc.

operating conditions. Addressing these challenges will not only advance the field of autonomous driving but also contribute significantly to improving road safety and traffic efficiency. By enhancing the accuracy and reliability of PVs' state estimation, intelligent connected vehicle systems can better anticipate and respond to dynamic driving environments, ultimately reducing the incidence of accidents and optimizing transportation logistics.

4.2 Related Works

Over the past few decades, there has been a growing research focus on ADAS and AVs due to their potential to significantly enhance road safety [1–3]. ADAS features like adaptive cruise control and emergency collision avoidance systems rely on real-time perception of the longitudinal motion of PVs to operate effectively. Additionally, the lateral motion state of PVs is crucial for AV control scenarios, such as when a PV performs a sudden lane change in front of the AV. In such situations, the AV must quickly assess the relative distance, vehicle velocity, and yaw rate of the PV to decide whether to brake or change lanes to avoid a collision. However, accurately measuring the lateral motion state of PVs poses a significant challenge for advanced sensors [4]. While some sophisticated lidar-like sensors can capture PVs' state information, their high cost prevents widespread adoption in mass-produced vehicles. Consequently, extensive research efforts have been dedicated to developing accurate methods for estimating PVs' motion states in recent years. These studies aim to leverage existing sensor data and advanced algorithms to reliably infer critical PV dynamics without the need for prohibitively expensive hardware. The pursuit of effective PV motion state estimation is driven by the need to enhance the operational safety and decision-making capabilities of AVs in complex traffic scenarios. By improving the accuracy of these estimations, researchers aim to enable AVs to react swiftly and appropriately to dynamic changes in their surroundings, thereby reducing the risk of accidents and optimizing traffic flow.

Predicting the longitudinal motion state of PVs using a car-following model is an effective method. A car-following model is a foundational concept in transportation engineering and traffic flow theory, essential for simulating how vehicles adjust their speed and position relative to PVs on roadways. These models, such as Gipps' model, Keller-Segel model, and the intelligent driver model, incorporate factors like driver reaction times, safety distances, and vehicle dynamics to predict how drivers respond to changes in traffic conditions. They are crucial for optimizing traffic management strategies, enhancing road safety, and developing ITS. Ongoing research focuses on refining these models to accommodate diverse traffic environments, integrating with emerging technologies like autonomous vehicles and connected infrastructure, and improving accuracy in predicting and managing traffic flow dynamics. In [5], the Gazis-Herman-Rothery model was constructed, which used the motion function of the ego vehicle (EV) relative to the PV to describe the acceleration change of the EV. Subsequently, a number of similar models have been reported in [6, 7]. This category of approach usually requires only the relative distance between the PV and the EV to be capable of estimating the motion state of PVs. However, these models assume that the vehicle moved along a straight line and that rotational motion is not considered. Thus, some curvilinear models are used to compensate for this deficiency. It was demonstrated in [8] that the constant turn rate and velocity (CTRV) model and the constant turn rate and acceleration (CTRA) model have higher estimation accuracy. The CTRV model and the CTRA model are both widely used in the field of state estimation and motion prediction for vehicles. The CTRV model assumes that the vehicle moves with a constant velocity and undergoes a constant rate of turn, making it suitable for scenarios where the vehicle's speed remains relatively steady while it navigates curved paths. The state vector in the CTRV model typically includes position, velocity, and heading angle.

On the other hand, the CTRA model extends the CTRV model by incorporating a constant longitudinal acceleration. This model is more complex than the CTRV model but allows for a more realistic representation of vehicle dynamics, especially in scenarios where vehicles accelerate or decelerate while turning. The state vector in the CTRA model includes position, velocity, heading angle, and possibly acceleration Both models are frequently employed in applications such as autonomous driving, trajectory prediction, and sensor fusion. They serve as fundamental tools for Kalman filtering and other estimation techniques, enabling accurate prediction of vehicle trajectories and states under various driving conditions. The choice between CTRV and CTRA models depends on the specific requirements of the application, particularly the need to account for changes in velocity or acceleration during maneuvers while maintaining computational efficiency and accuracy in state estimation. Both CTRV and CTRA presume that the yaw rate and the longitudinal vehicle velocity are independent of each other. Nevertheless, the vehicle velocity is usually highly coupled with the yaw rate, especially in high-speed driving situations.

To better represent the dynamic characteristics of the vehicle, a bicycle model was used to predict the lateral state of the PV [4]. However, additional prediction is needed for the longitudinal velocity in the bicycle model. Correspondingly, [9, 10] proposed a combined bicycle model and car-following model or CTRV to predict the motion state of the PV, respectively. These types of approaches usually consider that longitudinal velocity and yaw rate are decoupled. In addition, the proposed methods in [4, 9, 10] supposed the intrinsic parameters of the PV are available in advance. Although modeling and information acquisition are essential in vehicle state estimation, the estimation accuracy is also directly related to the estimation algorithm. Among these algorithms, Kalman-based approaches have been the mainstream research direction. EKF [11] and variable structure EKF [12, 13] were developed to predict the vehicle state. In addition, fault-tolerant EKF was also proposed to predict vehicle velocity and sideslip angle in the presence of missing data [14]. The UKF is a variant of the traditional Kalman filter that addresses nonlinearities in system dynamics and measurement models more effectively. Unlike the EKF, which linearizes the nonlinear functions through first-order Taylor expansion, the UKF approximates the distribution of state variables and their uncertainties using a set of carefully chosen sample points, known as sigma points. These sigma points are propagated through the nonlinear functions, capturing the mean and covariance of the transformed variables more accurately than linearization. The key principle behind the UKF is to approximate the probability distribution of the state variables using a minimal set of sigma points, typically chosen to capture the mean and covariance of the distribution accurately up to the second order. This approach avoids the potentially large errors introduced by linearization in highly nonlinear systems, making it particularly suitable for systems with non-Gaussian and highly nonlinear dynamics.

In practice, the UKF operates in two main steps: prediction and update. During the prediction step, sigma points are propagated through the state transition function to estimate the predicted state and covariance. In the update step, these predicted sigma points are mapped through the measurement function to obtain the predicted measurement mean and covariance. These predicted values are then combined with the actual measurements using Bayes' rule to update the state estimate and its uncertainty.

The advantages of the UKF include its ability to handle arbitrary nonlinearities without requiring analytical derivatives, which simplifies the implementation compared to EKF. Moreover, by using sigma points to represent the distribution, the UKF preserves the Gaussian nature of the posterior distribution under nonlinear transformations, ensuring robustness and accuracy in state estimation. Applications of the UKF span various fields, including robotics, autonomous navigation, aerospace, and biomedical engineering, where accurate estimation of state variables from noisy and nonlinear measurements is critical. Its flexibility and robustness make it a preferred choice in scenarios where traditional linearization methods like EKF may fail to provide accurate results. Therefore, UKF [15, 16] and interactive multi-model UKF [17] algorithms were developed to enhance the estimation accuracy.

Recently, The CKF is an advanced variant of the traditional Kalman filter designed to handle nonlinear state estimation problems more accurately. Unlike the EKF, which linearizes nonlinear functions using first-order Taylor expansion, the CKF approximates the integral of a nonlinear function using numerical integration techniques, specifically the cubature rule. In the CKF, the state and covariance estimates are propagated through the nonlinear functions by evaluating the weighted sum of function evaluations at carefully chosen cubature points within the state space. These cubature points are symmetrically distributed around the mean of the Gaussian distribution representing the state estimate, ensuring accurate representation of the nonlinear transformations without the limitations of linearization. The main advantage of the CKF lies in its ability to provide more accurate estimates of state variables in highly nonlinear systems compared to EKF, which can suffer from divergence or poor performance in such scenarios. By approximating the integral of the nonlinear function using numerical methods, the CKF preserves the Gaussian nature of the posterior distribution, maintaining robustness and stability in state estimation. In practice, the CKF operates similarly to the traditional Kalman filter with two main steps: prediction and update. During the prediction step, the cubature points are propagated through the state transition function to estimate the predicted state and covariance. In the update step, these predicted cubature points are mapped through the measurement function to obtain the predicted measurement mean and covariance, which are then used to update the state estimate based on actual measurements. Applications of the CKF are found in various fields such as robotics, navigation, finance, and signal processing, where accurate estimation of nonlinear systems from noisy measurements is crucial. Its ability to handle nonlinearities effectively while maintaining computational efficiency makes it a preferred choice in scenarios where traditional linearization-based methods like EKF may not perform optimally. The CKF was also used for state estimation [18]. Robust CKF [19] and extended square root CKF [20] are representative methods for estimating vehicle state.

A V2V system is integral to modern automotive technology, enabling direct communication between vehicles via dedicated short-range communication or cellular vehicle-to-everything technologies. This system consists of onboard sensors like radar and cameras for real-time data collection, communication modules for broadcasting vehicle status and receiving updates from nearby vehicles, and sophisticated data processing algorithms for information fusion and decision-making. V2V systems enhance safety by providing timely alerts about nearby vehicles' movements and potential hazards, improve traffic efficiency through cooperative driving strategies like adaptive cruise control and cooperative merging, support autonomous vehicles in navigating complex traffic scenarios, facilitate emergency vehicle prioritization, and promote scalability and interoperability across different vehicle types and manufacturers. Overall, V2V technology plays a pivotal role in advancing road safety, traffic management, and future mobility solutions. As V2V communication has become an indispensable application in ITS [21], it becomes possible to exchange some intrinsic parameters such as mass and cornering stiffness between vehicles. This also means that the PV can transmit its motion state directly to the EV through V2V communication. In recent studies, researchers have explored the potential of V2V communication in various applications. Zhou et al. [22] hypothesized that longitudinal speed data of a PV could be acquired through V2V communication, leading to the development of a lateral state estimator. Similarly, Wei et al. [23] utilized V2V communication to derive the state of PVs and designed an EV tracking control system. However, a significant limitation arises from the EV's inability to consistently receive real-time motion states of the PV within the sampling interval. Consequently, relying solely on V2V communication for transmitting comprehensive motion information of the PV proves unreliable in practice. This challenge underscores the need for robust solutions to ensure accurate and timely data exchange in dynamic traffic environments.

Although many exciting research outcomes have been made in vehicle state estimation, existing research on state estimation of PVs based on V2V communication requires periodic transmission of sensor data from the PV to the EV. Unfortunately, the periodic transmission may take up more network bandwidth than required. How to balance the estimation accuracy of PVs state and communication rate in the connected vehicle environment has rarely been considered. Previous research work considering limited in-vehicle network resources using event-triggered methods in vehicle control is a useful reference [24]. In addition, traditional studies on lateral state estimation of PVs assume that the longitudinal velocity of the PV is known or design an additional estimator to predict the longitudinal velocity. This results in the dynamical coupling characteristics of the vehicle not being fully considered. On the other hand, the ECKF [25] has been shown to have higher filtering accuracy and better numerical stability than the CKF. The use of better-performing filters is also an effective means to improve the estimation accuracy of PVs state. In addition to the challenges posed by nonlinearities, another critical issue in wireless communication is the occurrence of data packet loss. This phenomenon can significantly impact the reliability and efficiency of data transmission. Data packets may be lost due to various factors, including network congestion, signal interference, hardware malfunction, or deliberate cyberattacks [26]. Cyberattacks, in particular, pose a serious threat by targeting communication channels to disrupt data flow or compromise information integrity. Mitigating these risks requires robust protocols and security measures to ensure the continuity and integrity of wireless data transmission, especially in sensitive applications such as autonomous systems, real-time monitoring, and secure data exchange environments. Addressing these challenges is essential for enhancing the resilience and performance of wireless communication networks in diverse operational contexts. Once some key sensor data is lost, it will inevitably affect the estimation accuracy of the vehicle state. Furthermore, unknown system dynamics usually has a direct impact on system state estimation and control [27]. Therefore, the imprecise model parameters of the vehicle, such as the vehicle mass, can directly reduce the estimation performance of the vehicle state. Although many research efforts are devoted to the state estimation of connected vehicles, these studies seldom consider the effect of model parameters perturbation and packet loss simultaneously.

4.3 Event-Triggered State Estimation for Connected Vehicles with Data Loss

In this section, an event-triggered estimation framework by fusing an event-triggered mechanism with an ECKF based on a 3-DOF model is proposed for state estimation of the PVs. The 3-DOF vehicle model is a simplified representation used in vehicle dynamics to analyze and predict vehicle behavior by focusing on three primary components: longitudinal, lateral, and yaw motions. This model operates under assumptions such as planar motion, small angle approximations, and simplified tire behavior. It is particularly suitable for normal driving conditions, control system design, simulations, and autonomous vehicle algorithms. The 3-DOF model offers several advantages, including simplicity, computational efficiency, and the ability to provide valuable insights into vehicle behavior, making it a practical and widely used tool in automotive engineering for vehicle design, control systems, and autonomous driving technologies. The event-triggered mechanism is employed to determine the optimal moments for transmitting acceleration information from PVs to the EV. This mechanism addresses the limitation of the conventional ECKF, which assumes continuous and uninterrupted transmission of measurement data. By incorporating the event-triggered mechanism into the ECKF framework, an enhanced model known as the ETECKF is developed. The ETECKF is designed to perform state estimation of PVs more efficiently, by predicting their longitudinal and lateral states using the transmitted acceleration data and the 3-DOF model. To validate the effectiveness of the ETECKF, a comprehensive set of simulations and experiments are conducted, demonstrating its capability to improve state estimation accuracy and reliability under conditions where data transmission is not continuous. This approach enhances the robustness of the state estimation process, making it more suitable for real-world applications where communication interruptions are common.

The overall framework, as illustrated in Figure 4.1, consists of four key modules: acquisition of onboard sensor data, event-triggered mechanism, sensor data update, and the ETECKF algorithm. First, the acquisition module collects longitudinal and lateral acceleration data through the controller area network (CAN)

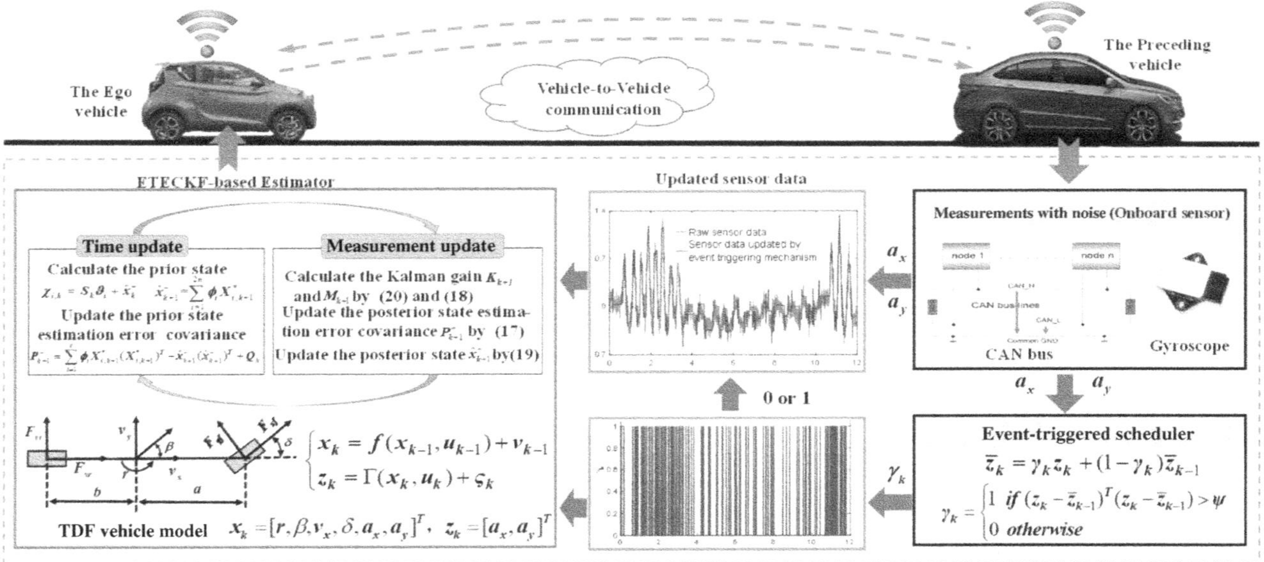

Figure 4.1 The event-triggered scheme for the PV state estimation.

bus. This raw sensor data provides the foundational information needed for further processing. Next, the event-triggered mechanism dynamically evaluates the incoming acceleration data and outputs a binary signal (0 or 1). This mechanism determines whether the current acceleration data should be used for updating based on predefined criteria, ensuring that only significant changes in acceleration are transmitted to the EV. In the sensor data update module, the raw acceleration data is adjusted according to the output of the event-triggered mechanism. When the output mechanism is 1, indicating a significant change, the current acceleration data is updated. If the output is 0, the previous value is retained, preventing unnecessary updates and reducing data transmission load. Finally, the updated acceleration data is fed into the ETECKF algorithm, which incorporates the 3-DOF model. This enhanced algorithm predicts the longitudinal and lateral states of the PVs based on the processed acceleration data. The integration of the event-triggered mechanism with the ETECKF algorithm ensures accurate and reliable state estimation of PVs, even in scenarios with intermittent data transmission. This structured approach, combining real-time data acquisition, dynamic event triggering, and advanced state estimation, provides a robust solution for maintaining accurate vehicle state information in real-world conditions.

4.3.1 Vehicle Model and Problem Statement

Considering real-time computing requirements and estimator design issues, a 3-DOF model (see Figure 4.2) is employed to describe the coupled dynamics characteristics of the PV.

The DOF model can be given in

$$\dot{r} = \frac{a^2k_f + b^2k_r}{I_z v_x} r + \frac{ak_f - bk_r}{I_z} \beta - \frac{ak_f}{I_z} \delta \tag{4.1}$$

$$\dot{\beta} = \left(\frac{ak_f - bk_r}{m{v_x}^2} - 1 \right) r + \frac{k_f + k_r}{mv_x} \beta - \frac{k_f}{mv_x} \delta \tag{4.2}$$

$$\dot{v}_x = r\beta v_x + a_x \tag{4.3}$$

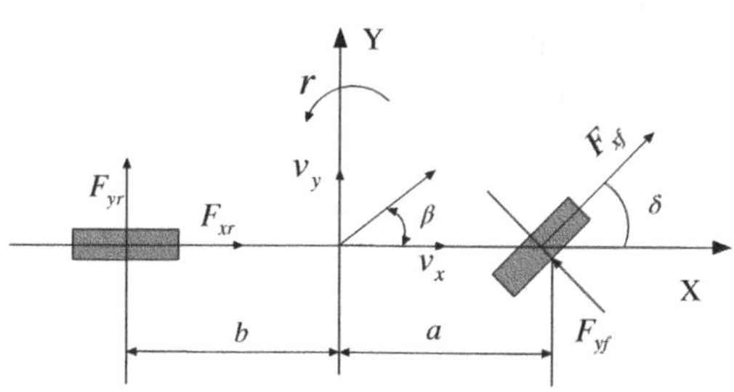

Figure 4.2 The 3-DOF model.

$$a_y = \frac{ak_f - bk_r}{mv_x} r + \frac{k_f + k_r}{m} \beta - \frac{k_r}{m} \delta \tag{4.4}$$

$$v_y = v_x \tan(\beta) \tag{4.5}$$

where a and b are distances from the center of gravity to front axle and rear axle, m is vehicle mass, a_x and a_y are longitudinal acceleration and lateral acceleration, δ is angle of front wheel, I_z is the inertia moment about the vehicle vertical axis, F_{yf} and F_{yr} are lateral forces of front axle and rear axle, F_{xf} and F_{xr} are longitudinal forces of front axle and rear axle, k_f and k_r are the front and rear axle cornering stiffnesses, r is yaw rate, β is sideslip angle, and v_x and v_y are longitudinal velocity and lateral velocity.

We estimate the longitudinal velocity, lateral velocity, and yaw rate of the PV based on the longitudinal and lateral acceleration of the PV obtained from V2V communication. To estimate vehicle state using discrete sensor data, Eq. (4.1–4.4) can be transformed into discrete-time state-space models.

$$\begin{cases} x_k = f(x_{k-1}, u_{k-1}) + v_{k-1} \\ z_k = \Gamma(x_k, u_k) + \varsigma_k \end{cases} \tag{4.6}$$

$$x_k = [r, \beta, v_x, \delta, a_x, a_y]^T, \quad z_k = [a_x, a_y]^T$$

$$\begin{bmatrix} r_k \\ \beta_k \\ v_{x,k} \\ \delta_k \\ a_{x,k} \\ a_{y,k} \end{bmatrix} = \begin{bmatrix} r_{k-1} + \left(\frac{a^2 k_f + b^2 k_r}{I_z v_{x,k-1}} r_{k-1} + \frac{ak_f - bk_r}{I_z} \beta_{k-1} - \frac{ak_f}{I_z} \delta_{k-1} \right) \Delta t \\ \beta_{k-1} + \left(\left(\frac{ak_f - bk_r}{mv^2_{x,k-1}} - 1 \right) r_{k-1} + \frac{k_f + k_r}{mv_{x,k-1}} \beta_{k-1} - \frac{k_f}{mv_{x,k-1}} \delta_{k-1} \right) \Delta t \\ v_{x,k-1} + (r_{k-1} \beta_{k-1} v_{x,k-1} + a_x) \Delta t \\ \delta_{k-1} \\ a_{x,k-1} \\ a_{y,k-1} \end{bmatrix} \tag{4.7}$$

$$\begin{bmatrix} a_{x,k} \\ a_{y,k} \end{bmatrix} = \begin{bmatrix} 0 & 0 & 0 & 1 & 0 \\ 0 & 0 & 0 & 0 & 1 \end{bmatrix} x_k \tag{4.8}$$

where k is the sampling instant, x_k is the state variables, u_k is the control input, z_k is the measurement output, and f and Γ are the process and measurement functions, respectively. Δt is the sampling interval, v_k is the system process noise, Q_k is its covariance, ς_k is the system measurement noise, R_k is its covariance, and v_k and ς_k are uncorrelated with each other. The initial sate x_0 is also independent of all noise signals.

Event-triggered data communication mechanisms can reduce data transmission while ensuring system performance [28]. Figure 4.3 presents an event-triggered mechanism for state estimation of PVs.

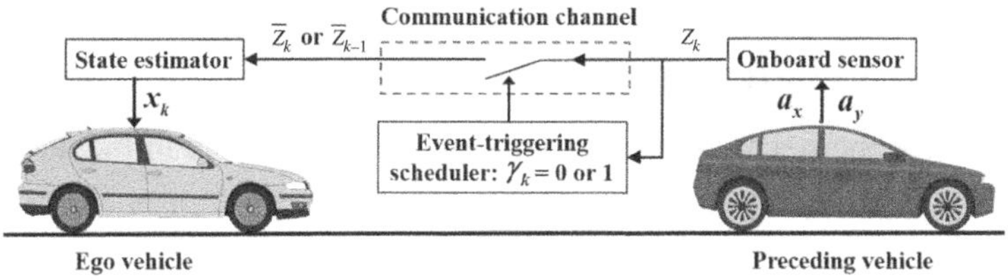

Figure 4.3 The event-triggered mechanism.

We use a variable γ_k to model measurement transmission. The γ_k can be expressed as

$$\gamma_k = \begin{cases} 1 & \text{if } (z_k - \bar{z}_{k-1})^T(z_k - \bar{z}_{k-1}) > \psi \\ 0 & \text{otherwise} \end{cases} \tag{4.9}$$

$$\bar{z}_k = \gamma_k z_k + (1 - \gamma_k)\bar{z}_{k-1} \tag{4.10}$$

where $\bar{z}_{k-1}$ represents the measurement previously transmitted by the sensor before time instant k, z_k is the current measurement output, $\bar{z}_k$ is the current transmitted measurement at time instant k, and ψ is the event-triggered threshold. Let γ_k be an indicator of whether the measurement data is transmitted. When $\gamma_k = 1$, the sensor data z_k of the PV is transmitted to the state estimator. And, $\gamma_k = 0$ means there is no data transmission.

4.3.2 Methodology

In this section, the ETECKF is derived to estimate the PV state under V2V communication. Some necessary lemmas are introduced first.

Lemma 4.1 [29] For any two vectors $x, y \in \mathbf{R}^n$, and $\sigma > 0$, one has

$$xy^T + yx^T \le \sigma xx^T + \sigma^{-1} yy^T \tag{4.11}$$

Lemma 4.2 [30] For any three matrices F, B, $H \in \mathbf{R}^{n\times n}$, if B, H are symmetric positive-definite, and $H - FBF^T > 0$, then

$$B^{-1} - F^T H^{-1} F > 0 \tag{4.12}$$

Lemma 4.3 [31] For a stochastic process $V_k(\xi_k)$ and real numbers $\underline{v}, \bar{v}, \eta > 0$, and $0 < \tau \le 1$ such that

$$\frac{1}{\bar{v}}\|\xi_k\|^2 \le V_k(\xi_k) \le \frac{1}{\underline{v}}\|\xi_k\|^2 \tag{4.13}$$

and

$$E[V_{k+1}(\xi_{k+1}) \mid \xi_k] \leq (1-\tau)V_k(\xi_k) + \eta \tag{4.14}$$

are fulfilled. Then the stochastic process is exponentially bounded in the mean square, i.e., we have

$$E\left[\|\xi_k\|^2\right] \leq \frac{\overline{v}}{\underline{v}} E\left[\|\xi_0\|^2\right](1-\tau)^k + \frac{\eta}{\underline{v}} \sum_{i=1}^{k-1}(1-\tau)^i \tag{4.15}$$

For $k > 0$.

Theorem 4.1 For the system described by Eq. (4.6) with an event-triggered mechanism, if there is no data packet loss and the following conditions

$$\begin{cases} E[v_k] = 0, & E[\varsigma_k] = 0, \quad E\left[v_k \varsigma_j^T\right] = 0, \\ E\left[v_k v_j^T\right] = Q_k \Omega_{k-j}, & E\left[\varsigma_k \varsigma_j^T\right] = R_k \Omega_{k-j}, \\ \Omega_{k-j} = 1(k=j); & \Omega_{k-j} = 0 \quad (k \neq j) \\ E\left[v_k x_0^T\right] = 0, & E\left[\varsigma_k x_0^T\right] = 0 \end{cases} \tag{4.16}$$

are fulfilled, then the estimation error covariance $\widetilde{P}_{k+1}^{+}$ is represented as

$$\begin{aligned} \widetilde{P}_{k+1}^{+} &= P_{k+1}^{-} - \gamma_{k+1} K_{k+1} P_{zz,k+1} K_{k+1}^T + (1-\gamma_{k+1})\left[(1+\sigma_1)\right. \\ &\quad \times (I - M_{k+1}\beta_{k+1}C_{k+1})P_{k+1}^{-}(I - M_{k+1}\beta_{k+1}C_{k+1})^T + (1+\sigma_2) \\ &\quad \times M_{k+1}R_{k+1}M_{k+1}^T + \left(1+\sigma_1^{-1}+\sigma_2^{-1}\right) M_{k+1}\psi M_{k+1}^T - \left.P_{k+1}^{-}\right] \end{aligned} \tag{4.17}$$

and the filter gain M_{k+1} is given by

$$\begin{aligned} M_{k+1} &= (1+\sigma_1)P_{k+1}^{-} C_{k+1}^T \beta_{k+1}^T \left[(1+\sigma_1)\beta_{k+1}C_{k+1}P_{k+1}^{-}\right. \\ &\quad \times C_{k+1}^T \beta_{k+1}^T + (1+\sigma_2)R_{k+1} + \left.\left(1+\sigma_1^{-1}+\sigma_2^{-1}\right)\psi I\right]^{-1} \end{aligned} \tag{4.18}$$

where $(\cdot)^T$ is matrix transpose, σ_1 and σ_2 are positive scalars, β_{k+1} is an unknown diagonal matrix representing the error incurred in neglecting the higher-order terms, Ω_{k-j} is the Kronecker delta function, C_{k+1} is a Jacobian matrix, $(\cdot)^{-1}$ is the inverse of a matrix, I is the identity matrix, K_{k+1} and M_{k+1} are the estimator gains, and P_{k+1}^{-} and $P_{zz,k+1}$ are the prior state estimation error covariance and predicted measurement covariance, respectively.

Proof: Similar to [32], the recursive filter with event-triggered data communication is expressed as follows

$$\widehat{x}_{k+1}^{+} = \widehat{x}_{k+1}^{-} + \gamma_{k+1}K_{k+1}\left(z_{k+1} - \widehat{z}_{k+1}^{-}\right) + (1-\gamma_{k+1})M_{k+1}\left(\overline{z}_k - \widehat{z}_{k+1}^{-}\right) \tag{4.19}$$

where $\widehat{z}_{k+1}^{-} = \Gamma\left(\widehat{x}_{k+1}^{-}, u_{k+1}\right)$, $\widehat{x}_{k+1}^{+}$ is a posteriori estimate of at time $k+1$, $\widehat{x}_{k+1}^{-}$ is a priori estimate of at time $k+1$, and the state variables $e_{k+1}^{+} = x_{k+1} - \widehat{x}_{k+1}^{+}$ and

$e^{-}_{k+1} = x_{k+1} - \hat{x}^{-}_{k+1}$ represent the system posterior and prior state estimation errors, respectively.

If $\gamma_{k+1} = 1$, the estimator gain matrix K_{k+1} is the same as ECKF.

$$K_{k+1} = P_{xz,k+1} P^{-1}_{zz,k+1} \tag{4.20}$$

$$P^{+}_{k+1} = P^{-}_{k+1} - K_{k+1} P_{zz,k+1} K^{T}_{k+1} \tag{4.21}$$

If $\gamma_{k+1} = 0$, Eq. (4.19) is given by

$$\hat{x}^{+}_{k+1} = \hat{x}^{-}_{k+1} + M_{k+1}\left(z_{k+1} - \hat{z}^{-}_{k+1}\right) + M_{k+1}(\bar{z}_k - z_{k+1}) \tag{4.22}$$

Linearizing $f(x_k, u_k)$ by performing first-order Taylor series expansion for $\hat{x}^{+}_{k}$.

$$\begin{cases} f(x_k, u_k) = f\left(\hat{x}^{+}_{k}, u_k\right) + \alpha_k A_k e^{+}_{k} \\ A_k = \frac{\partial f(x_k, u_k)}{\partial x_k}\Big|_{x_k = \hat{x}^{+}_{k}} \\ \alpha_k = diag(\alpha_{1,k}, \alpha_{2,k}, \dots, \alpha_{n,k}) \end{cases} \tag{4.23}$$

$$e^{-}_{k+1} = \alpha_k A_k e^{+}_{k} + v_k \tag{4.24}$$

where α_k is an unknown diagonal matrix representing the higher-order terms.

Linearizing $\Gamma(x_{k+1}, u_{k+1})$ by performing first-order Taylor series expansion for $\hat{x}^{-}_{k+1}$.

$$\begin{cases} \Gamma(x_{k+1}, u_{k+1}) = \Gamma\left(\hat{x}^{-}_{k+1}, u_{k+1}\right) + \beta_{k+1} C_{k+1} e^{-}_{k+1} \\ C_{k+1} = \frac{\partial \Gamma(x_{k+1}, u_{k+1})}{\partial x_{k+1}}\Big|_{x_{k+1} = \hat{x}^{-}_{k+1}} \\ \beta_{k+1} = diag(\beta_{1,k+1}, \beta_{2,k+1}, \dots, \beta_{n,k+1}) \end{cases} \tag{4.25}$$

Subtracting Eq. (4.22) from Eq. (4.6), the estimation error e^{+}_{k+1} is given by

$$e^{+}_{k+1} = e^{-}_{k+1} - M_{k+1}\left(z_{k+1} - \hat{z}^{-}_{k+1}\right) - M_{k+1}(\bar{z}_k - z_{k+1}) \tag{4.26}$$

Based on Eq. (4.25), the estimation error e^{+}_{k+1} is arranged as

$$e^{+}_{k+1} = (I - M_{k+1}\beta_{k+1}C_{k+1})e^{-}_{k+1} - M_{k+1}\varsigma_{k+1} - M_{k+1}(\bar{z}_k - z_{k+1}) \tag{4.27}$$

According to Eq. (4.27), P^{+}_{k+1} can be expressed as

$$\begin{aligned} P^{+}_{k+1} &= E\left[e^{+}_{k+1}\left(e^{+}_{k+1}\right)^{T}\right] \\ &= (I - M_{k+1}\beta_{k+1}C_{k+1})P^{-}_{k+1}(I - M_{k+1}\beta_{k+1}C_{k+1})^{T} + M_{k+1} \\ &\quad \times R_{k+1}M^{T}_{k+1} + M_{k+1}E\left\{(\bar{z}_k - z_{k+1})(\bar{z}_k - z_{k+1})^{T}\right\}M^{T}_{k+1} - \Omega_{k+1} \\ &\quad - \mathbf{Z}_{k+1} - (\mathbf{Z}_{k+1})^{T} + \Psi_{k+1} - (\Omega_{k+1})^{T} + (\Psi_{k+1})^{T} \end{aligned} \tag{4.28}$$

where $\mathbf{Z}_{k+1} = (I - M_{k+1}\beta_{k+1}C_{k+1})E\left\{e^{-}_{k+1}\varsigma^{T}_{k+1}\right\}M^{T}_{k+1}$ and $\Omega_{k+1} = (I - M_{k+1}\beta_{k+1}C_{k+1})E\left\{e^{-}_{k+1}(\bar{z}_k - z_{k+1})^{T}\right\}M^{T}_{k+1}$

$\Psi_{k+1} = M_{k+1}E\left\{\varsigma_{k+1}(\bar{z}_k - z_{k+1})^{T}\right\}M^{T}_{k+1}$.

Because e_k and ς_k are mutually independent, $\mathbf{Z}_{k+1}$ is equal to zero. Based on Theorem 4.1 and the event-trigger mechanism Eq. (4.11), we can derive the following inequalities:

$$\begin{aligned}&-\Omega_{k+1}-(\Omega_{k+1})^T \leq \sigma_1(I-M_{k+1}\beta_{k+1}C_{k+1})P_{k+1}^-\\&\times(I-M_{k+1}\beta_{k+1}C_{k+1})^T+\sigma_1^{-1}M_{k+1}\psi M_{k+1}^T\\&\Psi_{k+1}+(\Psi_{k+1})^T \leq \sigma_2 M_{k+1}R_{k+1}M_{k+1}^T+\sigma_2^{-1}M_{k+1}\psi M_{k+1}^T\end{aligned} \tag{4.29}$$

where σ_1 and σ_2 are positive numbers. Substituting Eq. (4.29) into (4.28), the upper bound $\overline{P}_{k+1}^+$ for P_{k+1}^+ is obtained

$$\begin{aligned}\overline{P}_{k+1}^+ &= (1+\sigma_1)(I-M_{k+1}\beta_{k+1}C_{k+1})P_{k+1}^-(I-M_{k+1}\beta_{k+1}C_{k+1})^T\\&+(1+\sigma_2)M_{k+1}R_{k+1}M_{k+1}^T+\left(1+\sigma_1^{-1}+\sigma_2^{-1}\right)M_{k+1}\psi M_{k+1}^T\end{aligned} \tag{4.30}$$

Based on Eqs. (4.21) and (4.30), the upper bound of covariance $\widetilde{P}_{k+1}^+$ is expressed as

$$\begin{aligned}\widetilde{P}_{k+1}^+ &= P_{k+1}^- - \gamma_{k+1}K_{k+1}P_{zz,k+1}K_{k+1}^T+(1-\gamma_{k+1})\left[(1+\sigma_1)\right.\\&\times(I-M_{k+1}\beta_{k+1}C_{k+1})P_{k+1}^-(I-M_{k+1}\beta_{k+1}C_{k+1})^T+(1+\sigma_2)\\&\left.\times M_{k+1}R_{k+1}M_{k+1}^T+\left(1+\sigma_1^{-1}+\sigma_2^{-1}\right)M_{k+1}\psi M_{k+1}^T-P_{k+1}^-\right]\end{aligned} \tag{4.31}$$

Take the partial derivative of $\widetilde{P}_{k+1}^+$ with respect to M_{k+1}, and let the derivative be 0:

$$\frac{\partial\left(\widetilde{P}_{k+1}^+\right)}{\partial M_{k+1}}=0 \tag{4.32}$$

Then the optimal filter gain M_{k+1} is expressed as

$$\begin{aligned}M_{k+1} &= (1+\sigma_1)P_{k+1}^-C_{k+1}^T\beta_{k+1}^T\left[(1+\sigma_1)\beta_{k+1}C_{k+1}P_{k+1}^-\right.\\&\left.\times C_{k+1}^T\beta_{k+1}^T+(1+\sigma_2)R_{k+1}+\left(1+\sigma_1^{-1}+\sigma_2^{-1}\right)\psi I\right]^{-1}\end{aligned} \tag{4.33}$$

This completes the proof. The entire iterative process of the ETECKF algorithm is given in Algorithm 4.1.

Next, we will analyze the stochastic boundedness of the ETECKF. To facilitate the analysis, the error expression is simplified based on the method reported in [33], namely,

$$z_{k+1}-\widehat{z}_{k+1}^- = \beta_{k+1}C_{k+1}e_{k+1}^- + \varsigma_{k+1} \tag{4.34}$$

Similarly, the following covariance can be derived by

$$\begin{cases}P_{k+1}^- = \alpha_k A_k \widetilde{P}_k^+ A_k^T \alpha_k + Q_k\\P_{xz,k+1} = P_{k+1}^- C_{k+1}^T \beta_{k+1}\\P_{zz,k+1} = \beta_{k+1}C_{k+1}P_{k+1}^- C_{k+1}^T\beta_{k+1}+R_{k+1}\end{cases} \tag{4.35}$$

Algorithm 4.1 The ETECKF Algorithm

Step 1: Set $\hat{x}_0 = E(x_0)\Pi_0 = E[(x_0 - \hat{x}_0)(x_0 - \hat{x}_0)^T]$

The embedded cubature points ϑ_i and the weights ϕ_i are, respectively, given by

$$\begin{cases} \phi_i = 1 - \frac{1}{2\rho^2} \quad i = 1; \quad \phi_i = \frac{1}{2^{n+1}\rho^2} \quad i = 1, 2, \ldots 2^n + 1 \\ \vartheta_i = [0]_i \quad i = 1; \quad \vartheta_i = e_i = [eye(n), -eye(n)]\sqrt{2}\rho_1, \quad i = 2, 3, \ldots 2n + 1 \\ s_l^+ = \{e_i + e_j, i, j = 1, 2, \ldots n, i \neq j\}, \quad s_l^- = \{e_i - e_j, i, j = 1, 2, \ldots n, i \neq j\} \\ \widehat{s}_i^+ = [s_l^+, -s_l^+] * \rho_1, \quad \widehat{s}_i^- = [s_l^-, -s_l^-]\sqrt{2}\rho 1, \quad \vartheta_i = \widehat{s}_i^+, \quad i = 2n + 2, \ldots 4n + 1 \\ \vartheta_i = \widehat{s}_i^-, \quad i = 4n + 2, \ldots 6n + 1, \quad \vartheta_i = e_i\rho_2, \quad i = 6n + 2, \ldots 8n + 1 \\ \vartheta_i = \widehat{s}_i^+\rho_2, \quad i = 8n + 2, \ldots 10n + 1, \quad \vartheta_i = \widehat{s}_i^-\rho_2, \quad i = 10n + 2, \ldots 2^n - 10n - 1 \end{cases}$$

Step 2: Time update

Singular value decomposition of P_k^+: $P_k^+ = U_k S_k V_k^T$

Compute the Cubature Points: $\chi_{i,k} = S_k \vartheta_i^+ \hat{x}_k^+$

Propagate the Cubature Points: $X_{i,k+1}^* = f(\chi_{i,k}, U_k)$

Estimate the predicted state $\hat{x}_{k+1}^-$ and P_{k+1}^-:

$$\hat{x}_{k+1}^- = \sum_{i=1}^{2n} \phi_i X_{i,k+1}^*, \quad P_{k+1}^- = \sum_{i=1}^{c} \phi_i X_{i,k+1}^* (X_{i,k+1}^*)^T - \hat{x}_{k+1}^- (\hat{x}_{k+1}^-)^T + Q_k$$

Step 3: Measurement update:

Singular value decomposition of P_{k+1}^-: $P_{k+1}^- = U_{k+1} S_{k+1} V_{k+1}^T$

Compute the Cubature Points: $\chi_{i,k+1} = S_{k+1}\vartheta_i + \hat{x}_{k+1}^-$

Propagate the Cubature Points: $Z_{i,k+1}^* = h(\chi_{i,k+1}, U_{k+1})$

Estimate the predicted measurement $\hat{z}_{k+1}^-$, the innovation covariance matrix $P_{zz,k+1}^-$, and the cross-covariance matrix $P_{xz,k+1}^-$

$$\hat{z}_{k+1}^- = \sum_{i=1}^{2n} \phi_i Z_{i,k+1}^* P_{zz,k+1}^- = \sum_{i=1}^{2n} \phi_i Z_{i,k+1}^* (Z_{i,k+1}^*)^T - \hat{z}_{k+1}^- (\hat{z}_{k+1}^-)^T + R_{k+1},$$

$$P_{xz,k+1}^- = \sum_{i=1}^{2n} \phi_i \chi_{i,k+1} (Z_{i,k+1}^*)^T - \hat{x}_{k+1}^- (\hat{z}_{k+1}^-)^T$$

Compute the value of γ_{k+1} using Eq. (4.9) and the filter gain matrices K_{k+1} or M_{k+1} through Eqs. (4.20) and (4.18).

Estimate the updated state $\hat{x}_{k+1}^+$ and P_{k+1}^+ through Eqs. (4.19) and (4.17).

Step 4: Repeat steps 1–3 for the next sample.

Theorem 4.2 For the system described by Eq. (4.6) with an event-triggered mechanism, If there exist real constants $\underline{f}$, $\overline{f}$, $\underline{h}$, $\overline{h}$, $\underline{\beta}$, $\overline{\beta}$, $\underline{\alpha}$, $\overline{\alpha} \neq 0$ and $\overline{p}$, $\underline{pq}$, $\overline{q}$, $\underline{r}$, $\overline{r} > 0$ such that the following inequalities are satisfied for $k > 0$.

$$\begin{cases} \underline{q}I \le Q_k \le \overline{q}I, \quad \underline{r}I \le R_k \le \overline{r}I, \quad \underline{f}^2 I \le A_k A_k^T \le \overline{f}^2 I, \\ \underline{h}^2 I \le C_k C_k^T \le \overline{h}^2 I, \quad \underline{\alpha}^2 I \le \alpha_k \alpha_k^T \le \overline{\alpha}^2 I, \\ \underline{\beta}^2 I \le \beta_k \beta_k^T \le \overline{\beta}^2 I, \quad \underline{p} I \le P_{k+1}^- \le \overline{p} I, \quad \underline{\delta} \le \delta \le \overline{\delta}. \end{cases} \tag{4.36}$$

Then the prior state estimation error e_{k+1}^- is exponentially bounded in the mean square.

Proof: Based on the assumption that $\underline{p}I \le P_{k+1}^- \le \overline{p}I$, which can be derived by Theorem 4.5 in [30]. By defining the Lyapunov function $V_{k+1}\left(e_{k+1}^-\right) = \left(e_{k+1}^-\right)^T \left(P_{k+1}^-\right)^{-1} e_{k+1}^-$, we have

$$\frac{1}{\overline{p}}\left\|e_{k+1}^-\right\|^2 \le V_{k+1}\left(e_{k+1}^-\right) \le \frac{1}{\underline{p}}\left\|e_{k+1}^-\right\|^2 \tag{4.37}$$

According to Eq. (4.19), the unified form of e_{k+1}^+ is derived as

$$e_{k+1}^+ = e_{k+1}^- - \gamma_{k+1} K_{k+1}\left(z_{k+1} - \hat{z}_{k+1}^-\right) - (1-\gamma_{k+1}) M_{k+1}\left(\overline{z}_k - \hat{z}_{k+1}^-\right) \tag{4.38}$$

Combining Eqs. (4.25) and (4.38), then

$$\begin{aligned} e_{k+1}^+ = e_{k+1}^- &- \gamma_{k+1} K_{k+1}\left(\beta_{k+1} C_{k+1} e_{k+1}^- + \varsigma_{k+1}\right) \\ &- (1-\gamma_{k+1}) M_{k+1}\left[\beta_{k+1} C_{k+1} e_{k+1}^- + \varsigma_{k+1} - (z_{k+1} - \overline{z}_k)\right] \end{aligned} \tag{4.39}$$

Combining Eqs. (4.24), (4.34), and (4.39) to obtain

$$\begin{cases} e_{k+1}^- = \prod_{1,k} + \prod_{2,k} + \prod_{3,k} + v_k \\ \prod_{1,k} = \widehat{\prod}_{1,k} e_k^- = \alpha_k A_k [I - \gamma_k K_k \beta_k C_k - (1-\gamma_k) M_k \beta_k C_k] e_k^- \\ \prod_{2,k} = \widehat{\prod}_{2,k} \varsigma_k = \alpha_k A_k [-\gamma_k K_k - (1-\gamma_k) M_k] \varsigma_k \\ \prod_{3,k} = \widehat{\prod}_{3,k} (\overline{z}_{k-1} - z_k) = \alpha_k A_k (1-\gamma_k) M_k (z_k - \overline{z}_{k-1}) \end{cases} \tag{4.40}$$

Taking the conditional expectation for Eq. (4.40), we have

$$\begin{aligned} E\left\{V_{k+1}\left(e_{k+1}^-\right) \middle| e_k^-\right\} &= E\left\{\left(\prod_{1,k} + \prod_{2,k} + \prod_{3,k} + v_k\right)^T \left(P_{k+1}^-\right)^{-1}\right. \\ &\left.\times \left(\prod_{1,k} + \prod_{2,k} + \prod_{3,k} + v_k\right) \middle| e_k^-\right\} = E\left\{\prod_{1,k}^T \left(P_{k+1}^-\right)^{-1} \prod_{1,k} \middle| e_k^-\right\} + \mu_k \end{aligned} \tag{4.41}$$

$$\begin{cases} \mu_k = \mu_{1,k} + \mu_{2,k} + \mu_{3,k} + \mu_{4,k} \\ \mu_{1,k} = E\left\{ \Pi_{2,k}^T \left(P_{k+1}^-\right)^{-1} \Pi_{2,k} + \Pi_{3,k}^T \left(P_{k+1}^-\right)^{-1} \right. \\ \qquad \left. \times \Pi_{3,k} + (v_k)^T \left(P_{k+1}^-\right)^{-1} (v_k) \middle| e_k^- \right\} \\ \mu_{2,k} = E\left\{ \Pi_{1,k}^T \left(P_{k+1}^-\right)^{-1} \Pi_{3,k} + \Pi_{3,k}^T \left(P_{k+1}^-\right)^{-1} \Pi_{1,k} \middle| e_k^- \right\} \\ \mu_{3,k} = E\left\{ \Pi_{2,k}^T \left(P_{k+1}^-\right)^{-1} \Pi_{3,k} + \Pi_{3,k}^T \left(P_{k+1}^-\right)^{-1} \Pi_{2,k} \middle| e_k^- \right\} \\ \mu_{4,k} = E\left\{ v_k^T \left(P_{k+1}^-\right)^{-1} \Pi_{3,k} + \Pi_{3,k}^T \left(P_{k+1}^-\right)^{-1} v_k \middle| e_k^- \right\} \end{cases} \tag{4.42}$$

According to Eq. (4.35), then

$$P_{k+1}^- = \alpha_k A_k \widetilde{P}_k^+ A_k^T \alpha_k + Q_k \geq \left(1 + \frac{\underline{q}}{\overline{\alpha}^2 \overline{f}^2 \overline{p}^2}\right) \alpha_k A_k \widetilde{P}_k^+ A_k^T \alpha_k \tag{4.43}$$

Define $N_k = [I - \gamma_k K_k \beta_k C_k - (1-\gamma_k) M_k \beta_k C_k]$, $\sum_k = N_k P_k^- N_k^T$.

Based on Theorem 4.2 in [30] and combining Eqs. (4.20), (4.21), (4.30), and (4.35), the upper bound of $\widetilde{P}_k^+$ at time k satisfies $\widetilde{P}_k^+ > \sum_k$. Thus, Eq. (4.43) can be rewritten as

$$P_{k+1}^- > \left(1 + \frac{\underline{q}}{\overline{\alpha}^2 \overline{f}^2 \overline{p}^2}\right) [\alpha_k A_k N_k] P_k^- [\alpha_k A_k N_k]^T \tag{4.44}$$

According to Lemma 4.2, it is obtained that

$$\begin{cases} [\alpha_k A_k N_k]^T \left(P_{k+1}^-\right)^{-1} [\alpha_k A_k N_k] < \left(1 - \frac{\underline{q}}{\overline{\alpha}^2 \overline{f}^2 \overline{p}^2 + \underline{q}}\right) \left(P_k^-\right)^{-1} \\ \widehat{\Pi}_{1,k}^T \left(P_{k+1}^-\right)^{-1} \widehat{\Pi}_{1,k} = [\alpha_k A_k N_k]^T \left(P_{k+1}^-\right)^{-1} [\alpha_k A_k N_k] \end{cases} \tag{4.45}$$

$$\begin{aligned} E\left\{ \prod_{1,k}^{T} \left(P_{k+1}^-\right)^{-1} \prod_{1,k} \middle| e_k^- \right\} &\leq (1-\tau) E\left\{ \left(e_k^-\right)^T \left(P_k^-\right)^{-1} \left(e_k^-\right)^T \right\} \\ &\leq (1-\tau) E\left\{ V_k\left(e_k^-\right) \right\} \end{aligned} \tag{4.46}$$

where $\tau = \frac{\underline{q}}{\overline{\alpha}^2 \overline{f}^2 \overline{p}^2 + \underline{q}}$. It can be clearly seen that $0 < \tau < 1$.

Both sides of μ_k are scalars, and we can take trace on the terms of the right-hand side of μ_k without changing its value.

$$
\begin{cases}
\mu_{1,k} \leq \frac{1}{\underline{p}} E\left\{ tr\left(\Pi_{2,k}^T \Pi_{2,k}\right) + tr\left(\Pi_{3,k}^T \Pi_{3,k}\right) + tr\left(v_k^T v_k\right) \middle| e_k^- \right\} \\
\leq \frac{1}{\underline{p}} E\left\{ \overline{\alpha}^2 \overline{f}^2 (\overline{k} + \overline{m})^2 tr\left(\varsigma_k^T \varsigma_k\right) + \overline{\alpha}^2 \overline{f}^2 \overline{m}^2 \overline{\delta} \; + tr\left(v_k^T v_k\right) \right\} \\
= \frac{1}{\underline{p}} (\overline{\alpha}^2 \overline{f}^2 (\overline{k} + \overline{m})^2 \overline{r} n_\varsigma + \overline{\alpha}^2 \overline{f}^2 \overline{m}^2 \overline{\delta} + \overline{q} n_v) \\
\overline{m} = \frac{(1+\sigma_1)\overline{p}\overline{c}\overline{\beta}}{(1+\sigma_1)\underline{c}^2 \underline{\beta}^2 \underline{p} + (1+\sigma_2)\underline{r} + (1+\sigma_1^{-1}+\sigma_2^{-1})\underline{\delta}} \\
\overline{k} = \frac{\overline{p}\overline{c}\overline{\beta}}{\underline{c}^2 \underline{\beta}^2 \underline{p} + \underline{r}}
\end{cases}
\tag{4.47}
$$

Using Lemma 4.1, $tr(FH) = tr(HF)$, and computing the trace of $\mu_{2,k}$, $\mu_{3,k}$, and $\mu_{4,k}$, we have

$$
\begin{cases}
\mu_{2,k} \leq \frac{1}{\underline{p}} E\left\{ \sigma_3 tr\left(\Pi_{3,k} \Pi_{3,k}^T\right) + \sigma_3^{-1} tr\left(\Pi_{1,k} \Pi_{1,k}^T\right) \right\} \\
= \frac{1}{\underline{p}} \left(\sigma_3 \overline{\alpha}^2 \overline{f}^2 \overline{m}^2 \overline{\delta} + \sigma_3^{-1} \overline{\alpha}^2 \overline{f}^2 (1 + \overline{k}\overline{c}\overline{\beta} + \overline{m}\overline{c}\overline{\beta})^2 \overline{p} \right)
\end{cases}
\tag{4.48}
$$

$$
\begin{cases}
\mu_{3,k} \leq \frac{1}{\underline{p}} E\left\{ \sigma_4 tr\left(\Pi_{3,k} \Pi_{3,k}^T\right) + \sigma_4^{-1} tr\left(\Pi_{2,k} \Pi_{2,k}^T\right) \right\} \\
= \frac{1}{\underline{p}} \left(\sigma_4 \overline{\alpha}^2 \overline{f}^2 \overline{m}^2 \overline{\delta} + \sigma_4^{-1} \overline{\alpha}^2 \overline{f}^2 (\overline{k} + \overline{m})^2 \overline{r} n_\varsigma \right)
\end{cases}
\tag{4.49}
$$

$$
\begin{cases}
\mu_{4,k} \leq \frac{1}{\underline{p}} E\left\{ \sigma_5 tr\left(\Pi_{3,k} \Pi_{3,k}^T\right) + \sigma_5^{-1} tr\left(v_k^T v_k\right) \right\} \\
= \frac{1}{\underline{p}} \left(\sigma_5 \overline{\alpha}^2 \overline{f}^2 \overline{m}^2 \overline{\delta} + \sigma_5^{-1} \underline{q} n_v \right)
\end{cases}
\tag{4.50}
$$

where σ_3, σ_4, σ_5, n_ς, and n_v are positive scalars. Now, we define

$$
\eta = \frac{\left(1 + \sigma_2^{-1}\right) \overline{q} n_v + (1 + \sigma_3 + \sigma_4 + \sigma_5) \overline{\alpha}^2 \overline{f}^2 \overline{m}^2 \overline{\delta}}{\underline{p}} + \frac{\left(1 + \sigma_4^{-1}\right) \overline{\alpha}^2 \overline{f}^2 (\overline{k} + \overline{m})^2 \overline{r} n_\varsigma + \sigma_3^{-1} \overline{\alpha}^2 \overline{f}^2 (1 + \overline{k}\overline{c}\overline{\beta} + \overline{m}\overline{c}\overline{\beta})^2 \overline{p}}{\underline{p}}
\tag{4.51}
$$

According to Eqs. (4.46) and (4.51), we have

$$
E\left\{ V_{k+1}\left(e_{k+1}^-\right) \middle| e_k^- \right\} \leq (1 - \tau) E\left\{ V_k\left(e_k^-\right) \right\} + \eta
\tag{4.52}
$$

Therefore, based on Lemma 4.3, e_{k+1}^- is bounded in the mean square sense.

4.3.3 Simulation and Experiment Tests

To verify the effectiveness of the ETECKF algorithm, a comprehensive validation process is undertaken, encompassing both simulation and real vehicle experiments. The superiority of the ETECKF algorithm is demonstrated by comparing its estimation results with those obtained from the conventional CKF and the standard ECKF. In the simulation tests, the CarSim software is utilized and installed on two separate computers to create a realistic testing environment. These computers simulate two distinct vehicle models: one representing the PV and the other representing the EV. Data exchange between the two computers is facilitated through a network cable, employing the user datagram protocol (UDP) module in SIMULINK software to ensure efficient and real-time communication. CarSim software generates detailed and accurate vehicle models, providing a robust platform for evaluating the performance of the ETECKF algorithm. The simulation setup allows for controlled testing conditions, enabling a thorough comparison of the ETECKF, CKF, and ECKF algorithms under various scenarios. By generating two vehicle models on separate computers, the setup mimics real-world interactions between PVs and EVs, ensuring the results are both relevant and applicable.

Additionally, real vehicle experiments are conducted to validate the simulation results and further demonstrate the practical applicability of the ETECKF algorithm. These experiments involve equipping actual vehicles with the necessary sensors and communication modules to implement the ETECKF framework. By comparing the estimation accuracy and reliability across different algorithms, the experiments underscore the advantages of the ETECKF in real driving conditions. Through this rigorous validation process, the effectiveness and superiority of the ETECKF algorithm are clearly illustrated, showcasing its potential to enhance vehicle state estimation in both simulated and real-world environments.

4.3.3.1 Simulation Results

The trajectory of the PV in the simulation test (ST) is shown in Figure 4.4. The vehicle state values from CarSim are defined as reference values (RVs).

The parameters utilized in the CKF, ECKF, and ETECKF are the same. The communication rate γ is calculated as [30], i.e. $\gamma = \frac{1}{N}\sum_{k=1}^{N}\gamma_k$. According to this formula, the communication rate is 0.96% when $\psi = 0.0001$. Different methods to estimate the motion state of the PV are depicted in Figs. 4.5–4.7. The root mean square errors (RMSE) estimated by different methods are listed in Table 4.2.

In Figure 4.5, the longitudinal velocity curve estimated using the conventional CKF shows a noticeable deviation from the RV, indicating less accuracy. In contrast, the longitudinal velocity predicted by the ECKF closely aligns with the RV, demonstrating superior estimation performance. The ETECKF also performs

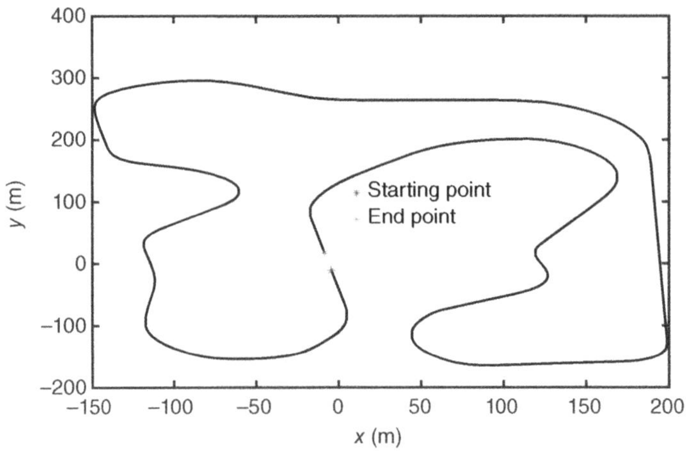

Figure 4.4 The trajectory of the PV in the ST.

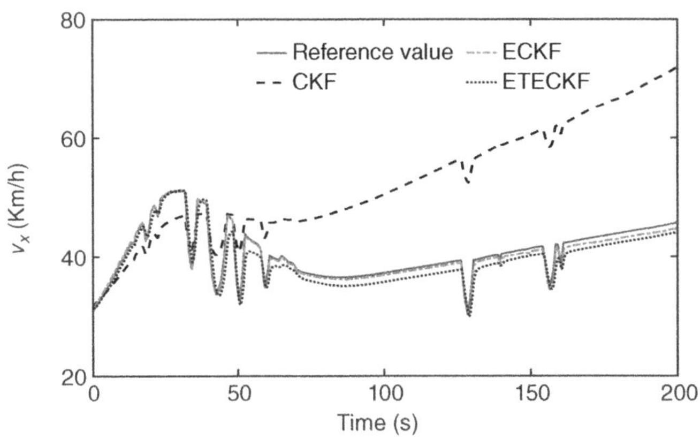

Figure 4.5 The curve of longitudinal velocity in the ST.

well, with its estimation closely matching that of the ECKF. Similarly, the lateral velocity and yaw rate estimation results depicted in Figs. 4.6 and 4.7 reinforce these findings. Both the ECKF and ETECKF algorithms outperform the CKF in these aspects, providing more accurate and reliable estimates. Table 4.1 further quantifies these results, clearly showing that among the three methods, the ECKF has the smallest estimation error, highlighting its superiority in terms of accuracy. Although the ETECKF experiences a slight degradation in performance due to the intermittent transfer of measurement data, its estimation accuracy remains within acceptable limits. Remarkably, even with 99.04% of the acceleration data not being transmitted, the ETECKF maintains satisfactory estimation

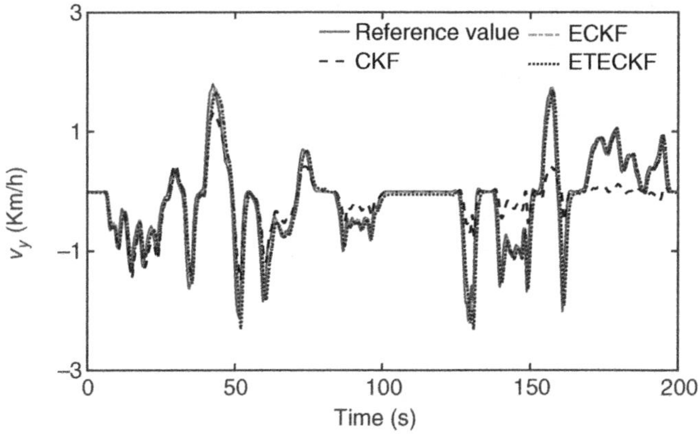

Figure 4.6 The curve of lateral velocity in the ST.

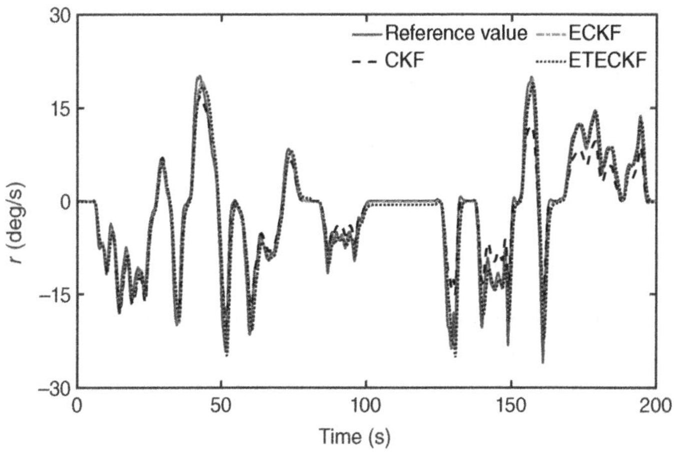

Figure 4.7 The curve of yaw rate in the ST.

performance. This demonstrates the robustness and reliability of the ETECKF algorithm, even under conditions of significant data loss. Overall, these results underscore the effectiveness of the ECKF and ETECKF algorithms in providing accurate vehicle state estimations, with the ECKF leading slightly in performance due to its continuous data transmission capabilities.

Figure 4.8 illustrates the variation of γ_k. Figures 4.9 and 4.10 provide a comparative analysis of the acceleration profiles used as inputs for the ETECKF algorithm versus the original acceleration profiles. A closer examination of the enlarged sections within these figures reveals that the acceleration signals fed into the ETECKF remain constant over certain time intervals. This constancy

Table 4.1 RMSE of ETECKF estimation under $\psi = 0.0001$.

Symbol	r	v_x	v_y
CKF	14.9423	0.4558	2.7099
ECKF	0.5702	0.2617	2.3655
ETECKF	1.6468	0.2992	2.6837

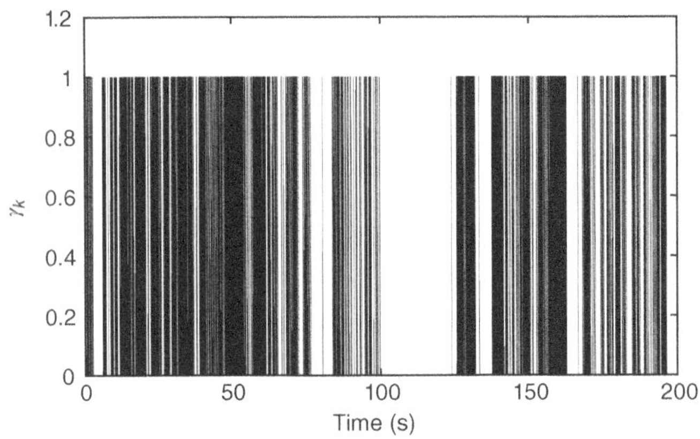

Figure 4.8 Event-triggered times in the ST.

indicates periods during which no new acceleration data is received, reflecting the event-triggered nature of the ETECKF algorithm. Specifically, during these intervals, the algorithm relies on previously received data to maintain state estimation, rather than incorporating new measurements. This behavior illustrates the algorithm's efficiency in managing data transmission, reducing the frequency of updates without compromising the overall estimation accuracy. By only triggering updates when significant changes in acceleration occur, the ETECKF minimizes unnecessary data transmission, thereby optimizing computational and communication resources. These findings underscore the ETECKF's capability to maintain accurate state estimation despite intermittent data inputs. The periods of constant acceleration input demonstrate the algorithm's robustness in handling scenarios where continuous data transmission is not feasible. This characteristic is particularly valuable in real-world applications where communication constraints or data loss might otherwise hinder performance. Overall, the comparison in Figs. 4.9 and 4.10 highlights the effective implementation of the event-triggered mechanism within the ETECKF, ensuring reliable and efficient vehicle state estimation.

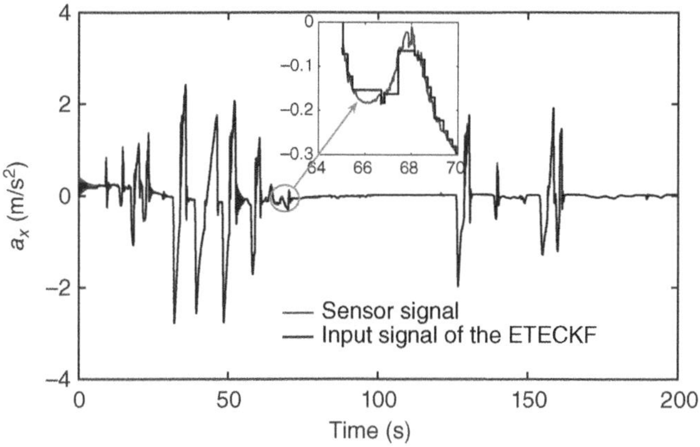

Figure 4.9 Comparison between the original longitudinal acceleration signal and the signal input to the ETECKF algorithm in the ST.

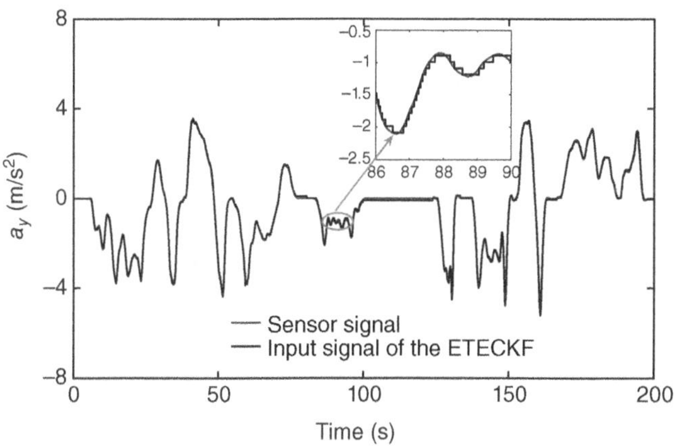

Figure 4.10 Comparison between the original lateral acceleration signal and the signal input to the ETECKF algorithm in the ST.

4.3.3.2 Real Vehicle Test Results

In a real vehicle test (RVT), an ARRIZO 5e equipped with an integrated navigation system (INS750D) is utilized as the PV, while a Chery eQ1 serves as the EV. These two vehicles, depicted in Figure 4.11, are employed to assess the effectiveness of the proposed method. The relevant sensor data is sampled at a high frequency of 100 Hz to ensure precise and detailed data collection. Data transmission between the two vehicles is facilitated through a wireless local area network using the UDP, which ensures efficient and reliable communication. The computer configuration for running the ETECKF algorithm and the MATLAB application is robust,

Figure 4.11 The two test vehicles.

featuring an AMD Ryzen 7 5800HS CPU and 16.0 GB of memory, as shown in Figure 4.12. This setup enables real-time processing and analysis, providing a suitable environment for evaluating the performance of the ETECKF algorithm under practical conditions. The use of high-frequency data sampling and advanced computational resources ensures that the test can accurately reflect the dynamics and interactions between the PV and EV, thereby validating the proposed method's applicability and effectiveness in real-world scenarios.

In the real vehicle experiments, the driver conducts lane-changing maneuvers to evaluate and verify the effectiveness of the ETECKF. Throughout these tests, the parameters used in the conventional CKF, the ECKF, and the ETECKF remain consistent to ensure a fair comparison. By maintaining the same parameter settings across all three algorithms, the experiments can accurately assess the performance improvements and robustness of the ETECKF in handling real-world dynamic driving scenarios, particularly during complex maneuvers such as lane changes. This rigorous testing process underscores the reliability and practical applicability of the ETECKF algorithm in enhancing vehicle state estimation and control. The communication rate γ is 37.55% when $\psi = 0.00012$. Different methods to estimate the motion state of the PV are depicted in Figs. 4.13–4.15. The RMSE estimated by different methods is listed in Table 4.2.

Consistent with the findings from simulation tests, both the ECKF and the ETECKF demonstrate superior estimation performance compared to the conventional CKF. Table 4.2 clearly illustrates that the ECKF exhibits the smallest estimation error among the three methods, highlighting its precision in vehicle state estimation. Although the performance of the ETECKF shows a slight degradation due to intermittent data transmission, its estimation accuracy remains within acceptable limits. Even with 62.45% of the measurement data not being

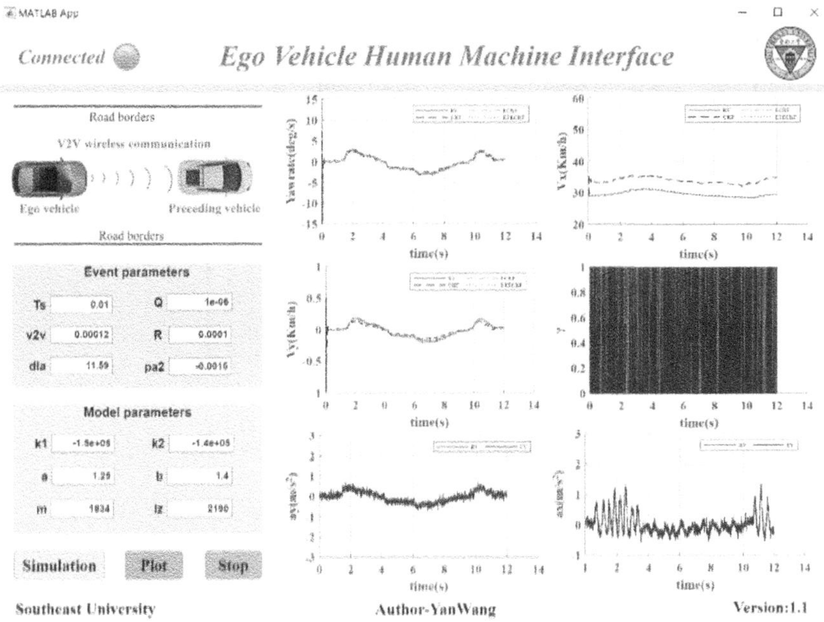

Figure 4.12 The app is used to configure the vehicle model and prediction algorithm parameters and to monitor the estimated results of the PV state.

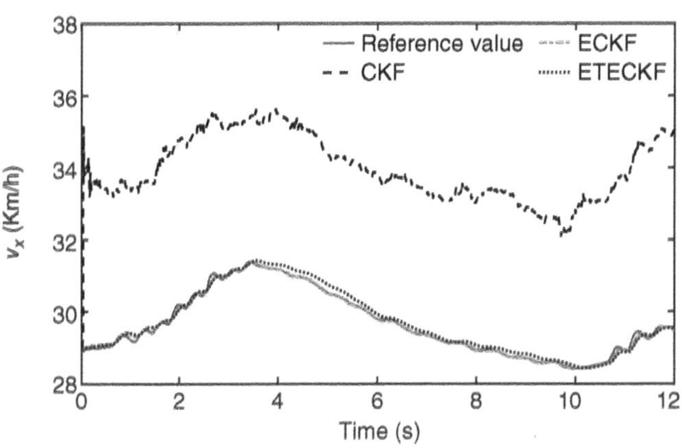

Figure 4.13 The curve of longitudinal velocity in the RVT.

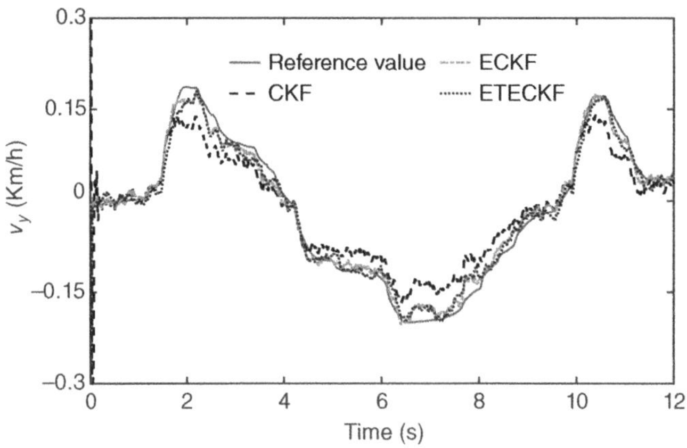

Figure 4.14 The curve of lateral velocity in the RVT.

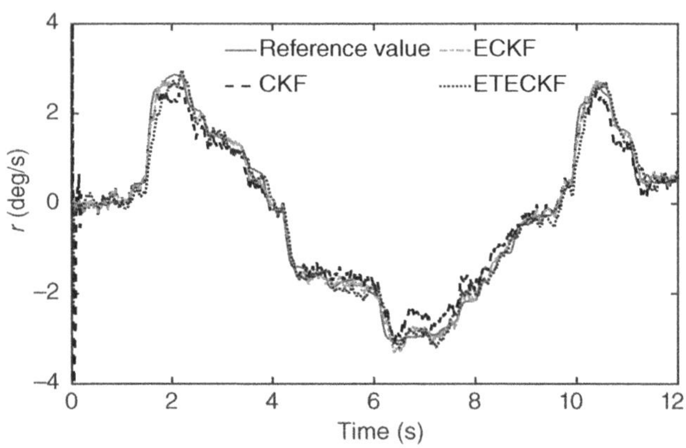

Figure 4.15 The curve of yaw rate in the RVT.

Table 4.2 RMSE of ETECKF estimation under $\psi = 0.00012$.

Symbol	r	v_x	v_y
CKF	0.3641	4.3191	0.0398
ECKF	0.1438	0.0574	0.0129
ETECKF	0.2470	0.1203	0.0171

transmitted, the ETECKF maintains satisfactory performance and consistently outperforms the CKF in real-world conditions. During the RVTs, it is observed that the estimated values of both ETECKF and CKF initially exhibit significant fluctuations. This behavior is attributed to several factors, including the inherent characteristics of the algorithms, the selection of parameters, and the dynamic updating of measurement data. These fluctuations stabilize over time as the algorithms adapt and refine their estimates based on continuous sensor inputs and data updates.

Figure 4.16 illustrates the variation of γ_k in the estimation process. Figures 4.17 and 4.18 present a detailed comparison of the acceleration profiles utilized as inputs for the ETECKF algorithm against the original acceleration profiles. Upon closer inspection of the enlarged sections within these figures, it becomes evident that the acceleration signal fed into the ETECKF remains unchanged over certain time intervals. This constancy indicates periods during which no new acceleration data is received, underscoring the event-triggered nature of the ETECKF algorithm. The consistent input signal illustrates the algorithm's capability to maintain state estimation without continuous updates, leveraging previously acquired data until significant changes in acceleration are detected. This approach optimizes computational resources and minimizes unnecessary data transmission, aligning with the algorithm's efficiency in managing communication bandwidth and processing requirements. These observations highlight the robustness of the ETECKF in handling scenarios where data transmission may be intermittent or limited. By effectively utilizing available data and triggering updates only when necessary, the ETECKF ensures reliable and accurate vehicle state estimation under varying real-world conditions.

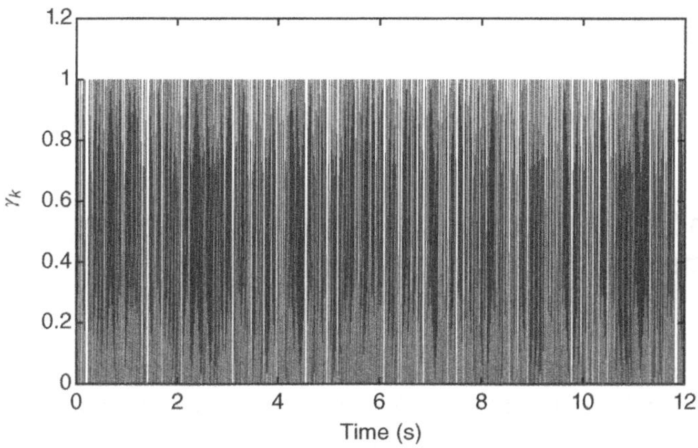

Figure 4.16 Event-triggered times in the RVT.

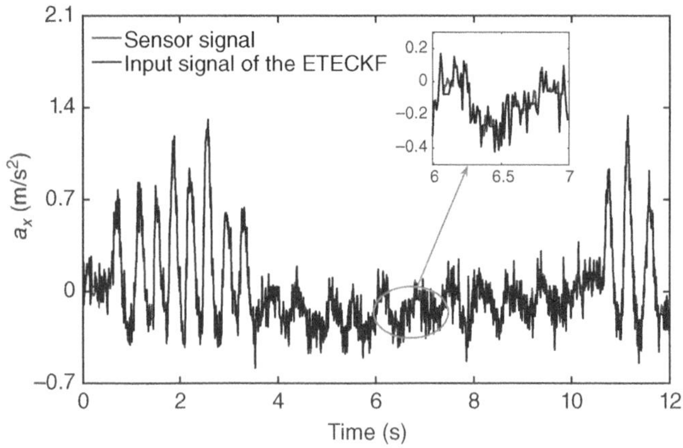

Figure 4.17 Comparison between the original longitudinal acceleration signal and the signal input to the ETECKF algorithm in the RVT.

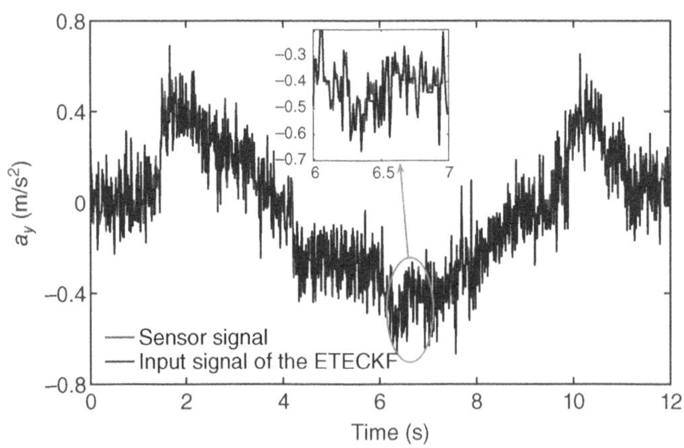

Figure 4.18 Comparison between the original lateral acceleration signal and the signal input to the ETECKF algorithm in the RVT.

4.4 Motion State Estimation of PVs with Unknown Model Parameters

In this section, we introduce an innovative event-triggered estimation framework tailored to address challenges such as packet loss and model parameter perturbations in predicting the state of PVs. The framework is designed to enhance the reliability and accuracy of state prediction in dynamic vehicular environments. The core of the proposed framework begins with deploying an event-triggered communication rule, which governs when the PV transmits its sensor data to the host vehicle (HV). This rule optimizes data transmission by sending updates only when significant changes in the PV's state occur, thereby conserving communication bandwidth and reducing computational load. Following this, an event-triggered cubature Kalman filter (ETCKF) is developed to perform the state estimation of the PV. Unlike traditional continuous-update filters, the ETCKF updates its estimates based on event-triggered rules, ensuring that computation occurs only when necessary data updates are received. This adaptive approach not only improves computational efficiency but also maintains accurate state estimation despite intermittent data availability. To further bolster the robustness of the ETCKF against uncertainties like model parameter variations, a strong tracking algorithm is integrated. This enhancement transforms the ETCKF into a strong tracking ETCKF (STETCKF). The STETCKF dynamically adjusts its estimation process to accommodate varying model parameters, thereby enhancing its ability to track the PV's state accurately over time. The theoretical convergence of the STETCKF is rigorously analyzed and proven, demonstrating its stability and reliability in real-world applications. Experimental validation further substantiates the effectiveness of the framework, conducted under controlled conditions to assess its performance in scenarios involving realistic levels of packet loss and model perturbations.

The entire estimation framework (as depicted in Figure 4.19) comprises five essential components, each playing a critical role in ensuring accurate and reliable state prediction for PVs in dynamic driving scenarios. These components include the CAN bus data collection module, event triggering mechanism, data loss simulation module, vehicle dynamics model, and STETCKF. To initiate the estimation process, the CAN bus data collection module acquires vital sensor information such as lateral acceleration and longitudinal velocity from the PV. These measurements serve as foundational data inputs for subsequent estimation procedures. The event-triggering mechanism operates next, evaluating the collected sensor data to determine the necessity of transmitting current sensor readings to the HV. This mechanism optimizes communication bandwidth by triggering data updates only when significant changes in the PV's state are detected, thereby enhancing efficiency and reducing unnecessary data transmission.

Subsequently, the data loss simulation module introduces controlled variability by applying a random variable following a Bernoulli distribution (ranging between 0 and 1) to the sensor data transmitted by the event trigger module. This simulation mimics real-world conditions of packet loss, ensuring the robustness of the estimation framework against communication interruptions. Finally, the STETCKF algorithm performs real-time state estimation of the PV based on the processed sensor data and the vehicle dynamics model. This advanced filtering technique

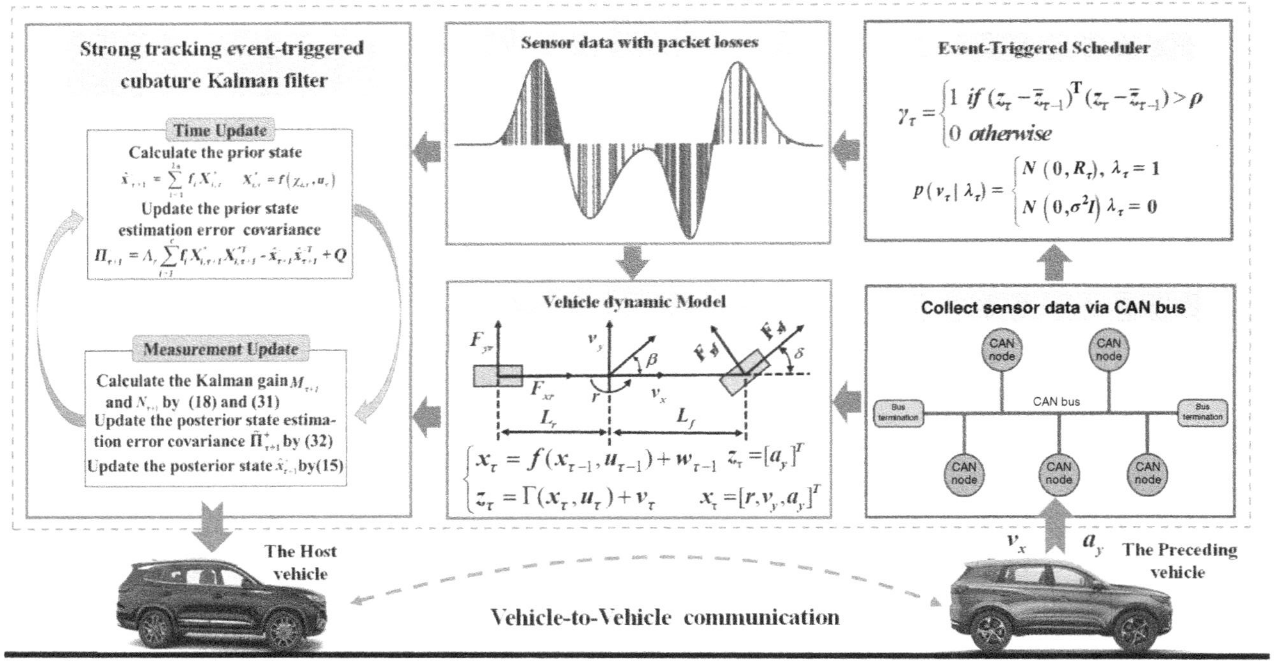

Figure 4.19 The estimation scheme based on STETCKF.

adjusts dynamically to accommodate uncertainties in model parameters, enhancing its capability to provide accurate and stable predictions of the PV's state over time. By integrating these components cohesively, the estimation framework not only addresses challenges like intermittent data transmission and model uncertainties but also validates its effectiveness through experimental validation. This approach ensures that the framework remains reliable and adaptable in diverse operational environments, offering significant advancements in vehicle state estimation technology for modern automotive applications.

4.4.1 Vehicle Model and Problem Statement

In this work, we choose the single-track model (see Figure 4.20) as the PV dynamics model. The single-track model, also known as the bicycle model, is a simplified representation of a vehicle's dynamics, widely used in the study of vehicle motion and control. This model reduces a four-wheeled vehicle to an equivalent two-wheeled system, capturing the essential behavior of the vehicle while significantly reducing computational complexity. In the single-track model, the vehicle's front and rear wheels are represented by single wheels located along the centerline, hence the term "single-track." This simplification allows for the effective analysis of lateral and longitudinal dynamics, crucial for understanding the vehicle's stability and control mechanisms. The single-track model incorporates key parameters such as the vehicle's mass, moment of inertia, wheelbase, and the positions of the center of gravity relative to the front and rear axles. These parameters help in formulating the equations of motion, which describe how the vehicle responds to steering inputs, external forces, and moments. The model typically includes equations for lateral dynamics (sideslip angle and yaw rate) and longitudinal dynamics (forward velocity and acceleration). One of the primary applications of the single-track model is in the development and testing of ADAS and autonomous driving algorithms. For instance, the model is used to design and validate systems like electronic stability control, traction control, and automated steering. Its relatively simple structure makes it suitable for real-time applications where computational efficiency is critical. The equations are given by

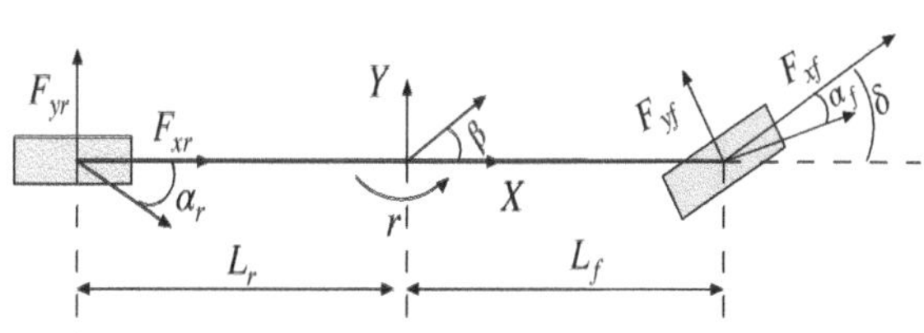

Figure 4.20 The single-track model.

$$\begin{cases} \dot{r} = \frac{L_f F_{yf} - L_r F_{yr}}{I_z} \\ \dot{v}_y = \frac{F_{yf} + F_{yr}}{m} - r v_x \end{cases} \tag{4.53}$$

$$\beta = \arctan\left(\frac{v_y}{v_x}\right) \tag{4.54}$$

$$a_y = \frac{L_f k_f - L_r k_r}{m v_x} r + \frac{k_f + k_r}{m} \beta - \frac{k_r}{m} \delta \tag{4.55}$$

where m is vehicle mass, I_z is the inertia moment about the vehicle vertical axis, δ is the angle of front wheel, F_{yf} and F_{yr} are lateral forces of front axle and rear axle, r is yaw rate, L_f and L_r are distances from the center of gravity to front axle and rear axle, β is sideslip angle, v_x and v_y are longitudinal velocity and lateral velocity, k_f and k_r are the front and rear tire cornering stiffnesses, and a_y is lateral acceleration.

The tire forces are given by

$$F_{yf} = 2k_f \alpha_f \tag{4.56}$$

$$F_{yr} = 2k_r \alpha_r \tag{4.57}$$

where α_f and α_r are the front and rear wheel sideslip angle and their calculation formulas are given by

$$\alpha_f = \delta - \beta - \frac{L_f r}{v_x} \tag{4.58}$$

$$\alpha_r = -\beta + \frac{L_r r}{v_x} \tag{4.59}$$

Rewrite the vehicle model into a state-space form, then we have

$$\begin{cases} x_\tau = f(x_{\tau-1}, u_{\tau-1}) + w_{\tau-1} \\ z_\tau = \Gamma(x_\tau, u_\tau) + v_\tau \end{cases} \tag{4.60}$$

$$x_\tau = [r, v_y, a_y]^T, \quad z_\tau = [a_y]^T, \quad u_\tau = [\delta, v_x]^T$$

$$\begin{bmatrix} r_\tau \\ v_{y,\tau} \\ a_{y,\tau} \end{bmatrix} = \begin{bmatrix} r_{\tau-1} + \left(\frac{L_f F_{yf} - L_r F_{yr}}{I_z}\right) \Delta t \\ v_{y,\tau-1} + \left(\frac{F_{yf} + F_{yr}}{m} - r_{\tau-1} v_{x,\tau-1}\right) \Delta t \\ a_{y,\tau-1} \end{bmatrix} \tag{4.61}$$

$$[a_{y,\tau}] = [0 \ \ 0 \ \ 1] x_\tau \tag{4.62}$$

Table 4.3 presents the meanings of the variables in the above equations.

where w_τ and v_τ are independent of each other. All noise signals are uncorrelated with the initial state x_0.

Table 4.3 The meanings of variables.

Symbol	Variables
τ	Sampling instant
w_τ	Process noise
x_τ	State variables
f	Process function
u_τ	Control input
R_τ	Covariance of measurement noise
z_τ	Measurement output
Δt	Sampling interval
Q_τ	Covariance of process noise
v_τ	Measurement noise
$\Gamma\Gamma$	Measurement function

The observability of a system is an indicator of how effectively its internal states can be calculated from knowing its inputs and outputs. We usually construct an observable matrix, and if it is a full-rank matrix, then the system is locally observable. Applying partial derivatives to the formulas in Section III A (4.4.1) and setting $v_x \neq 0$. It can be easily deduced that the rank of this observation matrix is 3, therefore the system is locally observable.

Figure 4.21 illustrates an event-triggered communication rule, a crucial mechanism for ensuring accurate vehicle state estimation while optimizing communication efficiency. The design principle of this rule revolves around assessing whether a significant change in the vehicle state has occurred by comparing the current measurement data with historical measurement data. This comparison helps determine the need for updating the vehicle's predicted state with fresh data. When the deviation between the current measurement and the historical data is minimal, indicating that the vehicle state has not changed drastically, the system deems it acceptable to use the historical measurement data for estimating the present vehicle (PV) state. This approach leverages the stability of the vehicle's dynamics, reducing the need for frequent data transmission and thus conserving communication bandwidth. By using historical data in such scenarios, the system maintains a balance between accuracy and efficiency, ensuring that minor fluctuations in the vehicle state do not trigger unnecessary data updates. Conversely, when the deviation is substantial, indicating a significant change in the vehicle state, the system recognizes the potential for considerable errors if it continues to rely on outdated data. In such cases, the event-triggered communication rule mandates the use of the latest measurement data to predict

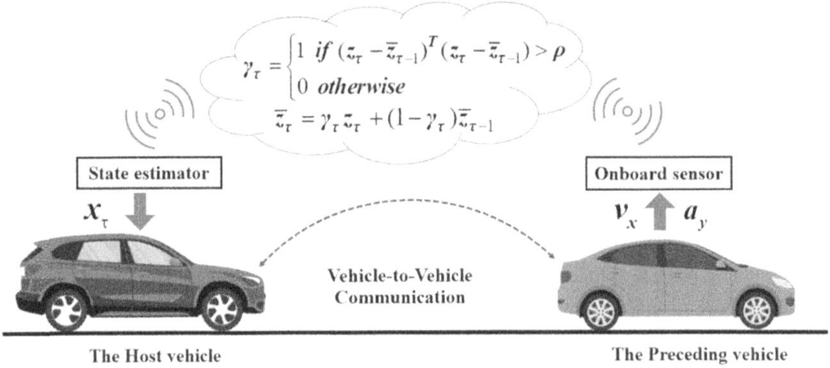

Figure 4.21 The event-triggered communication rule.

the PV state accurately. This timely update mechanism ensures that the vehicle's control systems and predictive models are based on the most current information, thereby minimizing the risk of errors that could compromise safety and performance. The event-triggered communication rule is particularly valuable in scenarios where communication resources are limited, and it is essential to optimize data transmission without sacrificing the accuracy of vehicle state estimation. By dynamically adjusting the reliance on historical versus current data based on the magnitude of state changes, this rule enhances the robustness and reliability of the vehicle's control systems. Furthermore, the implementation of this rule requires sophisticated algorithms capable of real-time data analysis and decision-making. These algorithms must efficiently process incoming data, compare it with historical records, and determine the appropriate course of action based on predefined thresholds for acceptable deviations. The continuous monitoring and adaptive response facilitated by the event-triggered communication rule play a critical role in ADAS and autonomous driving technologies, where precise and reliable vehicle state estimation is paramount.

γ_τ is an index of whether the sensor information of the PV is sent or not. The specific equation of γ_τ is as follows

$$\gamma_\tau = \begin{cases} 1 & if\ (z_\tau - \overline{z}_{\tau-1})^T (z_\tau - \overline{z}_{\tau-1}) > \rho \\ 0 & otherwise \end{cases} \tag{4.63}$$

$$\overline{z}_\tau = \gamma_\tau z_\tau + (1 - \gamma_\tau)\overline{z}_{\tau-1} \tag{4.64}$$

where ρ represents the event-triggered threshold, $\overline{z}_\tau$ indicates that the HV receives the sensor information from the PV at the current moment, $\overline{z}_{\tau-1}$ represents that the HV received the sensor information from the PV at the last moment, z_τ indicates the sensor information of the PV at the current moment, and if $\gamma_\tau = 1$, it means that the current sensor information of the PV will be sent to the HV for motion state estimation and vice versa.

After the event-triggering mechanism has worked, the communication rate will be reduced but there is still a possibility of losing the transmitted data. The momentary arrival of the observation at the moment τ can be described using the binary random variable λ_τ with the value 0 or 1. This description of packet loss is classic [34]. Furthermore, this expression is also commonly used for problems such as vehicle localization [35] and platoon control [36] in connected vehicle environments where data loss is considered. The following equation describes the probability density distribution of observation noise v_τ with respect to λ_τ.

$$p(v_\tau \mid \lambda_\tau) = \begin{cases} N(0, R_\tau) & \lambda_\tau = 1 \\ N(0, \sigma^2 I) & \lambda_\tau = 0 \end{cases} \tag{4.65}$$

If $\lambda_\tau = 0$, it indicates that data will be lost. This case is considered as a limiting case whose covariance is $\sigma^2 I$ and $\sigma \to \infty$. When $\lambda_\tau = 1$, it means that data will not be missing. The variance of the estimator is R_τ, I is an identity matrix. Additional details can be found in [34].

Remark 4.1 The packet loss only occurs with $\gamma_\tau = 1$. When packet loss occurs, i.e. $\lambda_\tau = 0$, the latest PV sensor data is sent to the HV, but the HV does not receive the latest data, it can only predict the PV state based on the historical data. Using a probabilistic formula to describe packet loss and embedding it into a filtering algorithm for iterative updating can effectively suppress the effects of packet loss.

4.4.2 Methodology

After establishing the event-triggered communication mechanism, the sensor information from the PV is transmitted to the HV only under specific conditions. This selective communication strategy implies that our estimator must rely on historical data to predict the PV state during periods when these conditions are not met. In such instances, using the traditional CKF for state estimation becomes problematic. The CKF operates under the assumption that sensor data is always available in real-time, corresponding to a 100% communication rate, typically achievable through uninterrupted wireless communication. When the communication mechanism triggers the transmission of sensor data, the CKF can be employed to estimate the PV state accurately. However, during intervals where only historical measurement data is accessible, the CKF's accuracy diminishes. This is because the CKF's derivation hinges on the continuous real-time availability of measurement data. Without this, the estimator cannot maintain its expected performance levels. To address this challenge, it becomes necessary to re-derive the filtering algorithm to handle scenarios where real-time data is intermittently unavailable. This new filter must effectively mitigate the impact of reduced real-time measurement data and packet loss on estimation accuracy. The goal is to ensure that the estimation remains robust even when the communication rate drops below 100%. The first step in deriving this new filter involves rewriting the state update equation to account for data loss in a recursive form. By incorporating data loss directly into the state update process, the new algorithm can adapt to the varying availability of sensor data. This approach ensures that the filter remains functional and accurate despite the interruptions in data transmission. The next step involves developing the mathematical framework and proving the effectiveness of the new filter. This entails deriving the necessary lemmas and proofs to establish the theoretical foundation of the filter. By demonstrating that the new algorithm can maintain high estimation accuracy even with sporadic data updates, we can validate its suitability for the event-triggered communication mechanism. In summary, transitioning from a traditional CKF to a newly derived filtering algorithm that accounts for intermittent data availability is crucial. This new filter must suppress the negative impacts of reduced real-time data and packet loss, ensuring that the HV can accurately predict the PV state despite communication challenges. The recursive state update equation and the subsequent theoretical derivation and proof form the backbone of this innovative approach, ultimately enhancing the robustness and reliability of vehicle state estimation in an event-triggered communication environment. The derived lemma and proof are as follows.

Theorem 4.3 For an event-triggered system considering data loss, if it can be rewritten into the state space form described in Eq. (4.60), the estimation error covariance $\widetilde{\Pi}^{+}_{\tau+1}$ is shown as

$$\begin{aligned}\widetilde{\Pi}^{+}_{\tau+1} &= \Pi^{-}_{\tau+1} - \lambda_{\tau+1}\gamma_{\tau+1}M_{\tau+1}\Pi_{zz,\tau+1}M^{T}_{\tau+1} + (1-\gamma_{\tau+1})\left[(1+a_1)\right.\\ &\times (I - N_{\tau+1}\theta_{\tau+1}C_{\tau+1})\Pi^{-}_{\tau+1}(I - N_{\tau+1}\theta_{\tau+1}C_{\tau+1})^{T} + (1+a_2)\\ &\times N_{\tau+1}R_{\tau+1}N^{T}_{\tau+1} + \left(1 + a_1^{-1} + a_2^{-1}\right)N_{\tau+1}\rho N^{T}_{\tau+1} - \left.\Pi^{-}_{\tau+1}\right]\end{aligned} \tag{4.66}$$

where $C_{\tau+1}$ is a Jacobian matrix, a_1 and a_2 are positive scalars, $\Pi^{-}_{\tau+1}$ is the prior error covariance, $M_{\tau+1}$ and $N_{\tau+1}$ are the estimator gains, $\theta_{\tau+1}$ is an undetermined diagonal matrix is introduced to model errors due to the first order linearization, and $\Pi_{zz,\tau+1}$ represents the predicted measurement covariance.

Proof: The state update with data loss is as follows

$$\begin{aligned}\hat{x}^{+}_{\tau+1} &= \hat{x}^{-}_{\tau+1} + \lambda_{\tau+1}\gamma_{\tau+1}M_{\tau+1}\left(z_{\tau+1} - \hat{z}^{-}_{\tau+1}\right)\\ &+ (1-\gamma_{\tau+1})N_{\tau+1}\left(\bar{z}_{\tau} - \hat{z}^{-}_{\tau+1}\right)\end{aligned} \tag{4.67}$$

where $\hat{z}^{-}_{\tau+1} = \Gamma\left(\hat{x}^{-}_{\tau+1}, u_{\tau+1}\right)$, $\hat{x}^{+}_{\tau+1}$ represents a posterior estimate and $\hat{x}^{-}_{\tau+1}$ is a priori estimate.

We define the posterior and prior estimation errors as $\widetilde{x}^{+}_{\tau+1}$ and $\widetilde{x}^{-}_{\tau+1}$, respectively, and the calculation formulas are as follows

$$\widetilde{x}^{+}_{\tau+1} = x_{\tau+1} - \widehat{x}^{+}_{\tau+1} \tag{4.68}$$

$$\widetilde{x}^{-}_{\tau+1} = x_{\tau+1} - \widehat{x}^{-}_{\tau+1} \tag{4.69}$$

If $\gamma_{\tau+1} = 1$, the gain matrix $M_{\tau+1}$ of CKF can be expressed as

$$M_{\tau+1} = \Pi_{xz,\tau+1}\left(\Pi^{-}_{zz,\tau+1}\right)^{-1} \tag{4.70}$$

$$\Pi^{+}_{\tau+1} = \Pi^{-}_{\tau+1} - M_{\tau+1}\Pi_{zz,\tau+1}M^{T}_{\tau+1} \tag{4.71}$$

If $\gamma_{\tau+1} = 0$, the Eq. (4.67) is given by

$$\widehat{x}^{+}_{\tau+1} = \widehat{x}^{-}_{\tau+1} + N_{\tau+1}(\overline{z}_{\tau} - z_{\tau+1}) + N_{\tau+1}\left(z_{\tau+1} - \widehat{z}^{-}_{\tau+1}\right) \tag{4.72}$$

Expanding $f(x_\tau, u_\tau)$ in the Taylor series, then we have

$$\begin{cases} f(x_\tau, u_\tau) = f\left(\widehat{x}^{+}_{\tau}, u_\tau\right) + \alpha_\tau A_\tau \widetilde{x}^{+}_{\tau} \\ A_\tau = \frac{\partial f(x_\tau,u_\tau)}{\partial x_\tau}\Big|_{x_\tau=\widehat{x}^{+}_{\tau}} \\ \alpha_\tau = diag(\alpha_{1,\tau}, \alpha_{2,\tau}, \dots, \alpha_{n,\tau}) \end{cases} \tag{4.73}$$

$$\widetilde{x}^{-}_{\tau+1} = \alpha_\tau A_\tau \widetilde{x}^{+}_{\tau} + w_\tau \tag{4.74}$$

where α_τ is an undetermined diagonal matrix that represents model errors due to the first-order linearization and n is the dimension of the system state.

Expanding $\Gamma(x_{\tau+1}, u_{\tau+1})$ in the Taylor series, then we have

$$\begin{cases} \Gamma(x_{\tau+1}, u_{\tau+1}) = \Gamma\left(\widehat{x}^{-}_{\tau+1}, u_{\tau+1}\right) + \theta_{\tau+1}C_{\tau+1}\widetilde{x}^{-}_{\tau+1} \\ C_{\tau+1} = \frac{\partial \Gamma(x_{\tau+1},u_{\tau+1})}{\partial x_{\tau+1}}\Big|_{x_{\tau+1}=\widehat{x}^{-}_{\tau+1}} \\ \theta_{\tau+1} = diag(\theta_{1,\tau+1}, \theta_{2,\tau+1}, \dots, \theta_{n,\tau+1}) \end{cases} \tag{4.75}$$

Subtracting Eqs. (4.72) from (4.67), the $\widetilde{x}^{+}_{\tau+1}$ is given by

$$\widetilde{x}^{+}_{\tau+1} = \widetilde{x}^{-}_{\tau+1} - N_{\tau+1}(\overline{z}_{\tau} - z_{\tau+1}) - N_{\tau+1}\left(z_{\tau+1} - \widehat{z}^{-}_{\tau+1}\right) \tag{4.76}$$

Based on Eq. (4.75), the $\widetilde{x}^{+}_{\tau+1}$ can be rewritten as

$$\widetilde{x}^{+}_{\tau+1} = (I - N_{\tau+1}\theta_{\tau+1}C_{\tau+1})\widetilde{x}^{-}_{\tau+1} - N_{\tau+1}v_{\tau+1} - N_{\tau+1}(\overline{z}_{\tau} - z_{\tau+1}) \tag{4.77}$$

According to Eq. (4.77), $\Pi^{+}_{\tau+1}$ is given by

$$\begin{aligned} \Pi^{+}_{\tau+1} &= E\left[\widetilde{x}^{+}_{\tau+1}\left(\widetilde{x}^{+}_{\tau+1}\right)^{T}\right] \\ &= (I - N_{\tau+1}\theta_{\tau+1}C_{\tau+1})\Pi^{-}_{k+1}(I - N_{\tau+1}\theta_{\tau+1}C_{\tau+1})^{T} + N_{\tau+1} \\ &\quad \times R_{\tau+1}N^{T}_{\tau+1} + N_{\tau+1}E\left\{(\overline{z}_{\tau} - z_{\tau+1})\,(\overline{z}_{\tau} - z_{\tau+1})^{T}\right\}N^{T}_{\tau+1} - \Omega_{\tau+1} \\ &\quad - Z_{\tau+1} - (Z_{\tau+1})^{T} + \Psi_{\tau+1} - (\Omega_{\tau+1})^{T} + (\Psi_{\tau+1})^{T} \end{aligned} \tag{4.78}$$

where

$$Z_{\tau+1} = (I - N_{\tau+1}\theta_{\tau+1}C_{\tau+1})E\left\{\widetilde{x}_{\tau+1}^{-}v_{\tau+1}^{T}\right\}N_{\tau+1}^{T}\Omega_{\tau+1} = (I\text{-}N_{\tau+1}\theta_{\tau+1}C_{\tau+1})$$
$$\times E\left\{\widetilde{x}_{\tau+1}^{-}(\overline{z}_{\tau} - z_{\tau+1})^{T}\right\}N_{\tau+1}^{T}$$
$$\Psi_{\tau+1} = N_{\tau+1}E\left\{v_{\tau+1}(\overline{z}_{\tau} - z_{\tau+1})^{T}\right\}N_{\tau+1}^{T}$$

$Z_{\tau+1}$ is equal to zero because $\widetilde{x}_{\tau}$ and v_{τ} are independent of each other. For further derivation, the following lemma [32] is described.

For any two vectors $x, y \in R^{n}$, and $a > 0$, one has

$$xy^{T} + yx^{T} \leq axx^{T} + a^{-1}yy^{T} \tag{4.79}$$

According to the above inequality, $\Omega_{\tau+1}$ and $\Psi_{\tau+1}$ can be rewritten as

$$-\Omega_{\tau+1} - (\Omega_{\tau+1})^{T} \leq a_{1}(I - N_{\tau+1}\theta_{\tau+1}C_{\tau+1})\Pi_{k+1}^{-}$$
$$\times (I - N_{\tau+1}\theta_{\tau+1}C_{\tau+1})^{T} + a_{1}^{-1}N_{\tau+1}\rho N_{\tau+1}^{T}$$
$$\Psi_{\tau+1} + (\Psi_{\tau+1})^{T} \leq a_{2}N_{\tau+1}R_{\tau+1}N_{\tau+1}^{T} + a_{2}^{-1}N_{\tau+1}\rho N_{\tau+1}^{T} \tag{4.80}$$

Combining Eqs. (4.80) and (4.78), the upper bound $\overline{\Pi}_{\tau+1}^{+}$ for $\Pi_{\tau+1}^{+}$ is given by

$$\overline{\Pi}_{\tau+1}^{+} = (1 + a_{1})(I - N_{\tau+1}\theta_{\tau+1}C_{\tau+1})\Pi_{k+1}^{-}(I - N_{\tau+1}\theta_{\tau+1}C_{\tau+1})^{T}$$
$$+ (1 + a_{2})N_{\tau+1}R_{\tau+1}N_{\tau+1}^{T} + \left(1 + a_{1}^{-1} + a_{2}^{-1}\right)N_{\tau+1}\rho N_{\tau+1}^{T} \tag{4.81}$$

Combining Eqs. (4.71) and (4.81), the new upper bound of $\widetilde{\Pi}_{\tau+1}^{+}$ can be rewritten as

$$\widetilde{\Pi}_{\tau+1}^{+} = \Pi_{\tau+1}^{-} - \gamma_{\tau+1}M_{\tau+1}\Pi_{zz,\tau+1}M_{\tau+1}^{T} + (1 - \gamma_{\tau+1})\left[(1 + a_{1})\right.$$
$$\times (I - N_{\tau+1}\theta_{\tau+1}C_{\tau+1})\Pi_{k+1}^{-}(I - N_{\tau+1}\theta_{\tau+1}C_{\tau+1})^{T} + (1 + a_{2})$$
$$\left.\times N_{\tau+1}R_{\tau+1}N_{\tau+1}^{T} + \left(1 + a_{1}^{-1} + a_{2}^{-1}\right)N_{\tau+1}\rho N_{\tau+1}^{T} - \Pi_{\tau+1}^{-}\right] \tag{4.82}$$

Take the partial derivative of $\widetilde{\Pi}_{\tau+1}^{+}$ with respect to $N_{\tau+1}$, and let the derivative be zero. Then the gain $N_{\tau+1}$ can be rewritten as

$$N_{\tau+1} = (1 + a_{1})\Pi_{\tau+1}^{-}C_{\tau+1}^{T}\theta_{\tau+1}^{T}\left[(1 + a_{1})\theta_{\tau+1}C_{\tau+1}\Pi_{\tau+1}^{-}\right.$$
$$\left.\times C_{\tau+1}^{T}\theta_{\tau+1}^{T} + (1 + a_{2})R_{\tau+1} + \left(1 + a_{1}^{-1} + a_{2}^{-1}\right)\rho I\right]^{-1} \tag{4.83}$$

Equation (4.83) can only use historical measurement data for filtering iterations. As long as our communication threshold ρ is set reasonably, its estimation accuracy can be guaranteed. Furthermore, taking into account the impact of packet loss, the upper bound $\widetilde{\Pi}_{\tau+1}^{+}$ is given by

$$\widetilde{\Pi}_{\tau+1}^{+} = \Pi_{\tau+1}^{-} - \lambda_{\tau+1}\gamma_{\tau+1}M_{\tau+1}\Pi_{zz,\tau+1}M_{\tau+1}^{T} + (1 - \gamma_{\tau+1})\left[(1 + a_{1})\right.$$
$$\times (I - N_{\tau+1}\theta_{\tau+1}C_{\tau+1})\Pi_{k+1}^{-}(I - N_{\tau+1}\theta_{\tau+1}C_{\tau+1})^{T} + (1 + a_{2})$$
$$\left.\times N_{\tau+1}R_{\tau+1}N_{\tau+1}^{T} + \left(1 + a_{1}^{-1} + a_{2}^{-1}\right)N_{\tau+1}\rho N_{\tau+1}^{T} - \Pi_{\tau+1}^{-}\right] \tag{4.84}$$

This completes the proof.

In the ETCKF algorithm, we address the effect of data loss but do not account for the impact of model parameter perturbations on estimation accuracy. Studies have shown that the strong tracking filter exhibits strong robustness to model parameter perturbations [37]. This robustness is achieved through adaptive processing by incorporating fading factors into the filtering process. The principle behind fading factors is to purposefully increase the variance of the predicted state vector. This is done by applying a matrix of fading factors to the predicted covariance matrix, thereby enhancing the filter's ability to track rapid changes in the system. The fading factors effectively allow the filter to "forget" older information at a controlled rate, which helps in adapting to changes and uncertainties in model parameters. By adjusting the fading factors online based on the a priori knowledge of the system's state model and relevant parameters, the filter maintains robustness against model parameter perturbations. Given the advantages of the strong tracking filter, we propose to combine it with the ETCKF algorithm to form the STETCKF. This combined approach aims to enhance the robustness of the ETCKF algorithm against model parameter perturbations, in addition to its existing capability to handle data loss. The STETCKF algorithm integrates the fading factor mechanism of the strong tracking filter with the event-triggered communication and state estimation framework of the ETCKF. This integration involves modifying the covariance prediction step of the ETCKF to include fading factors, which are adjusted dynamically based on the observed system behavior and measurement data. The online adjustment of fading factors ensures that the filter can adapt to changes in the model parameters, thereby improving the overall estimation accuracy and robustness.

Using a fading factor Λ_τ to update the $\Pi_{\tau+1}^{-}$, the specific equation is as follows

$$\Pi_{\tau+1}^{-} = \Lambda_\tau \sum_{i=1}^{c} \phi_i X_{i,\tau+1}^{*} X_{i,\tau+1}^{*T} - \hat{x}_{\tau+1}^{-} \hat{x}_{\tau+1}^{-T} + Q_\tau \tag{4.85}$$

where $\Lambda_\tau = diag\left[\wp_\tau^1, \wp_\tau^2, \cdots \wp_\tau^n\right]$. See Table 4.2 for the meaning of other variables.

$$\wp_\tau^k = \begin{cases} \Upsilon_k d_\tau & \Upsilon_k d_\tau \geq 1 \\ 1 & \Upsilon_k d_\tau < 1 \end{cases} \quad k = 1, 2, \ldots n \tag{4.86}$$

$$d_\tau = \frac{tr[\widetilde{N}_\tau]}{\sum_{k=1}^{n} \Upsilon_k \widetilde{M}_k} \tag{4.87}$$

$$\widetilde{N}_\tau = V_\tau - R_\tau - C_\tau Q_{\tau-1} C_\tau^T \tag{4.88}$$

$$\widetilde{M}_\tau = C_\tau A_\tau \Pi_\tau^{-} A_\tau^T C_\tau^T \tag{4.89}$$

$$V_\tau = \begin{cases} \varphi_1\varphi_1^T & \tau = 0 \\ \frac{\psi V_{\tau-1}+\varphi_\tau\varphi_\tau^T}{1+\psi} & \tau \geq 1 \\ \varphi_\tau = z_\tau - \hat{z}_\tau^- \end{cases} \tag{4.90}$$

where Υ_τ is a constant that is determined from a priori information about the state of the system, d_τ is the undetermined parameter, and n is the dimensionality of the state. $\psi \in [0, 1]$ is a forgetting factor, More details of the parameters can be found in [38]. Table 4.4 describes the iterative process of STETCKF.

Table 4.4 The STETCKF algorithm.

Estimation Algorithm
Step 1: Set $\hat{x}_0 = E(x_0)\ \Pi_0 = E\left[(x_0 - \hat{x}_0)(x_0 - \hat{x}_0)^T\right]$
The embedded cubature points ϑ_i and the weights ϕ_i are respectively given by
$\phi_i = \frac{1}{c},\quad \vartheta_i = \sqrt{\frac{c}{2}}\left[\begin{pmatrix}1\\0\\\vdots\\0\end{pmatrix}\cdots\begin{pmatrix}0\\0\\\vdots\\1\end{pmatrix}\begin{pmatrix}-1\\0\\\vdots\\0\end{pmatrix}\cdots\begin{pmatrix}0\\0\\\vdots\\-1\end{pmatrix}\right],\quad i = 1, 2, \ldots c,\quad c = 2n$
Step 2: Time update:
Singular value decomposition of Π_τ^+: $\Pi_\tau^+ = U_\tau S_\tau V_\tau^T$
Compute the Cubature Points: $\chi_{i,\tau} = S_\tau \vartheta_i + \hat{x}_\tau^+$
Propagate the $\chi_{i,\tau}$: $X_{i,\tau}^* = f(\chi_{i,\tau}, u_\tau)$
Estimate the $\hat{x}_{\tau+1}^-$ and $\Pi_{\tau+1}^-$:
$\hat{x}_{\tau+1}^- = \sum_{i=1}^{2n}\phi_i X_{i,\tau}^*\ \Pi_{\tau+1}^- = \Lambda_\tau \sum_{i=1}^{c}\phi_i X_{i,\tau+1}^* X_{i,\tau+1}^{*T} - \hat{x}_{\tau+1}^-\hat{x}_{\tau+1}^{-T} + Q$
Step 3: Measurement update:
Singular value decomposition of $\Pi_{\tau+1}^-$: $\Pi_{\tau+1}^- = U_{\tau+1}S_{\tau+1}V_{\tau+1}^T$
Compute the Cubature Points: $\chi_{i,\tau+1} = S_{\tau+1}\vartheta_i + \hat{x}_{\tau+1}^+$
Propagate the $\chi_{i,\tau+1}$: $Z_{i,\tau+1}^* = \Gamma(\chi_{i,\tau+1}, u_{\tau+1})$
Estimate the cross-covariance matrix $\Pi_{xz,\tau+1}^-$, the predicted measurement $\hat{z}_{\tau+1}^-$, and the innovation covariance matrix $\Pi_{zz,\tau+1}^-$
$\hat{z}_{\tau+1}^- = \sum_{i=1}^{2n}\phi_i Z_{i,\tau+1}^*\Pi_{xz,\tau+1}^- = \sum_{i=1}^{2n}\phi_i\chi_{i,\tau+1}Z_{i,\tau+1}^{*T} - \hat{x}_{\tau+1}^-\hat{z}_{\tau+1}^{-T}$
$\Pi_{zz,\tau+1}^- = \sum_{i=1}^{2n}\phi_i Z_{i,\tau+1}^* Z_{i,\tau+1}^{*T} - \hat{z}_{\tau+1}^-\hat{z}_{\tau+1}^{-T} + \lambda_{\tau+1}R_{\tau+1} + (1-\lambda_{\tau+1})\sigma^2 I$
Compute γ_τ using Eq. (4.63) and the filter gain $M_{\tau+1}$ and $N_{\tau+1}$ are calculated by Eqs. (4.70) and (4.83).
Update $\hat{x}_{\tau+1}^+$ and $\widetilde{\Pi}_{\tau+1}^+$ through Eqs. (4.67) and (4.83).
Step 4: At the next iteration loop, steps 1 to 3 will be repeated.

This section will evaluate the performance of the STETCKF and will give bounded conditions on covariance expectations.

Lemma 4.4 For two matrices F, $B \in \mathbf{R}^{n\times n}$, assuming F, B are symmetric positive-definite, then

$$(F+B)^{-1} > F^{-1} - F^{-1}BF^{-1} \tag{4.91}$$

Theorem 4.4 For an event-triggered system considering data loss, if it can be rewritten into the state space form described in Eq. (4.8), If there are real constants $\underline{f}, \bar{f}, \underline{h}, \bar{h}, \underline{\beta}, \bar{\beta}, \underline{\alpha}, \bar{\alpha} \neq 0, \bar{\Lambda} > 1, I \leq \Lambda_\tau \leq \bar{\Lambda} I$ and $\underline{q}, \bar{q}, \underline{r}, \bar{r} > 0$ such that these inequalities are met for $\tau > 0$.

$$\begin{cases} \underline{\alpha}^2 I \leq \alpha_\tau \alpha_\tau^T \leq \bar{\alpha}^2 I, \quad \underline{\theta}^2 I \leq \theta_\tau \theta_\tau^T \leq \bar{\theta}^2 I \\ \underline{h}^2 I \leq C_\tau C_\tau^T \leq \bar{h}^2 I, \quad \underline{f}^2 I \leq A_\tau A_\tau^T \leq \bar{f}^2 I \\ \underline{q} I \leq Q_\tau \leq \bar{q} I, \quad \underline{\rho} \leq \rho \leq \bar{\rho}, \quad \underline{r} I \leq R_\tau \leq \bar{r} I \end{cases} \tag{4.92}$$

If $\lambda > 1 - \frac{1}{\gamma \bar{\Lambda} \bar{\alpha}^2 \bar{f}^2}$ for every $\tau \geq 0$, then

$$\mathrm{E}\left[\tilde{\Pi}_{\tau+1}^{+}\right] \leq \mathrm{E}\left[\Pi_{\tau+1}^{-}\right] \leq \bar{\xi} I \tag{4.93}$$

Proof: For ease of analysis, the following covariates are simplified, then

$$\begin{cases} \Pi_{\tau+1}^{-} = \Lambda_\tau \alpha_\tau A_\tau \tilde{\Pi}_\tau^{+} A_\tau^T \alpha_\tau + Q_\tau \\ \Pi_{xz,\tau+1} = \Pi_{\tau+1}^{-} C_{\tau+1}^T \theta_{\tau+1} \\ \Pi_{zz,\tau+1} = \theta_{\tau+1} C_{\tau+1} \Pi_{\tau+1}^{-} C_{\tau+1}^T \theta_{\tau+1} + \lambda_{\tau+1} R_{\tau+1} + (1-\lambda_{\tau+1}) \sigma^2 I \end{cases} \tag{4.94}$$

Combining Eqs. (4.84) and (4.94), then

$$\begin{aligned} \tilde{\Pi}_{\tau+1}^{+} = {} & \Pi_{\tau+1}^{-} - \lambda_{\tau+1} \gamma_{\tau+1} \Pi_{\tau+1}^{-} C_{\tau+1}^T \theta_{\tau+1} \left(\theta_{\tau+1} C_{\tau+1} \Pi_{\tau+1}^{-} \right. \\ & \left. \times C_{\tau+1}^T \theta_{\tau+1} + R_{\tau+1}\right)^{-1} \theta_{\tau+1} C_{\tau+1} \Pi_{\tau+1}^{-} + (1-\gamma_{\tau+1}) \left[(1+a_1)\right. \\ & \times (I - N_{\tau+1} \theta_{\tau+1} C_{\tau+1}) \Pi_{\tau+1}^{-} (I - N_{\tau+1} \theta_{\tau+1} C_{\tau+1})^T + (1+a_2) \\ & \left. \times N_{\tau+1} R_{\tau+1} N_{\tau+1}^T + \left(1 + a_1^{-1} + a_2^{-1}\right) N_{\tau+1} \rho N_{\tau+1}^T - \Pi_{\tau+1}^{-}\right] \end{aligned} \tag{4.95}$$

Combining Eqs. (4.94) and (4.95), then

$$\begin{aligned} \mathrm{E}\left[\Pi_{\tau+1}^{-}\right] = {} & \mathrm{E}\left[\Lambda_\tau \alpha_\tau A_\tau \Pi_\tau^{-} A_\tau^T \alpha_\tau + Q_\tau - \lambda_\tau \gamma_\tau \Lambda_\tau \alpha_\tau A_\tau \Pi_\tau^{-} C_\tau^T \theta_\tau \right. \\ & \times \left(\theta_\tau C_\tau \Pi_\tau^{-} C_\tau^T \theta_\tau + R_\tau\right)^{-1} \theta_\tau C_\tau \Pi_\tau^{-} A_\tau^T \alpha_\tau + (1-\gamma_\tau) \Lambda_\tau \alpha_\tau A_\tau \\ & \times \left\{a_1 \Pi_\tau^{-} - (1+a_1)^2 \Pi_\tau^{-} C_\tau^T \left[(1+a_1) C_\tau \Pi_\tau^{-} C_\tau^T + (1+a_2) R_\tau \right.\right. \\ & \left.\left.\left. + \left(1 + a_1^{-1} + a_2^{-1}\right) \rho I\right]^{-1} C_\tau \Pi_\tau^{-}\right\} A_\tau^T \alpha_\tau\right] \end{aligned} \tag{4.96}$$

We first define $\gamma = \mathrm{E}[\gamma_\tau]$ as the average sensor communication rate and $\lambda = \mathrm{E}[\lambda_\tau]$. Then based on Lemma 4.2, Eq. (4.96) can be rewritten as

$$\begin{aligned}\mathrm{E}\left[\Pi_{\tau+1}^{-}\right] <&(1-\lambda\gamma)\Lambda_\tau\alpha_\tau A_\tau \mathrm{E}\left[\Pi_\tau^{-}\right]A_\tau^T\alpha_\tau + Q\\ &\lambda\gamma\Lambda_\tau\left(\alpha_\tau A_\tau C_\tau^{-1}\theta_\tau^{-1}R_\tau\theta_\tau^{-1}C_\tau^{-T}A_\tau^T\alpha_\tau\right)\\ &+(1-\gamma)\Lambda_\tau\alpha_\tau A_\tau\left\{-\mathrm{E}\left[\Pi_\tau^{-}\right]+C_\tau^{-1}\left[(1+a_2)R_\tau\right.\right.\\ &\left.\left.+\left(1+a_1^{-1}+a_2^{-1}\right)\rho I\right]C_\tau^{-T}\right\}A_\tau^T\alpha_\tau\end{aligned} \tag{4.97}$$

Combining Eqs. (4.92) and (4.97) to obtain

$$\begin{aligned}\mathrm{E}\left[\Pi_{\tau+1}^{-}\right] <& \gamma(1-\lambda)\Lambda_\tau\alpha_\tau A_\tau \mathrm{E}\left[\Pi_\tau^{-}\right]A_\tau^T\alpha_\tau\\ &+\left(\frac{\lambda\gamma\overline{\Lambda}(\overline{\alpha}^2\overline{f}^{-2}\overline{r})}{\underline{\theta}^2\underline{h}^2}+\frac{(1-\gamma)\overline{\Lambda}\overline{f}^{-2}\overline{\alpha}^2\left[(1+a_2)\overline{r}+\left(1+a_1^{-1}+a_2^{-1}\right)\overline{\rho}\right]}{\underline{h}^2}+\overline{q}\right)I\end{aligned} \tag{4.98}$$

$$\begin{aligned}\tilde{\xi}=\max&\left\{\|\Pi_1^{-}\|,\overline{q}+\frac{\lambda\gamma\overline{\Lambda}(\overline{\alpha}^2\overline{f}^{-2}r)}{\underline{\theta}^2\underline{h}^2}+\frac{(1-\gamma)\overline{\alpha}^2\overline{f}^{-2}(1+a_2)\overline{r}}{\underline{h}^2}\right.\\ &\left.+\frac{(1-\gamma)\overline{\Lambda}\overline{\alpha}^2\overline{f}^{-2}\left(1+a_1^{-1}+a_2^{-1}\right)\overline{\rho}}{\underline{h}^2}\right\}\end{aligned} \tag{4.99}$$

Then Eq. (4.98) recursively proved

$$\mathrm{E}\left[\Pi_\tau^{-}\right] < \tilde{\xi}\sum_{j=0}^{\tau-1}[\gamma(1-\lambda)\overline{\alpha}^2\overline{f}^{-2}\overline{\Lambda}]^jI \tag{4.100}$$

If $\mathrm{E}\left[\Pi_1^{-}\right] > 0$, then

$$\begin{aligned}\mathrm{E}\left[\Pi_2^{-}\right] &< \gamma(1-\lambda)\overline{\Lambda}\overline{\alpha}^2\overline{f}^{-2}\mathrm{E}\left[\Pi_1^{-}\right]\\ &+\left(\frac{\lambda\gamma\overline{\Lambda}(\overline{\alpha}^2\overline{f}^{-2}\overline{r})}{\underline{\theta}^2\underline{h}^2}+\frac{(1-\gamma)\overline{\Lambda}\overline{f}^{-2}\overline{\alpha}^2\left[(1+a_2)\overline{r}+\left(1+a_1^{-1}+a_2^{-1}\right)\overline{\rho}\right]}{\underline{h}^2}+\overline{q}\right)I\\ &< \gamma(1-\lambda)\overline{\Lambda}\overline{\alpha}^2\overline{f}^{-2}\tilde{\xi}I+\tilde{\xi}I\end{aligned} \tag{4.101}$$

According to Eqs. (4.100) and (4.101), then

$$\begin{aligned}\mathrm{E}\left[\Pi_{\tau+1}^{-}\right] &< \tilde{\xi}\sum_{j=0}^{\tau-1}[\gamma(1-\lambda)\overline{\Lambda}\overline{\alpha}^2\overline{f}^{-2}]^jI\\ &+\left(\frac{\lambda\gamma\overline{\Lambda}(\overline{\alpha}^2\overline{f}^{-2}\overline{r})}{\underline{\theta}^2\underline{h}^2}+\frac{(1-\gamma)\overline{\Lambda}\overline{f}^{-2}\overline{\alpha}^2\left[(1+a_2)r+\left(1+a_1^{-1}+a_2^{-1}\right)\overline{\rho}\right]}{\underline{h}^2}+\overline{q}\right)I\end{aligned}$$

$$
\begin{aligned}
&< \tilde{\xi} \sum_{j=0}^{\tau-1} [\gamma(1-\lambda)\overline{\Lambda}\overline{\alpha}^2\overline{f}^2]^j I + \tilde{\xi} I \\
&< \tilde{\xi} \sum_{j=0}^{\tau} [\gamma(1-\lambda)\overline{\Lambda}\overline{\alpha}^2\overline{f}^2]^j I \\
&< \overline{\xi} I
\end{aligned} \tag{4.102}
$$

where $\overline{\xi} = \tilde{\xi} \sum_{j=0}^{\tau} [\gamma(1-\lambda)\overline{\Lambda}\overline{\alpha}^2\overline{f}^2]^j I$. If $\lambda > 1 - \frac{1}{\gamma\overline{\Lambda}\overline{\alpha}^2\overline{f}^2}$, the filter will converge, and Eq. (4.93) is satisfied.

This completes the proof.

Remark 4.2 The sensor communication rate γ relies on the threshold ρ. The desired tradeoff between communication rate and estimation performance can be realized by appropriately adjusting ρ. In addition, it can be seen that STETCKF will converge as long as there is a lower bound on the packet delivery rate λ, i.e. λ is greater than $1 - \frac{1}{\gamma\overline{\Lambda}\overline{\alpha}^2\overline{f}^2}$. This means that the packet loss rate is less than $\frac{1}{\gamma\overline{\Lambda}\overline{\alpha}^2\overline{f}^2}$.

4.4.3 Simulation and Experiment Tests

Simulations and real vehicle experiments are conducted to validate the ETCKF and STETCKF algorithms. These experiments compare the estimation results of the conventional CKF with those of the ETCKF and STETCKF to demonstrate the superiority of the proposed algorithms. The CKF's estimation results serve as a benchmark, representing a scenario with a 100% communication rate. In contrast, the communication rate of the ETCKF and STETCKF is much lower due to the event-triggered communication mechanism. This difference in communication rates is critical for evaluating the robustness and effectiveness of the proposed algorithms under more realistic conditions where data transmission can be intermittent and subject to loss. The CKF has already been shown to provide good estimation accuracy in vehicle dynamics control. Therefore, if the estimation accuracy of the ETCKF and STETCKF approaches that of the CKF under a 100% communication rate, even when the communication rate is reduced and data is lost, it would effectively illustrate the robustness and efficacy of the STETCKF. To simulate packet loss, the primary vehicle sensor information transmitted through the event-triggered communication mechanism is multiplied by a random variable that follows a Bernoulli distribution between [0,1]. This processed data is then fed to the ETCKF and STETCKF, mimicking real-world communication challenges. Further verification of the strong tracking filtering effect involves intentionally setting the vehicle mass parameter in both the ETCKF and STETCKF to deviate from its true value. This deliberate perturbation tests the algorithms' ability to handle inaccuracies in model parameters, showcasing the adaptive capabilities of the strong tracking filter within the STETCKF framework. In the simulations, various driving scenarios are modeled to reflect different operational conditions, such as sudden maneuvers, changes in road friction, and varying speeds. These scenarios help assess how well the ETCKF and STETCKF algorithms can maintain accurate state estimation despite reduced communication rates and the presence of packet loss. The real vehicle experiments further substantiate these findings by applying the algorithms to actual vehicle data, ensuring that the simulations' results are transferable to real-world applications.

4.4.3.1 The Simulation Test

We set up a double-lane change scenario in CarSim, positioning two vehicles in a front-and-rear configuration on the road. The vehicle in front is defined as the PV, while the vehicle at the rear is defined as the HV. Communication between the two vehicles is established using Simulink, creating a robust simulation environment to test the proposed algorithms. The vehicle state from CarSim is determined as the RV, serving as the benchmark against which we measure the performance of the CKF, ETCKF, and STETCKF filter algorithms. All three algorithms use identical parameters to ensure a fair comparison. The packet loss rate is set at 38%, and the initial longitudinal velocity of the vehicles is 50 km/h, reflecting realistic driving conditions that include communication challenges. The computational setup for executing the proposed algorithms involves a personal computer equipped with an AMD 75800 CPU and 16.0 GB of RAM. The computation time for a single cycle of the STETCKF algorithm is 0.688 ms. This configuration is sufficient to handle the computational demands of the algorithms, ensuring that the simulations run smoothly and efficiently. Specifically, the computation time for a single cycle of the STETCKF algorithm is 0.688 ms, highlighting the algorithm's efficiency and suitability for real-time applications. During the double-lane change maneuver, the PV and HV navigate a predefined path, allowing us to observe the performance of the state estimation algorithms under dynamic conditions. The RVs provided by CarSim enable a precise assessment of how well each algorithm can track the true vehicle states. This setup is crucial for evaluating the algorithms' robustness and accuracy. The experiment focuses on comparing the estimation accuracy of the CKF, ETCKF, and STETCKF algorithms. The CKF operates under the assumption of a 100% communication rate, serving as a baseline for comparison. The ETCKF and STETCKF, however, must cope with the 38% packet loss rate, demonstrating their ability to function effectively under reduced communication conditions. The STETCKF algorithm, in particular, incorporates strong tracking filtering techniques to enhance its robustness against model parameter perturbations. By introducing fading factors that dynamically adjust based on the observed data, the STETCKF can adapt to changes in the vehicle's state and environment. As the vehicles execute the double-lane change maneuver, the algorithms continuously estimate the PV state.

Figures 4.22 and 4.23 illustrate the variations in longitudinal velocity and lateral acceleration, respectively. In these figures, the red solid line represents the original sensor data from the PV, which we define as the RV. This data is crucial for assessing the performance of our estimation algorithms. The sensor data undergoes an event-triggered processing mechanism before being input into the STETCKF. For simplicity, only the ETCKF is depicted in these figures, shown by the solid blue line, since both ETCKF and STETCKF share the same communication mechanism. In the magnified sections of these images, we observe that

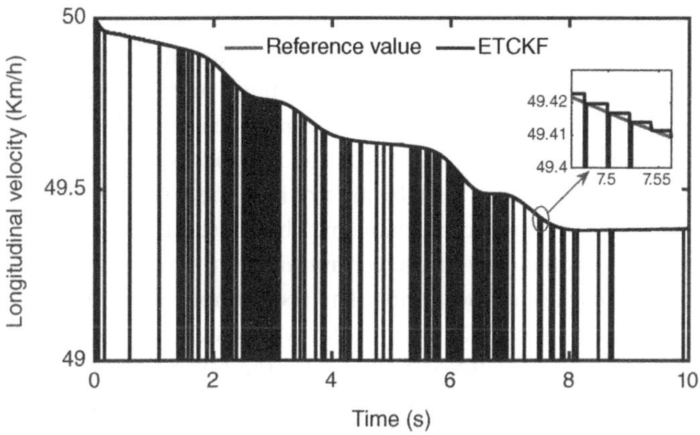

Figure 4.22 The longitudinal velocity curves in the case of DLC.

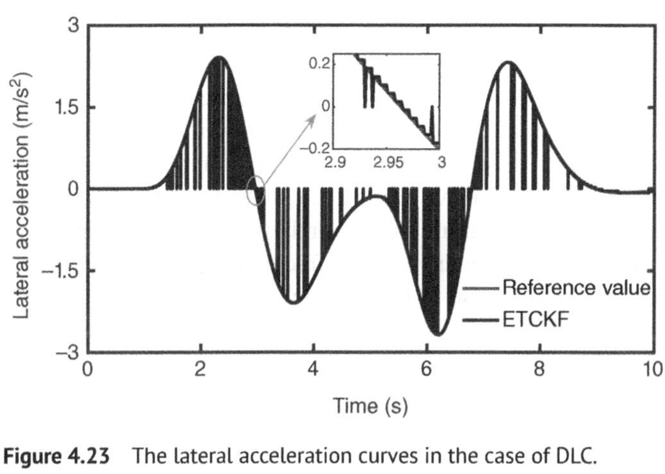

Figure 4.23 The lateral acceleration curves in the case of DLC.

the sensor signal remains constant for certain periods. This constancy indicates that no new sensor data is being received during these times, a direct result of the event-triggered mechanism, which only transmits data when significant changes occur. This behavior is expected and showcases the algorithm's ability to operate efficiently by reducing unnecessary data transmissions. Furthermore, Figure 4.23 shows instances where the data abruptly drops to zero. These sudden drops occur because real-time data is being sent, but packet loss leads to these zero values. This phenomenon is critical for understanding the robustness of our algorithms under conditions of data loss, as it simulates real-world scenarios where communication may be intermittent or unreliable. In Figure 4.22, the starting point of the vertical coordinate is set to 49 instead of zero. This adjustment is made to better reflect the fluctuations in longitudinal velocity, making it easier to observe the variations and performance of the ETCKF. Similar to the lateral acceleration, some points in the longitudinal velocity data also drop abruptly to zero, indicating packet loss at those moments. However, due to the adjusted starting point on the vertical axis, these drops are not as immediately apparent. This special note is included to avoid misunderstanding and to clarify the data presentation for readers. In conclusion, Figs. 4.22 and 4.23 provide a detailed visualization of how the ETCKF processes sensor data under the event-triggered mechanism. The constant periods in the sensor signal highlight the efficiency of the communication mechanism, while the zero values indicate packet loss scenarios. The special note about the vertical coordinate adjustment in Figure 4.22 ensures clarity in understanding the data representation. These figures collectively demonstrate the robustness and reliability of the ETCKF algorithm in maintaining accurate state estimation amidst communication challenges, underscoring its potential application in real-world autonomous driving scenarios.

In Figure 4.24, the data value of 1 indicates that the PV sensor data is being delivered to the HV at that specific time. Throughout the entire simulation process, the data is transmitted 593 times, corresponding to a communication rate of 5.93%. This communication rate is calculated as the percentage of the transmitted data relative to the total possible data points. The figure provides insight into the frequency and timing of these transmissions during different phases of the vehicle's operation. From the figure, we observe that during the first 1.6 seconds, the number of data transmissions from the PV is relatively low. However, as the vehicle begins to execute a turning maneuver, the frequency of data transmissions increases significantly. This increase in transmission frequency correlates with the higher variability and dynamic changes in the vehicle's state during the turning maneuver. It is important to note that the dense areas in Figure 4.24 do not represent continuous or 100% data transmission at those points. A closer inspection, by zooming in on these areas, reveals that 0 and 1 values still appear alternately, albeit more frequently. This pattern demonstrates that even during

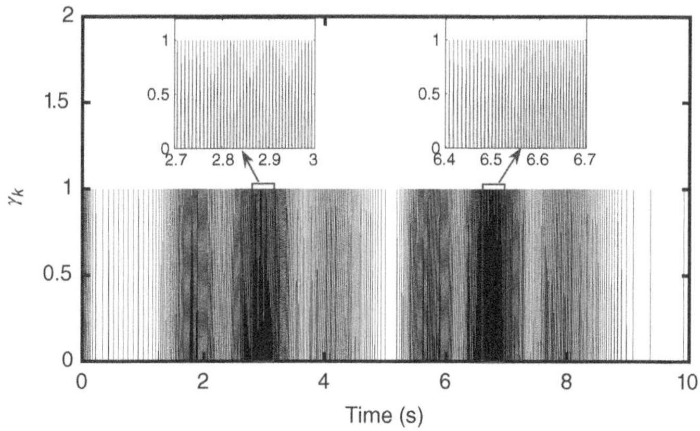

Figure 4.24 Event-triggered times in case of DLC.

periods of increased transmission, the event-triggered mechanism continues to selectively transmit data based on the significance of state changes rather than continuously streaming data. As the turning maneuver concludes, the frequency of data transmissions decreases rapidly. There are even periods where no data transmission occurs. This indicates that when the vehicle is in a stable operating state with minimal fluctuations, the need for frequent data updates diminishes. The event-triggered mechanism effectively reduces communication overhead by limiting transmissions to periods of significant state changes, thereby optimizing the use of communication resources. The raw sensor information from the PV is used as input for the CKF. Meanwhile, the data transmitted via the event-triggered mechanism, including the effects of packet loss, is input into the ETCKF and STETCKF. This setup allows for a direct comparison of how well each algorithm performs under realistic conditions of intermittent data transmission and packet loss. The comparison highlights the robustness and efficiency of the ETCKF and STETCKF algorithms in maintaining accurate state estimations despite a reduced communication rate. By selectively transmitting data only when significant changes occur, these algorithms can achieve reliable performance while minimizing communication bandwidth requirements. The STETCKF, with its additional strong tracking filtering, further enhances robustness by adapting to model parameter perturbations, providing more accurate and resilient state estimates even under challenging conditions.

The estimation results of different methods are displayed in Figs. 4.25–4.27. We use the root mean square error (RMSE) to evaluate the estimation performance of these methods, with the results detailed in Table 4.5. In Figure 4.25, the CKF with a 100% communication rate shows an estimated curve that closely follows the RV, indicating high accuracy under ideal conditions with no data loss.

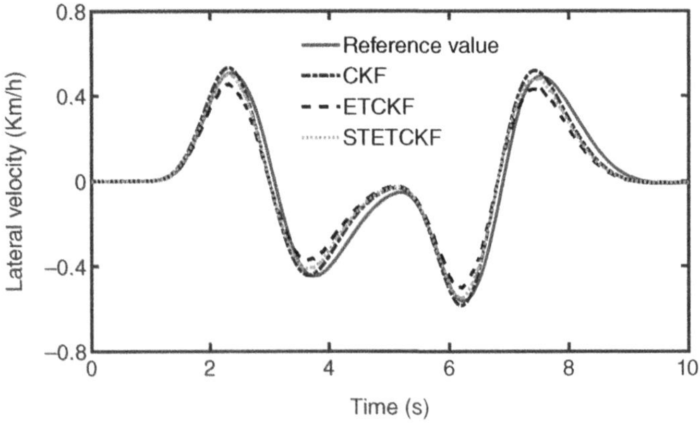

Figure 4.25 The lateral velocity in the case of DLC.

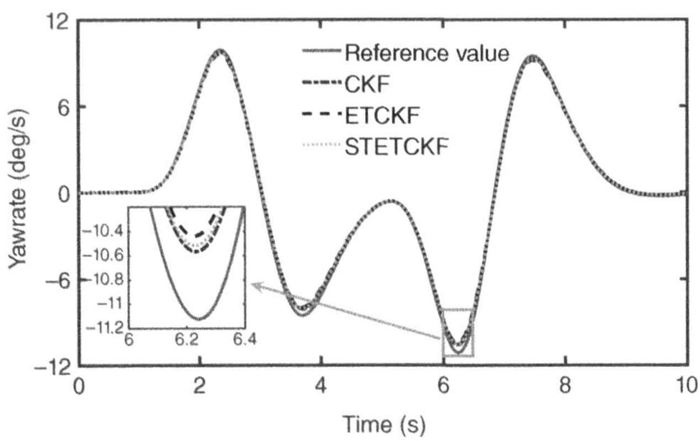

Figure 4.26 The yaw rate in the case of DLC.

However, when the communication rate drops to 5.93% and data loss occurs, the ETCKF and STETCKF estimation results remain commendably close to the RV. This demonstrates the robustness of these methods in handling communication rates and intermittent data. Specifically, the real vehicle's mass is 1270 kg, but in the ETCKF and STETCKF methods, the mass parameter is set to 1600 kg, intentionally introducing a discrepancy to test the robustness of the algorithms. The CKF maintains the correct mass parameter at 1270 kg. Figures 4.26 and 4.27 present the estimation results for yaw rate and sideslip angle, respectively. The STETCKF's estimated curves fit closely to the RVs, demonstrating its robustness to model parameter perturbation. The STETCKF method exhibits better estimation performance compared to the ETCKF, despite the incorrect mass

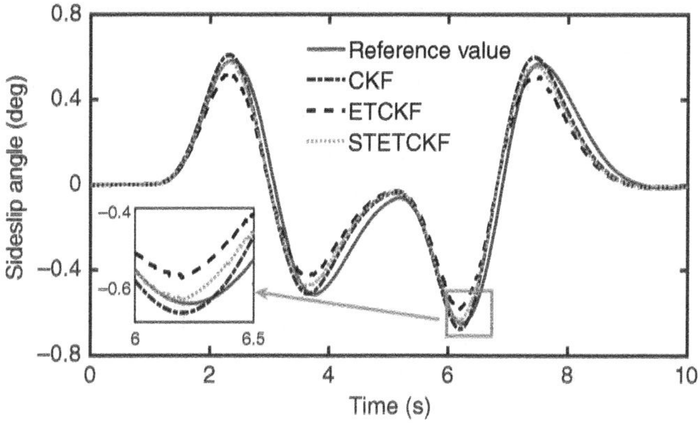

Figure 4.27 The sideslip angle in the case of DLC.

Table 4.5 RMSE of different methods in the case of DLC.

Symbol	r	v_y	β
CKF	0.2286	0.0479	0.0554
ETCKF	0.2754	0.0655	0.0757
STETCKF	0.2464	0.0509	0.0588

parameter. This is because the STETCKF is specifically designed to be robust to such perturbations, enabling it to maintain higher accuracy under challenging conditions. In Table 4.5, the RMSE values for each method are listed. The CKF, with no data loss and the correct model parameters, naturally achieves the smallest RMSE, showcasing its high estimation accuracy under ideal conditions. However, when data loss and model parameter perturbations are introduced, the RMSE of STETCKF, although increased relative to CKF, is still significantly lower than that of ETCKF. To further elaborate, the CKF's close adherence to the RV in Figure 4.25 underlines its efficiency when communication is uninterrupted. Conversely, the event-triggered nature of the ETCKF and STETCKF means they can effectively estimate states even with substantial data loss and lower communication rates. The ETCKF and STETCKF's ability to provide reliable state estimates with just a 5.93% communication rate is particularly noteworthy, as it suggests these methods can maintain performance in bandwidth-limited environments. The yaw rate estimations and sideslip angle estimations show that the STETCKF, in particular, maintains a closer fit to the RV compared to the ETCKF. The strong tracking capability of the STETCKF enables it to better adapt to the perturbations in model parameters, thereby providing more accurate state estimates under these conditions. Overall, the figures and RMSE values in Table 4.5 clearly illustrate the effectiveness and robustness of the STETCKF method. Even with a substantial reduction in communication rate and intentional discrepancies in model parameters, the STETCKF manages to achieve estimation results that are close to those of the CKF under ideal conditions.

4.4.3.2 The Real Vehicle Test

Figure 4.28 depicts the RVT scenario, which is designed to assess the effectiveness of our proposed algorithm under realistic conditions. In normal wireless communication environments, data packet loss tends to occur sporadically. This irregularity makes it challenging to systematically evaluate the robustness and reliability of estimation algorithms in the presence of communication disruptions. To address this challenge and provide a thorough validation of our algorithm, we opted to collect vehicle driving data on real roads and subsequently conduct offline tests. This approach allows us to simulate various levels of data packet loss and communication interruptions in a controlled manner, ensuring a comprehensive evaluation of the algorithm's performance. During these real-world driving sessions, the vehicle's sensor data was continuously recorded, capturing essential parameters such as longitudinal velocity, lateral acceleration, yaw rate, and sideslip angle. The collected data reflects the natural driving conditions and the inherent variability in sensor measurements, thereby offering a realistic basis for testing our algorithms. For vehicle data collection, we employ the ASENSING P-Box INS 570, a high-precision inertial navigation system that provides comprehensive data on vehicle dynamics. The collected data, initially recorded under the northeast coordinate system, is subsequently transformed

Figure 4.28 The two vehicles on the urban highway.

into vehicle state values under the vehicle coordinate system. This transformation is accomplished using the heading angle to ensure accurate alignment of the data with the vehicle's orientation and movement. These converted data serve as the ground truth values, also referred to as RVs, which are critical for validating the accuracy of our estimation algorithms. In the real vehicle experiment, we designed a specific test scenario to rigorously assess the performance of our algorithms under challenging conditions. The test involves having the driver perform a steering operation while accelerating, simulating a dynamic driving scenario with significant changes in vehicle state. The initial vehicle velocity is set at 11 km/h, providing a controlled starting condition for the experiment. To ensure consistency between our simulation tests and real vehicle experiments, we conduct the algorithm validation for offline experiments on a personal computer with the same configuration used for simulation tests. The computer is equipped with an AMD 75800 CPU and 16.0 GB of RAM, providing sufficient computational power to handle the complex calculations required by our algorithms. By using the same hardware setup, we maintain consistent individual cycle times across both simulation and RVT conditions, ensuring that our performance metrics are directly comparable. Once the data collection phase was completed, we introduced controlled packet loss scenarios during offline testing. By simulating different rates of data packet loss, we were able to rigorously test the resilience and accuracy of the STETCKF algorithms. This method allowed us to observe how well these algorithms maintain accurate state estimation despite intermittent data availability. Offline testing provides several advantages. It allows us to repeat the same scenarios multiple times with varying levels of packet loss, ensuring that our results are consistent and reliable. Furthermore, by using real-world data, we ensure that our findings are applicable to practical driving conditions, enhancing the validity of our conclusions. Throughout the test, we intentionally induce a high packet loss rate of 65%, representing a severe communication disruption scenario. This high level of packet loss allows us to evaluate the robustness of our estimation algorithms in maintaining accuracy despite substantial data loss. The validation process involves running the ETCKF and STETCKF algorithms on the collected and transformed vehicle data. The results are then compared against the RV to determine the estimation accuracy.

Figures 4.29 and 4.30 provide detailed depictions of the longitudinal velocity and lateral acceleration, respectively. Upon close examination of these images, particularly in the enlarged sections, we observe that the sensor signals remain constant over certain periods. This constancy indicates intervals during which no fresh sensor data is received, pointing to a temporary cessation of data transmission. Moreover, both figures reveal instances where the data values abruptly drop to zero. This sudden drop occurs because real-time data packets, although transmitted, are lost during these periods. Such occurrences are visually evident in the

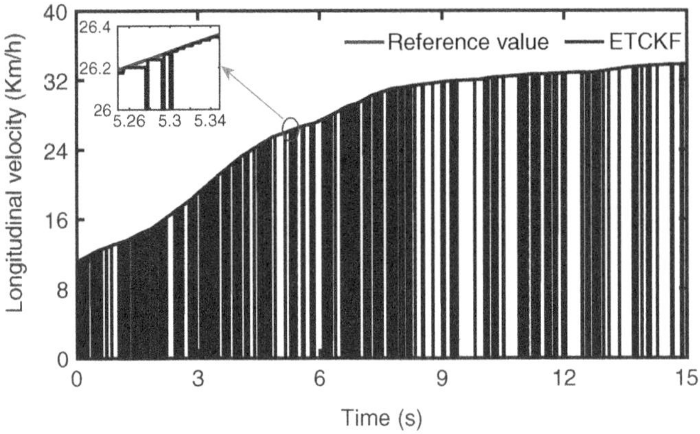

Figure 4.29 The longitudinal velocity curves in the real vehicle test.

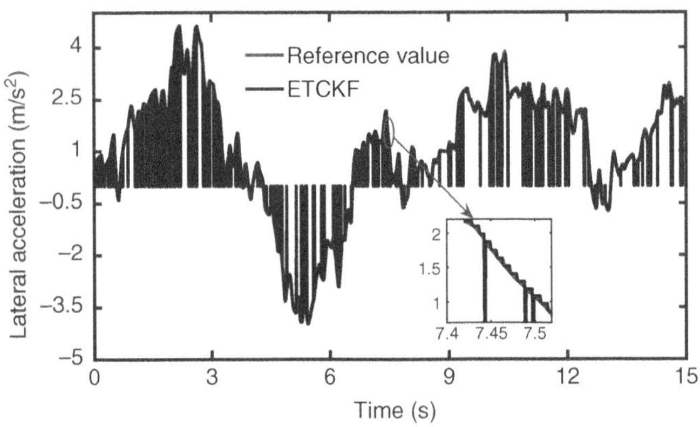

Figure 4.30 The lateral acceleration curves in the real vehicle test.

figures, where the signal flatlines at zero before resuming normal fluctuations. These zero-value points highlight the moments of data packet loss, which mimic real-world scenarios where wireless communication can be unreliable. Interestingly, these patterns of data constancy and abrupt zero drops closely resemble the curves observed in our simulation tests. In the simulations, we designed scenarios with controlled data packet loss to evaluate the robustness and performance of our estimation algorithms under intermittent data conditions. The similarity between the RVT results and the simulation outcomes underscores the validity of our simulation environment and its effectiveness in replicating real-world communication challenges.

The repeated constancy in sensor signals and the sudden drops to zero reflect the event-triggered communication mechanism's behavior. This mechanism prioritizes data transmission based on significant changes in vehicle state, thereby reducing the communication load during stable periods. However, it also means that when data transmission does occur, there is a higher likelihood of packet loss due to the increased data volume in those bursts. By comparing these RVT results with the simulation test curves, we can assess how well the ETCKF and STETCKF algorithms handle such communication interruptions. The algorithms are designed to maintain accurate state estimation even with substantial data loss, and the figures illustrate their performance in real-world conditions. The longitudinal velocity depicted in Figure 4.29 shows how the vehicle's speed fluctuates over time, with intervals of constant data and sudden drops mirroring the packet loss events. Similarly, the lateral acceleration in Figure 4.30 reveals how the vehicle's side-to-side movements are captured by the sensors, with the same patterns of data transmission gaps and packet loss.

In Figure 4.31, a comprehensive depiction of the test process reveals that data transmission occurs 1062 times throughout the experiment, equating to a communication rate of 7.08%. Notably, the initial phase of the test shows a notably high frequency of data transmission. This heightened frequency is primarily attributed to the significant fluctuations in vehicle state observed during the early stages of the experiment under these specific conditions. Similar to the observations in Figure 4.24, the curve density in Figure 4.31 appears more pronounced during the first 8 seconds of the test. This denser appearance results from the increased frequency of event triggering during periods of heightened vehicle state variability. It's crucial to note that while the density implies frequent triggering, it does not signify a constant 100% triggering rate at these points. By zooming into specific

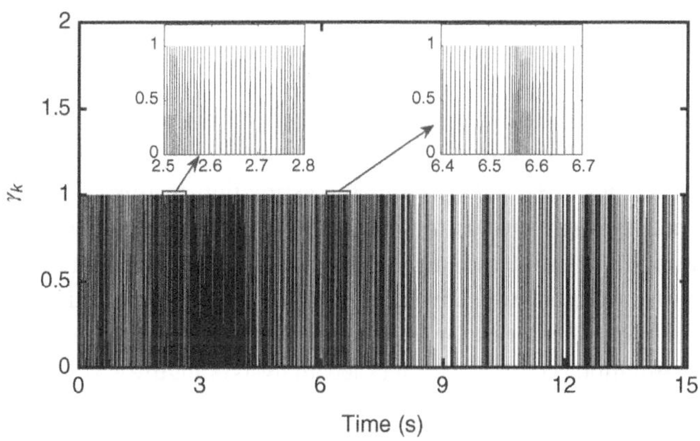

Figure 4.31 Event-triggered times in the real vehicle test.

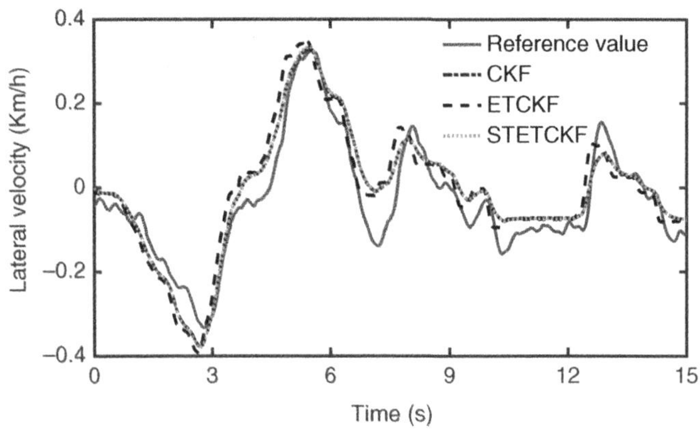

Figure 4.32 The lateral velocity in the real vehicle test.

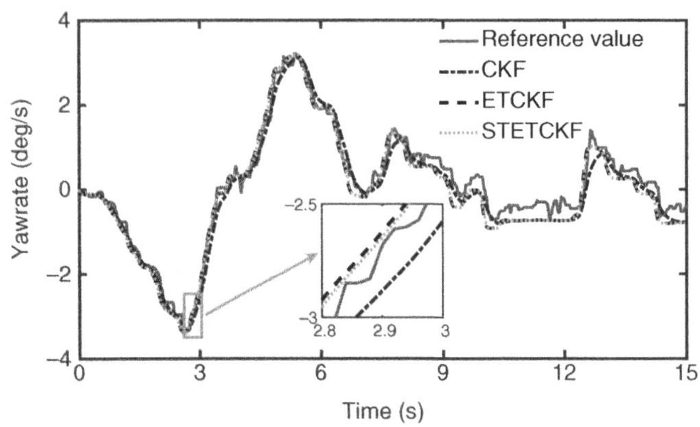

Figure 4.33 The yaw rate in the real vehicle test.

sections of Figure 4.31, it becomes evident that the data points alternate between 0 and 1, highlighting the intermittent nature of event-triggered data transmission. Moving forward, Figs. 4.32, 4.33 present the estimation results of the PV motion state obtained through various methods. These figures provide a visual comparison of how different estimation algorithms perform under the test conditions, offering insights into their respective accuracy and reliability. Additionally, Table 4.6 presents the RMSE metrics for each method, quantifying their performance in terms of estimation accuracy. The detailed analysis of Figure 4.31 and the subsequent estimation results in Figs. 4.32, 4.33 serve to evaluate the effectiveness of the ETCKF and STETCKF algorithms in real-world conditions with intermittent data transmission. These algorithms are specifically designed to adapt to varying

Table 4.6 RMSE of different methods in the case of real vehicle tests.

Symbol	r	v_y	β
CKF	0.2383	0.0493	0.1166
ETCKF	0.2751	0.0611	0.1509
STETCKF	0.2593	0.0505	0.1204

communication rates and data losses, ensuring robust state estimation even under challenging circumstances.

The real vehicle used in the experiment has a mass of 1790 kg, whereas for the ETCKF and STETCKF methods, the mass parameter is set to 1490 kg. This discrepancy in mass values between the real vehicle and the estimation algorithms is noted for consistency in comparing their performance. Figure 4.32 illustrates the estimation accuracy comparison between ETCKF and STETCKF. It is evident that the STETCKF algorithm outperforms ETCKF in terms of estimation accuracy. Specifically, the estimation accuracy of STETCKF closely approaches that of CKF. In Figs. 4.33 and 4.34, which depict the yaw rate and sideslip angle estimation results, respectively, the curves generated by STETCKF closely align with the reference curves, indicating robust performance in capturing these dynamic vehicle parameters. In Table 4.6, the RMSE values further quantify the performance differences among the estimation methods. The RMSE of STETCKF, while slightly higher than CKF due to the effects of model parameter adaptation and occasional data loss, is consistently lower than that of ETCKF. This confirms the

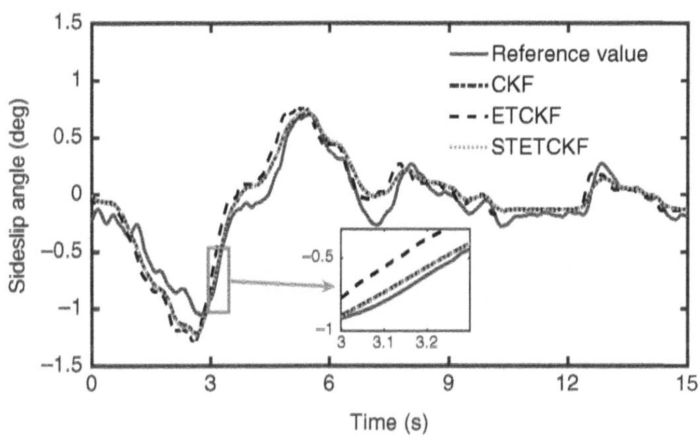

Figure 4.34 The sideslip angle in the real vehicle test.

Table 4.7 RMSE of different methods in the case of a double lane change.

Symbol	r	v_y	β
CKF	0.2146	0.0376	0.0632
ETCKF	0.2971	0.0742	0.0824
STETCKF	0.2275	0.0457	0.0684

Table 4.8 RMSE of different methods in the case of continuous steering.

Symbol	r	v_y	β
CKF	0.2315	0.0411	0.1012
ETCKF	0.2634	0.0645	0.1433
STETCKF	0.2403	0.0489	0.1183

superior accuracy and reliability of STETCKF in handling communication interruptions and model parameter uncertainties. To broaden the demonstration of our algorithm's applicability, we evaluated its performance under different working conditions. For instance, in a scenario involving a double lane change with an initial speed of 40 km/h and a 20% packet loss rate, Table 4.7 presents the RMSE values for each method. Similarly, under conditions of continuous steering at an initial speed of 20 km/h and a 50% packet loss rate, Table 4.8 lists the corresponding RMSE values. In Tables 4.7 and 4.8, the CKF consistently exhibits the smallest RMSE compared to the ETCKF and STETCKF. Specifically, the RMSE values of STETCKF are consistently better than those of ETCKF in both Tables 4.7 and 4.8. Despite occasional data loss and discrepancies in model parameters, STETCKF demonstrates robust performance in accurately estimating the motion state of the PV in both simulation and RVTs. These additional evaluations highlight the versatility and robustness of the STETCKF algorithm across various dynamic driving scenarios. By maintaining accurate state estimation despite fluctuations in communication reliability and vehicle dynamics, STETCKF demonstrates its suitability for real-world applications in autonomous driving and ADAS.

4.5 Summary and Future Research

In this chapter, we introduce an innovative event-triggered estimation framework that integrates an event-triggered mechanism with an ECKF based on a 3-DOF model. This framework is designed for state estimation of PVs in dynamic environments. Through extensive simulation and RVTs, our proposed prediction approach demonstrates its ability to effectively manage the trade-off between communication rate and estimation performance. The versatility of our method extends beyond PV state estimation; it can also be adapted for predicting the states of surrounding vehicles. This capability is crucial for enhancing the safety control of intelligent connected vehicles, thereby making significant strides in vehicle safety systems. While our experiments did not encounter data loss, we acknowledge that packet loss is a common occurrence in wireless communication systems. Our event-triggered mechanism proves effective in mitigating the impact of such packet loss incidents. Nevertheless, addressing data loss to further enhance estimation accuracy remains a critical area for future research. By focusing on strategies to handle data loss, future iterations of our framework aim to bolster estimation robustness and reliability under realistic wireless communication conditions. This direction underscores our commitment to advancing the applicability and resilience of our approach in real-world scenarios. Ultimately, our research contributes to the ongoing evolution of intelligent vehicle technologies, paving the way for safer and more efficient transportation systems.

Furthermore, we introduce an innovative event-triggered estimation scheme tailored to predict the motion state of PVs within the context of vehicle-to-vehicle communication, addressing the challenges posed by packet loss and uncertainties in model parameters. This approach includes the development of the novel STETCKF, which is validated for convergence through rigorous analysis of estimated error covariance. Through a comprehensive series of experimental tests, our findings consistently demonstrate that STETCKF achieves an optimal balance between estimation accuracy and communication efficiency. It effectively mitigates the detrimental effects of model parameter variations and data loss on estimation precision. This capability is crucial for ensuring reliable and robust state estimation in dynamic vehicular environments. One of the key advantages of STETCKF is its practical feasibility for deployment in real vehicles, characterized by its low computational resource requirements. This allows for the provision of real-time environmental information essential for informed decision-making in intelligent vehicle systems. In our current study, we acknowledge the simplifications made in the vehicle model, which may lead to reduced estimation accuracy under extreme operational conditions. Moving forward, our research agenda includes enhancing the performance and adaptability of the proposed method. This involves refining the vehicle model to better reflect real-world complexities and optimizing the estimation algorithm to further strengthen its robustness across diverse driving scenarios. In summary, our proposed event-triggered estimation framework, embodied by STETCKF, represents a significant step toward achieving reliable and efficient motion state prediction for PVs. It underscores our dedication to pushing the boundaries of vehicle intelligence and contributing to the evolution of smart transportation systems worldwide.

References

1 Wang Nidamanuri, J., Nibhanupudi, C., Assfalg, R., and Venkataraman, H. (2021). A progressive review-emerging technologies for ADAS driven solutions. *IEEE Transactions on Intelligent Vehicles*, early access, October 2021 doi: 10.1109/TIV.2021.3122898.

2 van Hoek, R., Ploeg, J., and Nijmeijer, H. (2021). Cooperative driving of automated vehicles using B-Splines for trajectory planning. *IEEE Transactions on Intelligent Vehicles* 6 (3): 594–604.

3 Wang, Y., Hu, J., Wang, F. et al. (2022). Tire road friction coefficient estimation: Review and research perspectives. *Chinese Journal of Mechanical Engineering* 35 (1): 1–11.

4 Wang, Y., Zhou, Z., Wei, C. et al. (2018). Host-target vehicle model-based lateral state estimation for preceding target vehicles considering measurement delay. *IEEE Transactions on Industrial Informatics* 14 (9): 4190–4199.

5 Hajidavalloo, M.R., Li, Z., Chen, D. et al. (2022). A mechanical system inspired microscopic traffic model: modeling, analysis, and validation. *IEEE Transactions on Intelligent Vehicles*, early access, January 2022. doi: 10.1109/TIV.2022.3146313.

6 Tajalli, M. and Hajbabaie, A. (2021). Traffic signal timing and trajectory optimization in a mixed autonomy traffic stream. *IEEE Transactions on Intelligent Transportation Systems*, early access, February 2021. doi: 10.1109/TITS.2021.3058193.

7 Liu, Y., Zhou, J., Tian, D. et al. (2021). Joint communication and computation resource scheduling of a UAV-assisted mobile edge computing system for platooning vehicles. *IEEE Transactions on Intelligent Transportation Systems*, June 2021 doi: 10.1109/TITS.2021.3082539.

8 Schubert, R., Richter, E., and Wanielik, G. (2008). Comparison and evaluation of advanced motion models for vehicle tracking. In *Proceedings of 11th International Conference on Information Fusion*, pp. 1–6.

9 Liu, H., Wang, P., Lin, J. et al. (2021). Real-time longitudinal and lateral state estimation of preceding vehicle based on moving horizon estimation. *IEEE Transactions on Vehicular Technology* 70 (9): 8755–8768.

10 Zhou, Z., Zhou, G., Wang, Y. et al. (2021). PTV longitudinal-lateral state estimation considering unknown control inputs and uncertain model parameters. *IEEE Transactions on Vehicular Technology* 70 (5): 4366–4376.

11 Doumiati, M., Victorino, A.C., Charara, A., and Lechner, D. (2011). Onboard real-time estimation of vehicle lateral tire-road forces and sideslip angle. *IEEE/ASME Transactions on Mechatronics* 16 (4): 601–614.

12 Li, L., Song, J., Li, H., and Zhang, X. (2011). A variable structure adaptive extended Kalman filter for vehicle slip angle estimation. *International Journal of Vehicle Design* 56 (1): 161–185.

13 Li, L., Jia, G., Ran, X. et al. (2014). A variable structure extended Kalman filter for vehicle sideslip angle estimation on a low friction road. *Vehicle System Dynamics* 52 (2): 280–308.

14 Wang, Y., Xu, L., Zhang, F. et al. (2021). An adaptive fault-tolerant EKF for vehicle state estimation with partial missing measurements. *IEEE/ASME Transactions on Mechatronics* 26 (3): 1318–1327.

15 Antonov, S., Fehn, A., and Kugi, A. (2011). Unscented Kalman filter for vehicle state estimation. *Vehicle System Dynamics* 49 (9): 1497–1520.

16 Heidfeld, H., Schünemann, M., and Kasper, R. (2020). UKF-based State and tire slip estimation for a 4WD electric vehicle. *Vehicle System Dynamics* 58 (10): 1479–1496.

17 Wang, Y., Lv, C., Yan, Y. et al. (2022). An integrated scheme for coefficient estimation of tire road friction with mass parameter mismatch under complex driving scenarios. *IEEE Transactions on Industrial Electronics* 69 (12): 13337–13347.

18 Arasaratnam, I. and Haykin, S. (2009). Cubature Kalman filters. *IEEE Transactions on Automatic Control* 54 (6): 1254–1269.

19 Wang, Y., Geng, K., Xu, L. et al. (2022). Estimation of sideslip angle and tire cornering stiffness using fuzzy adaptive robust cubature Kalman filter. *IEEE Transactions on Systems, Man, and Cybernetics: Systems* 52 (3): 1451–1462.

20 Song, R. and Fang, Y. (2021). Vehicle state estimation for INS/GPS aided by sensors fusion and SCKF-based algorithm. *Mechanical Systems and Signal Processing* 150.

21 Shieh, W., Hsu, C.J., Lin, C., and Wang, T. (2018). Investigation of vehicle positioning by infrared signal-direction discrimination for short-range vehicle-to-vehicle communications. *IEEE Transactions on Vehicular Technology* 67 (12): 11563–11574.

22 Zhou, Z., Wang, Y., Du, H. et al. (2020). Sub-full model-based heterogeneous sensor fusion for lateral state estimation of preceding target vehicles. *IEEE/ASME Transactions on Mechatronics* 25 (3): 1335–1345.

23 Wei, S., Zou, Y., Zhang, X. et al. (2019). An integrated longitudinal and lateral vehicle following control system with radar and vehicle-to-vehicle communication. *IEEE Transactions on Vehicular Technology* 68 (2): 1116–1127.

24 Zhang, J., Zhang, B., Zhang, N. et al. (2021). A novel robust event-triggered fault tolerant automatic steering control approach of autonomous land vehicles under in-vehicle network delay. *International Journal of Robust and Nonlinear Control*. https://doi.org/10.1002/rnc.53.

25 Zhang, Y., Huang, Y., Li, N., and Zhao, L. (2015). Embedded cubature Kalman filter with adaptive setting of free parameter. *Signal Processing* 114 (3): 112–116.

26 Wei, H., Zhang, K., Zhang, H., and Shi, Y. (2024). Resilient and constrained consensus against adversarial attacks: A distributed MPC framework. *Automatica* 160: 111417.

27 Wang, N. and Li, X.-J. (2021). Optimal output synchronization control of a class of complex dynamical networks with partially unknown system dynamics. *IEEE Transactions on Systems, Man, and Cybernetics: Systems* 51 (2): 822–832. https://doi.org/10.1109/TSMC.2018.2882827.

28 Shi, D., Chen, T., and Shi, L. (2014). An event-triggered approach to state estimation with multiple point and set-valued measurements. *Automatica* 50 (6): 1641–1648.

29 Hu, J., Wang, Z., Gao, H., and Stergioulas, L.K. (2012). Extended Kalman filtering with stochastic nonlinearities and multiple missing measurements. *Automatica* 48 (9): 2007–2015.

30 Kluge, S., Reif, K., and Brokate, M. (2010). Stochastic stability of the extended Kalman filter with intermittent observations. *IEEE Transactions on Automatic Control* 55 (2): 514–518.

31 Reif, K., Gunther, S., Yaz, E., and Unbehauen, R. (1999). Stochastic stability of the discrete-time extended Kalman filter. *IEEE Transactions on Automatic Control* 44 (4): 714–728.

32 Li, L., Yu, D., Xia, Y., and Yang, H. (2017). Event-triggered UKF for nonlinear dynamic systems with packet dropout. *International Journal of Robust and Nonlinear Control* 27 (18): 4208–4226.

33 Liu, X., Li, L., Zhen, L. et al. (2017). Stochastic stability condition for the extended Kalman filter with intermittent observations. *IEEE Transactions on Circuits and Systems II: Express Briefs* 64 (3): 334–338.

34 Sinopoli, B., Schenato, L., Franceschetti, M. et al. (2004). Kalman filtering with intermittent observations. *IEEE Transactions on Automatic Control* 49 (9): 1453–1464.

35 Zhu, H., Mi, J., Li, Y. et al. (2022). VB-Kalman based localization for connected vehicles with delayed and lost measurements: Theory and experiments. *IEEE/ASME Transactions on Mechatronics* 27 (3): 1370–1378.

36 Yan, Y., Du, H., Han, Q.L., and Li, W. (2023). Discrete multi-objective switching topology sliding mode control of connected autonomous vehicles with packet loss. *IEEE Transactions on Intelligent Vehicles* 8 (4): 2926–2938.

37 Jwo, D., Yang, C., Chuang, C., and Lee, T. (2013). Performance enhancement for ultra-tight GPS/INS integration using a fuzzy adaptive strong tracking unscented Kalman filter. *Nonlinear Dynamics* 73 (1): 377–395.

38 Zhou, D.H. and Frank, P.M. (1996). Strong tracking Kalman filtering of nonlinear time-varying stochastic systems with colored noise: application to parameter estimation and empirical robustness analysis. *International Journal of Control* 65: 295–307.

5

Tire–Road Friction Coefficient Estimation with Parameters Mismatch and Data Loss

5.1 Introduction

Accurate knowledge of the TRFC is essential for optimizing driver maneuvers, thereby enhancing the safety of intelligent vehicles. The TRFC plays a critical role in determining how well a vehicle can adhere to the road surface, influencing braking, acceleration, and cornering performance. This makes it a pivotal factor in vehicle safety systems, especially for autonomous and semi-autonomous driving technologies. Unfortunately, due to technological and cost constraints, TRFC cannot be directly measured by in-vehicle sensors. The lack of direct measurement capabilities poses a significant challenge for automotive engineers and researchers striving to develop more sophisticated and safer vehicles. To tackle this challenge, numerous researchers have proposed a variety of approaches over time. These approaches range from empirical models based on extensive road testing to advanced algorithms that infer TRFC from available sensor data, such as wheel speed, acceleration, and steering angle. Despite significant advancements in TRFC estimation, the impact of mass parameter mismatch in vehicle models on estimation accuracy has rarely been explored. This oversight can have profound implications, as the mass of the vehicle directly influences its dynamic behavior. The mass of a vehicle directly influences its longitudinal, lateral, and vertical dynamics. It affects how a vehicle accelerates, brakes, and handles corners, all of which are critical for accurate TRFC estimation. Unlike other vehicle parameters, the mass parameter can fluctuate widely due to the loading and unloading of goods and passengers. This variability necessitates careful consideration of the mass parameter's influence on TRFC estimation. For instance, a fully loaded vehicle behaves quite differently from an empty one, and failing to account for this can lead to significant errors in TRFC estimation. Additionally, many existing hybrid methods aim to enhance TRFC estimation accuracy through the weighted summation of results from individual models. However, each model independently updates its estimation based solely on its

State Estimation of Multi-Agent Vehicle-Road Interaction Systems, First Edition.
Yan Wang, Guodong Yin, and Chao Huang.
© 2026 The Institute of Electrical and Electronics Engineers, Inc. Published 2026 by John Wiley & Sons, Inc.

previous prediction, failing to fully leverage the benefits of multi-modal data fusion. This independent operation means that the potential synergies between different data sources and models are not fully exploited. Consequently, the accuracy improvements achieved by these methods are often limited, especially in complex driving scenarios that involve simultaneous braking and steering. Complex driving scenarios, such as navigating sharp turns while braking or accelerating on slippery surfaces, present significant challenges for TRFC estimation. In such conditions, the vehicle's dynamic behavior can change rapidly, and accurate TRFC estimation becomes crucial for maintaining stability and control. Existing methods often struggle in these scenarios because they do not incorporate real-time adjustments for changes in vehicle mass and do not fully integrate data from multiple sources. In conclusion, while significant progress has been made in TRFC estimation, the challenge of mass parameter variability remains largely unaddressed. A holistic approach that incorporates real-time mass parameter adjustments and multi-modal data fusion holds the promise of significantly enhancing TRFC estimation accuracy.

In addition, some parameters in these estimation algorithms are set in advance, and their performance heavily relies on the experience and intuition of the engineers and researchers who design them. This reliance on predefined parameters can be problematic, especially considering the complexity and variability of real-world driving conditions. As driving conditions change, the weights assigned to different models within the estimation algorithm need to be dynamically adjusted to maintain optimal performance. On the other hand, it should be noted that most existing estimation approaches ignore the effect of missing measurements on estimation accuracy. In reality, sensor data loss is a common phenomenon. The signals from onboard sensors, such as wheel speed sensors, accelerometers, and gyroscopes, are not always fully transmitted to the estimator due to various reasons, such as electromagnetic interference, hardware malfunctions, or connectivity issues. When sensor data is missing or corrupted, the accuracy of TRFC estimation can be significantly compromised, leading to potential safety risks. Most studies today focus on model-based identification methods. These methods typically rely on a well-defined mathematical representation of the vehicle dynamics and require prior knowledge of precise noise statistical characteristics to achieve optimal estimation results. Model-based approaches, while robust in certain controlled environments, can struggle when faced with the unpredictable nature of real-world driving. They often assume that the noise characteristics are stationary and well-known, which is rarely the case in practice. Data-driven methods, on the other hand, are still in their early stages. These approaches leverage large datasets and machine learning algorithms to infer TRFC from sensor data. However, significant challenges exist in both data collection and the generalization capability of these models.

Collecting comprehensive and high-quality data that captures the full range of possible driving conditions is a monumental task. Moreover, ensuring that the models trained on this data can generalize well to unseen conditions is another major hurdle. Data-driven methods often require vast amounts of labeled data to train accurately, and even then, they might not perform well in scenarios that differ from the training data. Furthermore, both model-based and data-driven approaches require real-time access to accurate sensor data. Real-time data processing is critical for making timely and accurate TRFC estimations, which are necessary for immediate decision-making in intelligent vehicle systems. Any delay or inaccuracy in data transmission can lead to suboptimal performance or even failure of the estimation algorithm. This need for real-time, accurate data further complicates the development of reliable TRFC estimation systems. In summary, while existing TRFC estimation methods have made significant strides, there is still much work to be done to improve their robustness and accuracy under varying and unpredictable real-world conditions. Developing adaptive, data-resilient algorithms and enhancing the generalization capabilities of data-driven models are crucial steps toward achieving more reliable TRFC estimation, ultimately leading to safer and more efficient intelligent vehicle systems.

In this chapter, an integrated scheme for TRFC estimation is proposed by combining strong tracking UKF and interactive multiple model UKF. Real-time experiments are implemented on a mass-produced vehicle to demonstrate the feasibility and effectiveness of the proposed method. Next, a fault-tolerant estimation scheme is proposed to estimate TRFC in the case of missing measurements. Experiments with different working conditions are performed to demonstrate the validity of the fault-tolerant estimation framework. Finally, a model-based learning approach, incorporating ETCKF EKFNet, is proposed for identifying TRFC. Multiple virtual experiment results demonstrate that the estimation performance of the model-based learning framework outperforms conventional EKF and UKF.

5.2 Related Works

Traffic accidents usually result in a large number of casualties and economic losses [1]. The development of ITS and intelligent vehicle technology can effectively reduce traffic accidents [2, 3]. Several advanced navigation [4], vehicle chassis control [5–8], and platoon control [9, 10] techniques have been proposed to enhance vehicle safety. These innovations aim to provide better control and coordination of vehicles on the road, thereby improving overall traffic safety. In recent years, the distributed drive electric vehicle (DDEV) has become a hot research topic. It has more control degrees of freedom than conventional electric vehicles, which makes the vehicle stability control system more applicable and reduces the risk of instability [11]. As tires are the only components connecting the vehicle to the ground, motion control or stability control of vehicles ultimately translates into the control of motor torque and braking torque. The TRFC directly limits the maximum tire forces available for the vehicle. Furthermore, many advanced driver assistance systems or high-level autonomous vehicles require dynamic adjustments to longitudinal and lateral control to enhance vehicle safety based on TRFC. For example, path tracking [12], active collision avoidance [13], and electronic stability control systems [14]. However, one significant challenge is that in-vehicle sensors do not directly measure TRFC. To address this, there are two popular methods for obtaining TRFC: direct measurement with specialized sensors and estimation using low-cost sensors. The former method, usually referred to as an off-board sensor-based approach, involves using external sensors to directly measure the TRFC. These sensors can provide highly accurate data but are often expensive and impractical for widespread use in consumer vehicles. The latter method, known as a model-based approach, relies on in-vehicle sensors to estimate TRFC. This approach is more cost-effective and practical for mass adoption. It uses data from various low-cost sensors, such as wheel speed sensors, accelerometers, and gyroscopes, combined with mathematical models to infer the TRFC. While this method is more accessible, it faces challenges in accuracy and reliability, especially under varying and unpredictable driving conditions. Both methods have their advantages and limitations. Off-board sensor-based approaches offer high accuracy but are not feasible for all vehicles due to their high cost and complexity. Model-based methods, on the other hand, provide a more economical solution but require sophisticated algorithms to handle the inherent uncertainties and variabilities in sensor data. In conclusion, while significant progress has been made in vehicle safety technology, accurately estimating TRFC remains a critical challenge. Both off-board sensor-based and model-based approaches offer viable solutions, each with its own set of trade-offs. Continued research and development in this area are essential for advancing intelligent vehicle technologies and ultimately reducing traffic accidents and their associated costs.

These off-board sensors typically include advanced cameras [15], accelerometers [16], and magnetometers [17], among others. Leng et al. [18] designed a visual fusion method to identify the TRFC, integrating data from multiple cameras to enhance the accuracy of their estimations. Similarly, Yu et al. [19] utilized image data to identify TRFC using a backpropagation neural network, demonstrating the potential of machine learning in processing visual information for this purpose. However, these vision-based approaches have a strong dependence on environmental visibility. Factors such as poor lighting, fog, rain, or snow can significantly impair the performance of camera-based systems, limiting their reliability in various driving conditions. Recognizing this limitation, researchers have proposed several methods based on mechanical response to estimate TRFC. For instance, Zou et al. [16] presented a methodology to determine TRFC by utilizing tire vibration. Their approach involves analyzing the vibrations transmitted through the tire as it interacts with the road surface, offering a more direct measure of the frictional forces at play. This method provides a robust alternative to vision-based systems, especially in conditions where visibility is compromised. Furthermore, the concept of the intelligent tire, equipped with advanced sensors, has been explored as another means to identify TRFC [20]. Intelligent tires incorporate sensors such as accelerometers, strain gauges, and pressure sensors within the tire structure itself. These sensors can capture real-time data on tire deformation, vibration, and other mechanical responses, providing valuable insights into the tire–road interaction. The data collected from intelligent tires can be processed using advanced algorithms to estimate TRFC with high accuracy. This approach not only enhances the robustness of TRFC estimation under various driving conditions but also enables continuous monitoring of tire performance and health. The integration of intelligent tires into modern vehicles represents a significant advancement in vehicle safety systems, offering a more comprehensive understanding of tire–road dynamics.

These off-board sensors usually include advanced cameras [15], accelerometers [16], magnetometers [17], etc. Leng et al. [18] designed a visual fusion method to identify TRFC. Yu et al. [19] identified TRFC based on image data using a backpropagation neural network. These vision-based approaches have a strong dependence on environmental visibility. To address the problem, several approaches on the basis of mechanical response have been proposed successively. Zou et al. [16] presented a methodology to determine the TRFC utilizing tire vibration. Furthermore, the intelligent tire with advanced sensors was also used to identify TRFC [20]. In summary, while vision-based methods for TRFC estimation have shown promise, their dependence on environmental visibility poses a significant challenge. Mechanical response-based approaches, such as those utilizing tire vibration and intelligent tire technology, offer viable alternatives that can operate effectively regardless of visibility conditions. By combining data from various sensors and employing advanced processing techniques, these methods can provide more reliable and accurate TRFC estimations, ultimately contributing to safer and more efficient intelligent transportation systems.

However, the frequent impact and harsh conditions experienced by sensors within tires make them susceptible to damage. The constant interaction with the road surface, exposure to extreme temperatures, and mechanical stress can lead to sensor malfunction or failure. Consequently, the durability and longevity of these sensors are significant concerns. These challenges, along with technical and cost constraints, limit the extensive application of off-board sensor-based approaches in everyday vehicles. Moreover, off-board sensor-based methods primarily detect the type of road surface but often fall short of providing accurate, real-time TRFC values. These methods might identify whether the road is dry, wet, icy, or covered with gravel, but they struggle to quantify the exact friction coefficient. This limitation reduces their effectiveness in applications that require precise TRFC data for optimal vehicle control and safety. To address these shortcomings, model-based estimation methods have gained increasing attention. Unlike off-board sensor-based approaches, model-based methods do not rely on external sensors embedded in tires. Instead, they combine vehicle dynamics models and tire models with sophisticated estimation algorithms to infer the TRFC. This approach leverages data from existing vehicle sensors, such as wheel speed sensors, accelerometers, and gyroscopes, to estimate the friction coefficient. Model-based estimation methods typically involve creating detailed mathematical representations of the vehicle and tire behavior. These models account for various factors, including tire deformation, vehicle speed, steering angle, and acceleration. By inputting sensor data into these models, the system can estimate the TRFC based on the observed vehicle dynamics. This process often involves complex algorithms, such as KFs, particle filters, or machine learning techniques, to refine the estimates and account for uncertainties in the data.

This class of methods is usually divided into three categories: longitudinal dynamics-based, lateral dynamics-based, and fusion estimation. These categories reflect the different aspects of vehicle dynamics that are leveraged to estimate the TRFC. Below are some of the major research efforts and advancements in each category. For straight driving conditions, the longitudinal dynamic model is typically used for estimating TRFC [21]. This model focuses on the forces and movements along the direction of travel. Rajamani et al. [22] introduced an approach that employs the longitudinal slip slope to estimate TRFC, enhancing the performance of the method by fusing information from the GPS. This integration of GPS data helps to refine the accuracy of the estimation by providing additional context about the vehicle's speed and trajectory. Another popular research direction involves identifying TRFC based on longitudinal tire force information. For instance, Xia et al. [23] designed a holistic estimator based on the Burckhardt tire model to predict TRFC. This model incorporates various parameters that characterize the tire–road interaction, allowing for a comprehensive estimation of friction. Similarly, Sharifzadeh et al. [24] used RLS to identify TRFC according to tire slip information. Although RLS is a powerful tool, it typically has only one degree of freedom to modify the adaptation of the algorithm, which can limit its flexibility and accuracy. To address this limitation, many scholars have turned to Kalman filtering techniques to estimate TRFC.

KFs provide a robust framework for integrating multiple sources of information and handling uncertainties in the data. Krisztian et al. [25] developed an EKF to predict TRFC, which allows for more accurate and adaptive estimations by considering the nonlinearities in the vehicle dynamics. Additionally, Castillo et al. [26] developed an identification methodology that integrates fuzzy logic and Kalman filtering, combining the strengths of both approaches to improve estimation accuracy. A similar estimation method that leverages these advanced filtering techniques was also reported in [27]. Lateral dynamics-based methods focus on the forces and movements perpendicular to the direction of travel. These methods are particularly useful for estimating TRFC during cornering and other maneuvers where lateral forces play a significant role. By analyzing the lateral slip and tire force information, researchers can develop models that accurately estimate TRFC under various driving conditions.

For cornering driving conditions, the lateral dynamic state of the vehicle is critical for accurate TRFC estimation. This dynamic state is first obtained using state-of-the-art estimation algorithms or sensors, which capture the lateral forces and movements experienced by the vehicle during cornering. Once this lateral dynamic data is acquired, various estimation methods are employed to predict the TRFC. One approach involves using an analytical model based on the aligning torque signal to identify TRFC [28]. This model leverages the torque generated by the lateral forces acting on the tires to estimate the friction coefficient, providing valuable insights into the tire–road interaction during cornering. Hu et al. [29] developed an EKF based on lateral vehicle dynamics to estimate TRFC. The EKF is well-suited for handling the nonlinearities in vehicle dynamics, making it an effective tool for this application. A similar EKF-based estimator was also reported successively [30], demonstrating the robustness and reliability of EKF in TRFC estimation. The UKF offers even higher estimation accuracy compared to the EKF. This filter method captures the statistical properties of the system more accurately by using a deterministic sampling technique. Consequently, the UKF has been applied to further improve the estimation accuracy of TRFC [29], making it a preferred choice in scenarios requiring high precision. In addition to Kalman-based methods, particle filters have been designed to estimate TRFC, particularly in situations where the noise is non-Gaussian [31, 32]. Particle filters can handle complex, non-linear, and non-Gaussian problems by using a set of random samples (particles) to represent the posterior distribution of the estimated parameters. This makes them highly versatile and capable of providing accurate TRFC estimates under a wide range of conditions. However, the practical application of particle filters poses significant challenges. The computational time required for particle filters is considerable, making real-time implementation difficult. Additionally, particle filters suffer from the degeneracy problem, where after several iterations, a large portion of the particles may have negligible weights, leading to a loss of diversity in the sample set. This issue can degrade the estimation accuracy and reliability [33]. To mitigate these challenges, researchers are exploring various strategies to optimize the performance of particle filters. Techniques such as resampling and parallel computing are being investigated to reduce computation time and prevent degeneracy. Despite these challenges, particle filters remain a promising approach for TRFC estimation, especially in scenarios where traditional Kalman-based methods may fall short. In conclusion, estimating TRFC for cornering driving conditions involves leveraging lateral dynamic data through advanced estimation algorithms and sensors. Analytical models based on aligning torque, EKF, and UKF methods provide robust solutions for TRFC estimation. Particle filters offer additional capabilities in handling non-Gaussian noise, although their practical application requires overcoming significant computational challenges. Continued research and development in these areas are essential for improving the accuracy and reliability of TRFC estimation, ultimately enhancing vehicle safety and performance in intelligent transportation systems.

Nevertheless, the above studies primarily considered either vehicle longitudinal or lateral dynamics in isolation, which can result in a significant underestimation of the TRFC [34]. To address this limitation and improve estimation accuracy, some hybrid estimation methods have been proposed that integrate both longitudinal and lateral dynamics. Choi et al. [34] introduced an approach to estimate TRFC based on a coupled longitudinal and lateral tire model using a linearized RLS method. This method effectively combines the two dynamic aspects, providing a more comprehensive understanding of tire–road interaction and improving the reliability of TRFC estimation. Ren et al. [35] designed a hybrid estimator that leverages the UKF to identify TRFC. By integrating longitudinal and lateral dynamics, this hybrid approach utilizes the strengths of UKF in handling nonlinearities and uncertainties in vehicle dynamics, resulting in more accurate TRFC estimates. Li et al. [36] presented a signal fusion method to obtain TRFC. This method combines signals from various sensors and integrates information from both longitudinal and lateral dynamics, enhancing the overall accuracy and robustness of the TRFC estimation. Chen et al. [37] utilized a MSE weighted fusion approach to observe TRFC by leveraging longitudinal and lateral tire forces. This technique weights the contributions of different sensor inputs based on their estimated accuracy, resulting in a more precise and reliable TRFC estimation. Similarly, another method for estimating TRFC using tire forces was described in [38], demonstrating the importance of integrating multiple sources of dynamic information for accurate TRFC estimation. Zareian et al. [39] proposed using a multilayer perceptron neural network to predict TRFC. This neural network model incorporates tire forces, vehicle velocity, and yaw rate as inputs, providing a data-driven approach to TRFC estimation that can capture complex relationships between these variables. Chen et al. [40] introduced a frequency domain data fusion method to estimate TRFC, based on information from the steering and drive systems. This method analyzes the frequency characteristics of the sensor data, offering a novel perspective on integrating longitudinal and lateral dynamics for TRFC estimation.

Recently, with the continuous enhancement of deep learning algorithms, data-driven methods have been increasingly employed for TRFC estimation. These methods offer the advantage of bypassing the need for complex vehicle or tire models, instead relying on the power of neural networks and other machine learning techniques to derive insights directly from data. Mirmohammad Sadeghi et al. [41] demonstrated this approach by collecting four distinct features and building a multilayer neural network to estimate TRFC. Their method showed that, with appropriate feature selection and network architecture, it is possible to achieve accurate TRFC predictions without delving into the intricacies of physical modeling. Expanding on this concept, a spatial-temporal convolutional neural network (ST-CNN) was proposed to predict TRFC [42]. This approach leverages the ability of convolutional neural networks to process spatial data and their temporal extensions to capture the dynamic changes over time. By analyzing patterns in both space and time, the ST-CNN can provide more robust and accurate TRFC estimates. However, the complexity of such network structures can significantly increase computational costs, posing challenges for real-time applications. Despite these advancements, data-driven approaches face several challenges. One of the main issues is the interpretability of the models. Unlike traditional methods grounded in physical laws and vehicle dynamics, neural networks often operate as "black boxes," making it difficult to understand how they arrive at their predictions. This lack of transparency can be a barrier to their acceptance and trustworthiness, particularly in safety-critical applications such as vehicle control systems.

5.3 TRFC Estimation with Mass Parameter Mismatch Under Complex Driving Scenarios

In this section, a novel integrated estimation scheme that combines the STUKF with the IMM-UKF is proposed. The primary objective of this scheme is to enhance the accuracy of TRFC estimation, particularly in scenarios involving vehicle mass parameter mismatch. The STUKF is employed to effectively estimate longitudinal and lateral axle forces, accounting for variations in vehicle mass parameters. By utilizing different axle force information, the scheme applies the UKF to determine both longitudinal and lateral TRFC. This approach leverages the UKF's capability to handle nonlinearities and uncertainties in the estimation process, thereby improving the precision of TRFC estimates. Furthermore, to further enhance estimation accuracy, the scheme incorporates an interactive multiple model (IMM) approach. This method dynamically selects and fuses multiple TRFC estimates derived from different models or conditions. By adapting to varying driving conditions and sensor inputs in real-time, the IMM-UKF ensures robust TRFC estimation across a range of scenarios. Overall, the proposed scheme integrates advanced filtering techniques (STUKF and IMM-UKF) to achieve more accurate and reliable TRFC estimation, crucial for enhancing vehicle dynamics control and safety in intelligent transportation systems.

The overall architecture is shown in Fig. 5.1. It has four main modules, which are measurements with noise, vehicle and tire models, vehicle axle forces estimation, and TRFC estimation. Sensor data acquired via the CAN bus includes essential parameters such as yaw rate, steering wheel angle, longitudinal acceleration, and lateral acceleration. Additionally, the front wheel angle information is determined through a lookup table based on the steering wheel angle data, enhancing the accuracy of vehicle dynamics modeling. In the context of the proposed scheme, the STUKF assumes a pivotal role in estimating vehicle axle forces. By leveraging the vehicle dynamics model and sensor measurements, the STUKF aims to mitigate the impact of mass parameter variations on estimation accuracy. This approach ensures that the estimated axle forces reflect the true dynamics of the vehicle under varying operational conditions. Simultaneously, the IMMUKF dynamically identifies TRFC by integrating longitudinal and lateral axle forces information. This adaptive approach allows the IMMUKF to select and fuse TRFC estimates from multiple models or conditions in real-time. By adjusting to changes in driving conditions and sensor inputs, the IMMUKF enhances the robustness and accuracy of TRFC estimation, crucial for optimizing vehicle control systems and enhancing safety in intelligent transportation environments. Overall, the integration of advanced filtering techniques (STUKF and IMMUKF) within the architecture facilitates a comprehensive approach to vehicle dynamics estimation. This not only improves the accuracy of axle force and TRFC estimation but also enhances the overall performance and reliability of intelligent vehicle systems in dynamic and challenging driving scenarios.

5.3.1 Vehicle Model and Problem Statement

Considering real-time computing requirements and estimator design issues, a bicycle model (see Fig. 5.2) is used to describe the vehicle's dynamic characteristics. The vehicle bicycle model is a simplified representation used in vehicle dynamics to predict and analyze the lateral motion of vehicles during maneuvers such as cornering. It conceptualizes a vehicle as a point mass with two main reference points: the center of gravity and a point on the ground where lateral

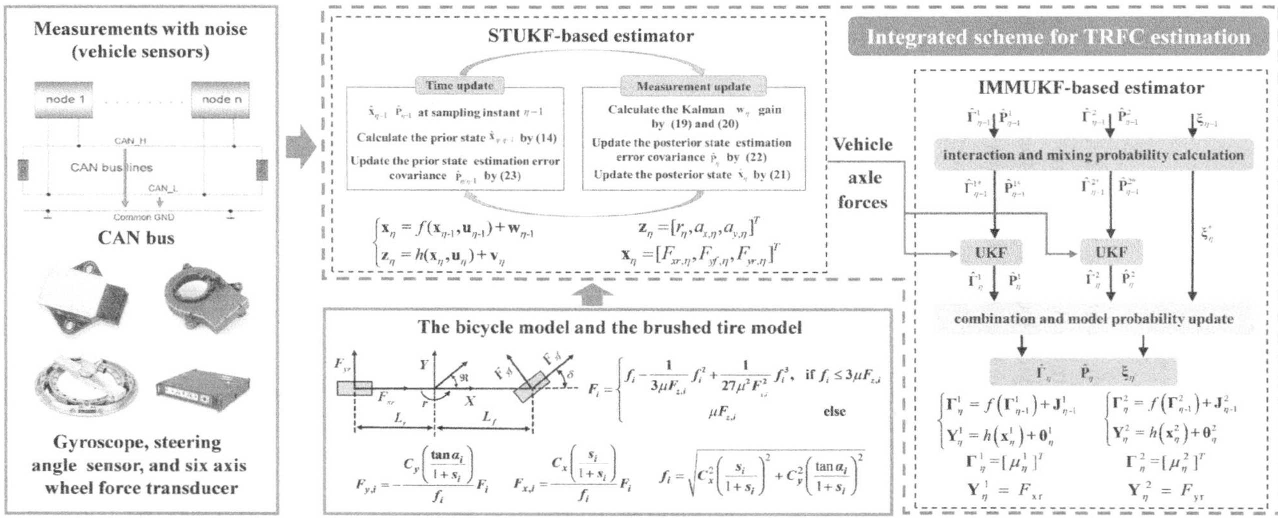

Figure 5.1 The integrated scheme for TRFC estimation.

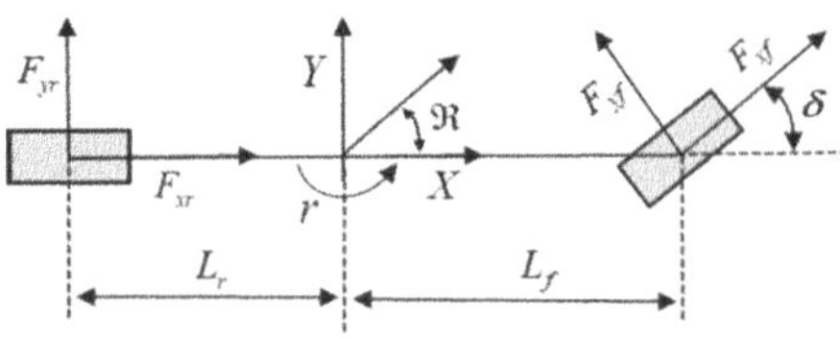

Figure 5.2 The bicycle model.

forces act, typically located at the center of the rear axle. This model operates in a 2D plane, separating longitudinal and lateral motions for ease of analysis. Key variables include longitudinal velocity, lateral velocity, yaw rate, and sideslip angle, which describe the vehicle's state. Equations derived from rotational dynamics and tire mechanics govern the lateral dynamics, relating lateral forces to yaw rate and sideslip angle. The model is instrumental in understanding vehicle handling characteristics like understeer and oversteer, essential for developing stability control systems. While the bicycle model simplifies vehicle dynamics, it does not capture all complexities, such as tire dynamics and road surface interactions, leading to its supplementation by more advanced multi-body vehicle models in practical applications. Despite its limitations, the vehicle bicycle model remains pivotal in optimizing vehicle control algorithms and advancing technologies in automotive safety and autonomous driving. To obtain a tractable estimation problem using in-vehicle sensors, a number of assumptions are adopted. The steering angles of the front right wheel and the left wheel are the same. The left and right wheels have the same stiffness. The effect of air resistance is neglected. Since the estimated TRFC needs to utilize only axial force information rather than individual tire force information, the transfer of elastic loads is also omitted. Furthermore, this study focuses on the rear-wheel-drive condition, and the front axle longitudinal force is ignored relative to the rear axle longitudinal force. The dynamic equations of this vehicle model are described as follows:

$$\dot{r} = \frac{F_{yf}L_f \cos\delta - L_r F_{yr}}{I_z} \tag{5.1}$$

$$a_x = \frac{F_{xr} - F_{yf}\sin\delta}{m} \tag{5.2}$$

$$a_y = \frac{F_{yf}\cos\delta + F_{yr}}{m} \tag{5.3}$$

The normal forces in the front and rear axles are expressed as follows:

$$F_{zf} = \frac{mgL_r - ma_x h}{L_f + L_r} \tag{5.4}$$

$$F_{zr} = \frac{mgL_f + ma_x h}{L_f + L_r} \tag{5.5}$$

where L_f and L_r are distances from the center of gravity to the front axle and rear axle; m is vehicle mass; a_x and a_y are longitudinal acceleration and lateral acceleration; δ is angle of front wheel; I_z is the inertia moment about the vehicle vertical axis; F_{yf} and F_{yr} are lateral forces of front axle and rear axle; F_{xf} and F_{xr} are longitudinal forces of front axle and rear axle; r is yaw rate; $\Re$ is sideslip angle; and h is height of the center of gravity.

In this work, we use the combined longitudinal and lateral brushed tire model to calculate the tire forces, which are expressed as follows:

$$F_{x,i} = \frac{C_x\left(\frac{s_i}{1+s_i}\right)}{f_i}F_i \tag{5.6}$$

$$F_{y,i} = -\frac{C_y\left(\frac{\tan\alpha_i}{1+s_i}\right)}{f_i}F_i \tag{5.7}$$

$$F_i = \begin{cases} f_i - \frac{1}{3\mu F_{z,i}}f_i^2 + \frac{1}{27\mu^2 F_{x,i}^2}f_i^3, & \text{if } f_i \le 3\mu F_{z,i} \\ \mu F_{z,i} & \text{else} \end{cases} \tag{5.8}$$

$$f_i = \sqrt{C_x^2\left(\frac{s_i}{1+s_i}\right)^2 + C_y^2\left(\frac{\tan\alpha_i}{1+s_i}\right)^2} \tag{5.9}$$

where μ is TRFC; C_x and C_y is the tire longitudinal and lateral stiffness coefficients; $F_{z,i}$ is vertical tire forces, its value is half of the vertical force of the corresponding axle; where $i = 1, 2, 3, 4$, which correspond to the left-front, right-front, left-rear, and right-rear wheels, respectively. s_i is the longitudinal slip ratio and α_i is the wheel sideslip angle.

5.3.2 Methodology

To estimate axle forces using discrete measurements, the vehicle model, i.e., eqs. (5.1)–(5.3) can be converted to a discrete-time state-space model.

$$\begin{cases} \mathbf{x}_\eta = f(\mathbf{x}_{\eta-1}, \mathbf{u}_{\eta-1}) + \mathbf{w}_{\eta-1} \\ \mathbf{z}_\eta = h(\mathbf{x}_\eta, \mathbf{u}_\eta) + \mathbf{v}_\eta \end{cases} \tag{5.10}$$

$$\mathbf{x}_\eta = [F_{xr,\eta}, F_{yf,\eta}, F_{yr,\eta}]^T$$

$$\mathbf{z}_\eta = [r_\eta, a_{x,\eta}, a_{y,\eta}]^T \quad \mathbf{u} = [\delta]^T$$

The state-space model can be calculated as

$$\begin{bmatrix} F_{xr,\eta} \\ F_{yf,\eta} \\ F_{yr,\eta} \end{bmatrix} = \begin{bmatrix} F_{xr,\eta-1} \\ F_{yf,\eta-1} \\ F_{yr,\eta-1} \end{bmatrix} \tag{5.11}$$

$$\begin{bmatrix} r_\eta \\ a_{x,\eta} \\ a_{y,\eta} \end{bmatrix} = \begin{bmatrix} r_{\eta-1} + [F_{yf,\eta-1}L_f \cos\delta - L_r F_{yr,\eta-1}]\Delta t/I_z \\ a_{x,\eta-1} + [F_{xr,\eta-1} - F_{yf,\eta-1}\sin\delta]\Delta t/m \\ a_{y,\eta-1} + [F_{yf,\eta-1}\cos\delta + F_{yr,\eta-1}]\Delta t/m \end{bmatrix} \tag{5.12}$$

where $\mathbf{z}$ is the measurement vector, $\mathbf{x}$ is the state vector, $\mathbf{u}$ is the input vector, $f(\cdot)$ is the state transition function, $h(\cdot)$ is the output function, $\mathbf{v}$ is process noise, $\mathbf{w}$ is measurement noise, and Δt is sampling interval. Most of the existing research works on vehicle axle force estimation have ignored the effect of model parameter perturbations, which may decrease estimation accuracy. To address this problem, we use STUKF to predict vehicle axle forces. The STUKF is an advanced variant of the UKF designed to improve state estimation accuracy in scenarios where significant nonlinearities and uncertainties are present. It addresses the limitations of the standard UKF by dynamically adjusting the filtering process to better track states in the face of model uncertainties and measurement noise. The STUKF builds upon the foundation of the UKF, which is itself an extension of the traditional KF tailored for nonlinear systems. Unlike the EKF, which linearizes nonlinear functions through Taylor series expansion, the UKF approximates the probability distribution of states using a set of carefully chosen sigma points. This approach avoids the linearization errors associated with EKF, making it more suitable for highly nonlinear systems. The core principle of the STUKF lies in its ability to adaptively adjust the sigma points used in the prediction and update stages of the UKF. By dynamically updating the spread and distribution of sigma points based on the evolving covariance matrix of the state estimate, the STUKF enhances its resilience to model uncertainties and measurement noise. This adaptive mechanism ensures that the filter can accurately track the true state of the system even under challenging conditions. In summary, the STUKF represents a significant advancement in nonlinear state estimation, offering robust performance in complex systems characterized by nonlinear dynamics and uncertain measurements. Its adaptive sigma point selection mechanism ensures accurate and reliable state estimation, making it a valuable tool in various fields, including aerospace, robotics, and autonomous systems. The iterative steps of the STUKF are as follows.

1) Initialization:
   The initial mean of $\mathbf{x}$ and its covariance matrix $\mathbf{P}$

$$\hat{\mathbf{x}}_0 = E(\mathbf{x}_0) \tag{5.13}$$

$$\mathbf{P}_0 = E\left[(\mathbf{x}_0 - \hat{\mathbf{x}}_0)(\mathbf{x}_0 - \hat{\mathbf{x}}_0)^T\right] \tag{5.14}$$

2) Time Update
   The weight ϕ_c^i, ϕ_m^i, and sigma sampling points $\boldsymbol{\tau}_{\eta-1}^i$ are given by

$$\begin{cases} \boldsymbol{\tau}_{\eta-1}^0 = \hat{\mathbf{x}}_{\eta-1} \\ \boldsymbol{\tau}_{\eta-1}^i = \hat{\mathbf{x}}_{\eta-1} + \sqrt{n+\lambda}\left(\sqrt{\mathbf{P}_{\eta-1}}\right)_i & i = 1, 2, \ldots n \\ \boldsymbol{\tau}_{\eta-1}^i = \hat{\mathbf{x}}_{\eta-1} - \sqrt{n+\lambda}\left(\sqrt{\mathbf{P}_{\eta-1}}\right)_i & i = n+1, \ldots 2n \end{cases} \tag{5.15}$$

$$\begin{cases} \phi_m^0 = \lambda/(n+\lambda), \phi_c^0 = \lambda/(n+\lambda) + 1 + \beta - \alpha^2 \\ \phi_m^i = \phi_c^i = \lambda/(2(n+\lambda)), \quad i = 1, 2, \dots 2n \end{cases} \tag{5.16}$$

where n is the dimension of $\mathbf{x}$, the specific formulas for λ, β, and α can be found in [43].

The propagated sigma points can be calculated by

$$\boldsymbol{\tau}_{\eta/\eta-1}^{*(i)} = f\left(\boldsymbol{\tau}_{\eta-1}^{(i)}, \mathbf{u}_{\eta-1}\right) \tag{5.17}$$

The prior state $\hat{\mathbf{x}}_{\eta/\eta-1}$ and corresponding state covariance matrix $\mathbf{P}_{\eta/\eta-1}$ are updated using eqs. (5.18) and (5.19), respectively.

$$\hat{\mathbf{x}}_{\eta/\eta-1} = \sum_{i=0}^{2n} \phi_m^i \boldsymbol{\tau}_{\eta/\eta-1}^{*(i)} \tag{5.18}$$

$$\mathbf{P}_{\eta/\eta-1} = \sum_{i=0}^{2n} \phi_c^i \left(\boldsymbol{\tau}_{\eta/\eta-1}^{*(i)} - \hat{\mathbf{x}}_{\eta/\eta-1}\right)\left(\boldsymbol{\tau}_{\eta/\eta-1}^{*(i)} - \hat{\mathbf{x}}_{\eta/\eta-1}\right)^T + \mathbf{Q}_{\eta-1} \tag{5.19}$$

where $\mathbf{Q}$ is the covariance matrix of the process noise.

3) Measurement Update

The new sigma points $\boldsymbol{\tau}_\eta^i$ are given by

$$\begin{cases} \boldsymbol{\tau}_\eta^0 = \hat{\mathbf{x}}_{\eta/\eta-1} \\ \boldsymbol{\tau}_\eta^i = \hat{\mathbf{x}}_{\eta/\eta-1} + \sqrt{n+\lambda}\left(\sqrt{\mathbf{P}_{\eta/\eta-1}}\right)_i & i = 1, 2, \dots n \\ \boldsymbol{\tau}_\eta^i = \hat{\mathbf{x}}_{\eta/\eta-1} - \sqrt{n+\lambda}\left(\sqrt{\mathbf{P}_{\eta/\eta-1}}\right)_i & i = n+1, \dots 2n \end{cases} \tag{5.20}$$

The propagated sigma points can be calculated by

$$\mathbf{z}_{\eta/\eta-1}^{*(i)} = h\left(\boldsymbol{\tau}_\eta^{(i)}, \mathbf{u}_\eta\right) \tag{5.21}$$

The estimated output $\hat{\mathbf{z}}_{\eta/\eta-1}$ and its covariance matrix $\mathbf{P}_{z,\eta}$ are updated using eqs. (5.22) and (5.23), respectively.

$$\hat{\mathbf{z}}_{\eta/\eta-1} = \sum_{i=0}^{2n} \phi_m^i \mathbf{z}_{\eta/\eta-1}^{*(i)} \tag{5.22}$$

$$\mathbf{P}_{z,\eta} = \sum_{i=0}^{2n} \phi_c^i \left(\mathbf{z}_{\eta/\eta-1}^{*(i)} - \hat{\mathbf{z}}_{\eta/\eta-1}\right)\left(\mathbf{z}_{\eta/\eta-1}^{*(i)} - \hat{\mathbf{z}}_{\eta/\eta-1}\right)^T + \mathbf{R}_\eta \tag{5.23}$$

where $\mathbf{R}$ is the covariance matrix of the measurement noise.

The covariance matrix $\mathbf{P}_{xz,\eta}$ between $\hat{\mathbf{x}}_{\eta/\eta-1}$ and $\hat{\mathbf{z}}_{\eta/\eta-1}$ is given by

$$\mathbf{P}_{xz,\eta} = \sum_{i=0}^{2n} \phi_c^i \left(\boldsymbol{\tau}_{\eta/\eta-1}^{*(i)} - \hat{\mathbf{x}}_{\eta/\eta-1}\right)\left(\mathbf{z}_{\eta/\eta-1}^{*(i)} - \hat{\mathbf{z}}_{\eta/\eta-1}\right)^T \tag{5.24}$$

The gain matrix $\mathbf{w}_\eta$ can be calculated using eq. (5.25), and the posterior state $\hat{\mathbf{x}}_\eta$ and its covariance matrix $\mathbf{P}_\eta$ can be calculated using eqs. (5.26 and 5.27).

$$\mathbf{w}_\eta = \mathbf{P}_{xz,\eta} \mathbf{P}_{z,\eta}^{-1} \tag{5.25}$$

$$\hat{\mathbf{x}}_\eta = \hat{\mathbf{x}}_{\eta/\eta-1} + \mathbf{w}_\eta(\mathbf{z}_\eta - \hat{\mathbf{z}}_{\eta/\eta-1}) \tag{5.26}$$

$$\mathbf{P}_\eta = \mathbf{P}_{\eta/\eta-1} - \mathbf{w}_\eta \mathbf{P}_{z,\eta} \mathbf{w}_\eta^T \tag{5.27}$$

In the above UKF, the effect of the model parameter perturbations on the prediction of vehicle axle forces is not considered. The strong tracking algorithm has robustness against model parameter uncertainties [44]. The core principle of STA involves two main components: adaptation of the filter gain and adjustment of the covariance matrix during the filtering process. These adaptations are guided by the current estimation error covariance, ensuring that the filter can accurately track system states even in the presence of significant uncertainties and measurement noise. By dynamically adjusting these parameters, STA maintains robustness and stability across a wide range of operating conditions. The STA is a robust estimation technique used primarily in nonlinear and non-Gaussian environments to improve the performance of filters such as the KF and its variants. Unlike traditional KF approaches that assume Gaussian noise and linear dynamics, STA adapts dynamically to varying conditions, enhancing the accuracy and reliability of state estimation in complex systems. Therefore, the STUKF is presented on the basis of the combination of a strong tracking algorithm and UKF.

Thus, a fading factor σ_η is employed to dynamically adjust the prior state covariance with the following equation.

$$\mathbf{P}_{\eta/\eta-1} = \sigma_\eta \sum_{i=0}^{2n} \phi_c^i \left(\boldsymbol{\tau}_{\eta/\eta-1}^{*(i)} - \hat{\mathbf{x}}_{\eta/\eta-1}\right)\left(\boldsymbol{\tau}_{\eta/\eta-1}^{*(i)} - \hat{\mathbf{x}}_{\eta/\eta-1}\right)^T + \mathbf{Q}_{\eta-1} \tag{5.28}$$

where $\sigma_\eta = diag\left[\Omega_\eta^1, \Omega_\eta^2, \cdots \Omega_\eta^n\right]$

$$\Omega_\eta^t = \begin{cases} \Upsilon_t d_\eta & \Upsilon_t d_\eta \geq 1 \\ 1 & \Upsilon_t d_\eta < 1 \end{cases} \quad t = 1, 2, \ldots n \tag{5.29}$$

$$N_\eta = V_\eta - R_\eta - B_\eta Q_{\eta-1} B_\eta^T \tag{5.30}$$

$$M_\eta = B_\eta A_\eta P_{\eta/\eta-1} A_\eta^T B_\eta^T \tag{5.31}$$

$$d_\eta = \frac{tr[N_\eta]}{\sum_{t=1}^{n} \Upsilon_t M_\eta} \tag{5.32}$$

$$A_\eta = \frac{\partial f(\mathbf{x}_\eta, \mathbf{u}_\eta)}{\partial \mathbf{x}_\eta}, \quad B_\eta = \frac{\partial h(\mathbf{x}_\eta, \mathbf{u}_\eta)}{\partial \mathbf{x}_\eta} \tag{5.33}$$

$$V_\eta = \begin{cases} \varphi_1 \varphi_1^T & \eta = 0 \\ \dfrac{\rho V_{\eta-1} + \varphi_\eta \varphi_\eta^T}{1+\rho} & \eta \geq 1 \end{cases} \tag{5.34}$$

where Υ_t is a constant that is determined based on the system state prior information, if there is a large change in the system state then a larger Υ_t needs to be selected; $\rho \in [0, 1]$ is the forgetting factor. d_η is the undetermined parameter. The reasonable selection of the above parameters helps to improve the stability of the STFUKF.

To demonstrate the validity of the IMM-UKF, we need at least two different estimation models. Therefore, we only use the vehicle rear axle forces to identify TRFC in this study. After obtaining vehicle rear axle forces, we estimate TRFC based on longitudinal and lateral rear axle forces, respectively. Two discrete-time state-space models are given by

$$\begin{cases} \mathbf{\Gamma}_\eta^1 = f\left(\mathbf{\Gamma}_{\eta-1}^1\right) + \mathbf{J}_{\eta-1}^1 \\ \mathbf{Y}_\eta^1 = h\left(\mathbf{\Gamma}_\eta^1\right) + \mathbf{\theta}_\eta^1 \end{cases} \tag{5.35}$$

$$f\left(\mathbf{\Gamma}_{\eta-1}^1\right) = \mathbf{\Gamma}_{\eta-1}^1 h\left(\mathbf{\Gamma}_\eta^1\right) = \mathbf{\Gamma}_\eta^1 F_{xr}$$

$$\begin{cases} \mathbf{\Gamma}_\eta^2 = f\left(\mathbf{\Gamma}_{\eta-1}^2\right) + \mathbf{J}_{\eta-1}^2 \\ \mathbf{Y}_\eta^2 = h\left(\mathbf{\Gamma}_\eta^2\right) + \mathbf{\theta}_\eta^2 \end{cases} \tag{5.36}$$

$$f\left(\mathbf{\Gamma}_{\eta-1}^2\right) = \mathbf{\Gamma}_{\eta-1}^2 h\left(\mathbf{\Gamma}_\eta^2\right) = \mathbf{\Gamma}_\eta^2 F_{yr}$$

where $\mathbf{J}_\eta^i$ and $\mathbf{\theta}_\eta^i$ are the process noise and measurement noise, $\mathbf{Y}_\eta^i$ is measurement vector; $\mathbf{\Gamma}_\eta^i$ is the state vector, where $i = 1, 2$, which correspond to the longitudinal dynamics model and the lateral dynamics model.

We use two UKFs to estimate the longitudinal and lateral TRFCs, respectively, and the iterative process of UKF is the same as eqs. (5.13)–(5.27). The TRFC estimation based on the longitudinal dynamics model has good results in vehicle acceleration or deceleration maneuvers, but it is not suitable for vehicle steering conditions. Similarly, the TRFC identification based on the lateral dynamics model has some limitations in vehicle acceleration or deceleration conditions. It is a trend to integrate information from different models because there is no perfect vehicle dynamics model suitable for all driving conditions [45]. Therefore, an integrated method based on UKF and IMM methods is proposed to estimate TRFC. The IMM approach represents a sophisticated method in state estimation, tailored to handle the complexities of dynamic systems with nonlinear behaviors, variable operating modes, and uncertainties. Unlike traditional single-model approaches such as the KF, which struggle with diverse and changing system dynamics, IMM leverages a collection of distinct models, each representing different hypotheses about system behavior. These models are dynamically selected based on real-time measurements and predictions, allowing the estimator to adaptively switch between them to accurately track the system's evolving state. At the heart of IMM is the probabilistic weighting of models, where each model's likelihood is continuously updated according to its agreement with observed data. This adaptive model selection ensures robust performance across various conditions, mitigating the impact of uncertainties and noise inherent in real-world environments. Moreover, IMM incorporates an interaction mechanism that enables models to exchange information, facilitating collaborative estimation and further enhancing accuracy. IMM finds wide application in fields such as aerospace, autonomous vehicles, robotics, and finance, where systems exhibit complex dynamics and operational variability. In aerospace, for instance, IMM is utilized to predict aircraft trajectory amidst changing flight conditions and sensor inaccuracies. In autonomous vehicles, it aids in precise localization and motion planning by dynamically adjusting to diverse driving scenarios. Similarly, in robotics and finance, IMM improves the estimation of complex system states, supporting decision-making processes under uncertain and dynamic conditions. The versatility and robustness of IMM make it a cornerstone of modern estimation theory, offering superior performance compared to single-model methods in capturing the true underlying state of dynamic systems. By integrating diverse models and adapting dynamically to changing environments, IMM ensures reliable and accurate state estimation critical for advancing technology and enhancing safety across various domains.

The probability of each UKF estimation model can be dynamically adjusted based on the information on the prediction variance of each model. Figure 5.3 shows the overall process of the IMM-UKF method, which is composed of three modules: interaction and mixing probability calculation, UKF, and combination and model probability update.

$\hat{\mathbf{\Gamma}}^i_{\eta-1}$ and $\hat{\mathbf{P}}^i_{\eta-1}$ are the mean and covariance of estimation results based on longitudinal or lateral dynamics models at time $\eta - 1$. $\xi_{\eta-1}$ is the weight coefficient vector for different estimation models at time $\eta - 1$. $\hat{\mathbf{\Gamma}}^{i*}_{\eta-1}$ and $\hat{\mathbf{P}}^{i*}_{\eta-1}$ are the mean and covariance estimation results after the interaction and mixing probability

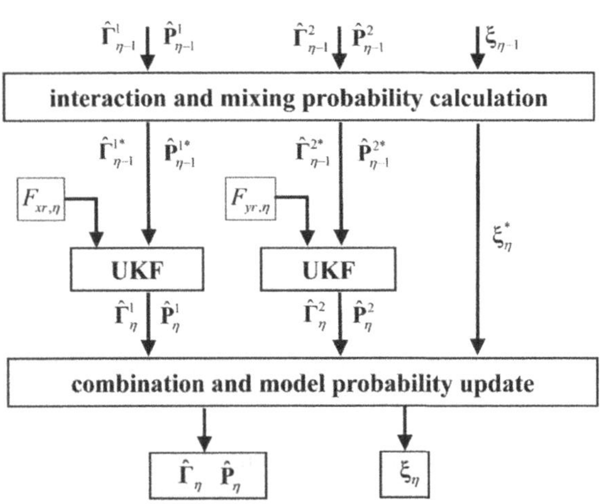

Figure 5.3 The IMM-UKF estimation method.

calculation. $\boldsymbol{\xi}_{\eta}^{*}$ is the weight coefficient vector after the interaction and mixing probability calculation. $\hat{\boldsymbol{\Gamma}}_{\eta}^{i}$ and $\hat{\mathbf{P}}_{\eta}^{i}$ are the mean and covariance estimation results using the UKF method. $\hat{\boldsymbol{\Gamma}}_{\eta}$, $\hat{\mathbf{P}}_{\eta}$, and $\boldsymbol{\xi}_{\eta}$ are mean, covariance, and the weight of the final estimation results, respectively. The IMM-UKF method combines the longitudinal dynamics-based estimation model (M^1) and the lateral dynamics-based estimation model (M^2). Therefore, this system is a discrete set that is composed of two models. The system can be expressed as

$$M = \{M^1, M^2\} \tag{5.37}$$

In this study, we assume that the system is a first-order Markov chain system and its transition matrix $\overline{P}$ is time-invariant. The transition matrix $\overline{P}$ is given by

$$\overline{P} = \begin{bmatrix} p_{11} \cdots p_{1j} \\ \cdots \quad \cdots \quad \cdots \\ p_{i1} \cdots p_{ij} \end{bmatrix} \tag{5.38}$$

$$p_{ij} = P\left(M_{\eta}^{j} \mid M_{\eta-1}^{i}\right) \quad i = 1,2 \;; j = 1,2 \tag{5.39}$$

where p_{ij} is the transition probability from the estimation model i to the estimation model j.

1) Interacting and Mixing

   The estimation results of each model from the previous step are combined with the mixing weight. The mixing weight $\xi_{\eta-1}^{i|j}$ and mixing probabilities ρ_j can be expressed as

$$\xi_{\eta-1}^{i|j} = \frac{1}{\rho_j} p_{ij} \xi_{\eta-1}^{i} \tag{5.40}$$

$$\rho_j = \sum_{i=1}^{2} p_{ij} \xi_{\eta-1}^{i} \tag{5.41}$$

   where $\xi_{\eta-1}^{i}$ is the probability of the model i.

   The mixing mean and covariance for each model i are given by

$$\hat{\boldsymbol{\Gamma}}_{\eta-1}^{i*} = \sum_{i=1}^{2} \boldsymbol{\xi}_{\eta}^{i|j} \hat{\boldsymbol{\Gamma}}_{\eta-1}^{i} \tag{5.42}$$

$$\hat{\mathbf{P}}_{\eta-1}^{i*} = \sum_{i=1}^{2} \boldsymbol{\xi}_{\eta}^{i|j} \left[\hat{\mathbf{P}}_{\eta-1}^{i} + \left(\hat{\boldsymbol{\Gamma}}_{\eta-1}^{i} - \hat{\boldsymbol{\Gamma}}_{\eta-1}^{i*}\right)\left(\hat{\boldsymbol{\Gamma}}_{\eta-1}^{i} - \hat{\boldsymbol{\Gamma}}_{\eta-1}^{i*}\right)^{T}\right] \tag{5.43}$$

2) Prediction

   Based on the mixing mean and covariance, the UKF of each model i predicts and updates the model mean $\hat{\boldsymbol{\Gamma}}_{\eta}^{i}$ and covariance $\hat{\mathbf{P}}_{\eta}^{i}$. The likelihood of the prediction for each model i at time η is given by

$$\Lambda_{\eta}^{i} = \frac{\exp\left[-\frac{1}{2}\left(\psi_{\eta}^{i}\right)^{T}\left(\pi_{\eta}^{i}\right)^{-1}\psi_{\eta}^{i}\right]}{\sqrt{\left|2\pi_{\eta}^{i}\right|}} \tag{5.44}$$

The innovation sequence ψ_{η}^{i} and innovation covariance π_{η}^{i} of the UKF can be expressed as

$$\psi_{\eta}^{i} = \mathbf{Y}_{\eta}^{i} - h\left(\mathbf{\Gamma}_{\eta}^{i}\right) \tag{5.45}$$

$$\pi_{\eta}^{i} = E\left[\psi_{\eta}^{i}\left(\psi_{\eta}^{i}\right)^{T}\right] \tag{5.46}$$

The probability of each model i at time η is

$$\xi_{\eta}^{i} = \frac{1}{\rho}\Lambda_{\eta}^{i}\rho_{i} \tag{5.47}$$

$$\rho = \sum_{i=1}^{2}\Lambda_{\eta}^{i}\rho_{i} \tag{5.48}$$

3) Combination

The final predicted mean $\hat{\mathbf{\Gamma}}_{\eta}$ and covariance $\hat{\mathbf{P}}_{\eta}$ can be calculated as

$$\hat{\mathbf{\Gamma}}_{\eta} = \sum_{i=1}^{2}\xi_{\eta}^{i}\hat{\mathbf{\Gamma}}_{\eta}^{i} \tag{5.49}$$

$$\hat{\mathbf{P}}_{\eta} = \sum_{i=1}^{2}\xi_{\eta}^{i}\left[\hat{\mathbf{P}}_{\eta}^{i} + \left(\hat{\mathbf{\Gamma}}_{\eta}^{i} - \hat{\mathbf{\Gamma}}_{\eta}\right)\left(\hat{\mathbf{\Gamma}}_{\eta}^{i} - \hat{\mathbf{\Gamma}}_{\eta}\right)^{T}\right] \tag{5.50}$$

5.3.3 Experiment Tests

A C-class vehicle equipped with a six-axis wheel force transducer was utilized to evaluate the efficacy of the proposed method under controlled conditions. The relevant measurement data was sampled at a frequency of 100 Hz. Tests were conducted on two distinct surfaces: dry asphalt and ice-snow roads, each characterized by different friction coefficients. The dry asphalt surface exhibited coefficients ranging between 0.7 and 0.8, whereas the ice-snow surface had coefficients ranging from 0.1 to 0.2 [46]. During the tests, lane change maneuvers served as the control inputs. These maneuvers were executed by an experienced driver to ensure consistency and accuracy in vehicle response. To simulate realistic conditions, the vehicle mass parameter used in the estimation method was set 10% higher than the nominal value. This adjustment aimed to evaluate the robustness and accuracy of the STUKF under varying mass conditions. Real-world vehicle tests were conducted under these controlled parameters to validate the effectiveness and performance of the STUKF method in accurately estimating TRFCs across different road surfaces and maneuvering scenarios.

5.3.3.1 The Test on the Dry Asphalt Road

The first test of the experiment involved steering maneuvers while the vehicle speed slowly changed. For the axial force estimation, the parameters utilized in the STUKF are selected as follows: $\mathbf{Q} = \text{diag}[40, 4, 4]$, $\alpha = 0.5$, $\beta = 2$, $\Upsilon = 1$, $\rho = 0.93$, $\mathbf{R} = \text{diag}[0.06, 0.06, 0.06]$. In addition, the vehicle mass parameter is 1972.3 kg, more than 10% of its true value. The FWSA, LA, and LV are shown in Fig. 5.4. The estimation results of vehicle axle forces are presented in Figs. 5.5, 5.6, and 5.7. The RMSE estimated by different methods are listed in Table 5.1.

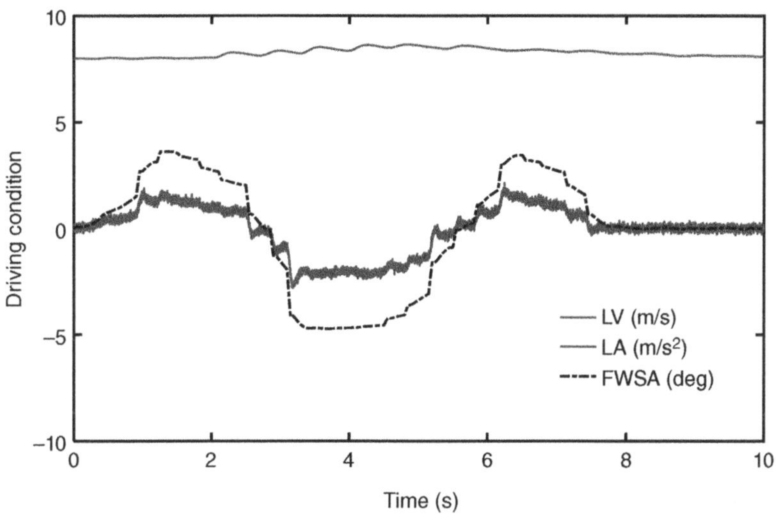

Figure 5.4 The change of FWSA, LA, and LV on dry asphalt roads.

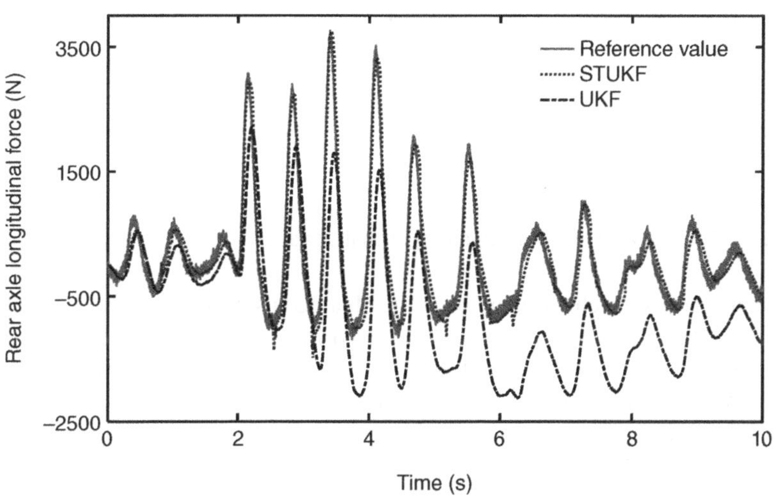

Figure 5.5 The rear axle longitudinal force on dry asphalt road.

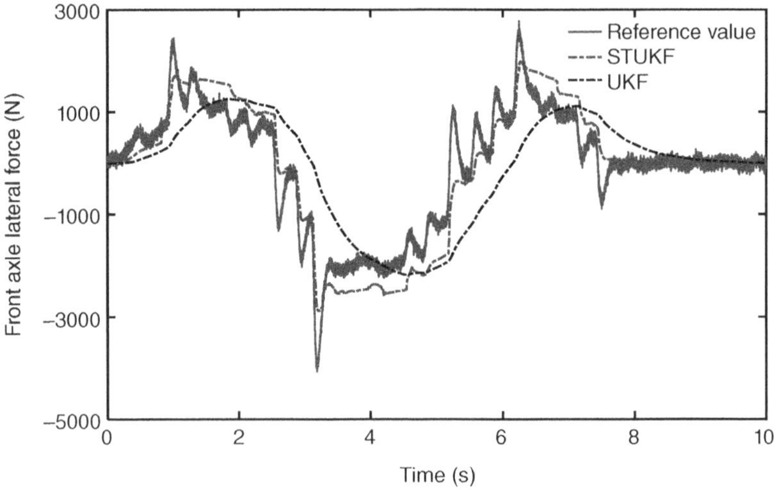

Figure 5.6 The front axle lateral force on dry asphalt road.

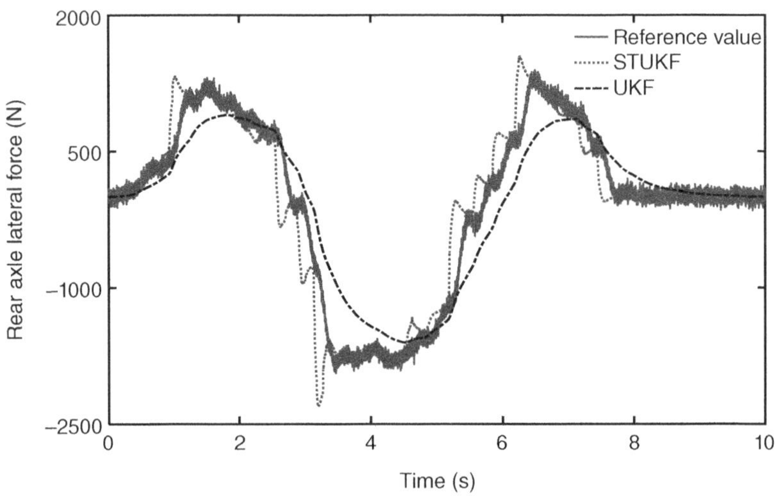

Figure 5.7 The rear axle lateral force on dry asphalt road.

Reference values of vehicle axial forces are obtained using a six-axis wheel force transducer. The effectiveness of the STUKF algorithm in estimating vehicle axle forces surpasses that of the UKF, as demonstrated in empirical testing. Particularly notable is STUKF's ability to enhance estimation accuracy even in scenarios where the vehicle mass parameter deviates from its nominal value. To assess the robustness of the STUKF algorithm, we introduced mass parameter perturbations within

Table 5.1 RMSE of vehicle axle forces.

RMSE	F_{xr}	F_{yf}	F_{yr}
UKF	**1223.4**	931.6	**343.4**
STUKF	**244.8**	448.2	295.7

a specified range, set at no more than 20% of its true value, which is $[0.8m, 1.2m]$. This range was chosen to simulate real-world conditions where vehicle mass might vary due to factors like cargo load or passenger occupancy. To thoroughly evaluate its performance, we conducted 100 offline tests using real vehicle test data, the results of which are depicted in Fig. 5.8. Figure 5.8(a) illustrates the distribution of mass perturbation values across the 100 experiments, demonstrating a randomly selected range of deviations. Figure 5.8(b)–(d) show the RMSE of the estimated axial forces relative to the reference values for both STUKF and UKF across these experiments. These figures provide a comparative view of how each method performs under varying mass perturbation conditions. From the analysis of these results, it is evident that STUKF consistently achieves superior estimation accuracy compared to UKF across all experiments. Moreover, STUKF exhibits robustness in the face of different mass parameter perturbations, maintaining stable and accurate estimations under varied conditions. This robust performance underscores the algorithm's capability to adapt and provide reliable estimates of vehicle axle forces despite uncertainties in vehicle mass. In conclusion, the empirical findings validate the effectiveness and robustness of the STUKF algorithm in real-world applications. By outperforming UKF and demonstrating resilience to mass parameter variations, STUKF proves to be a promising advancement in state estimation techniques for enhancing vehicle dynamics and safety systems. Future research may explore further applications and refinements of STUKF to address additional complexities in dynamic vehicle environments.

Figure 5.9 shows the results of identifying the TRFC based on the axial force obtained from STUKF estimation. The effectiveness of TRFC estimation is demonstrated by comparing IMM-UKF with the MSE weighted fusion method [37] and each single-model-based UKF. The parameters utilized in the IMM-UKF are selected as follows: $\mathbf{Q} = 1^{-9}, \mathbf{R} = 500, \overline{P} = [0.99, 0.01; 0.01, 0.99]$. The remaining IMM-UKF unlisted parameters are the same as those in the UKF algorithm used for axial force estimation. The parameters used in this study for the comparative fusion method (MSE) are the same as in [37], except for $p_{0-lon} = 0.003$. The other parameters of the MSE-weighted fusion method and the meaning of all the parameters of the method can be found in [37]. The legend text "UKF-lon" indicates that the prediction of TRFC using UKF is based on the longitudinal dynamics

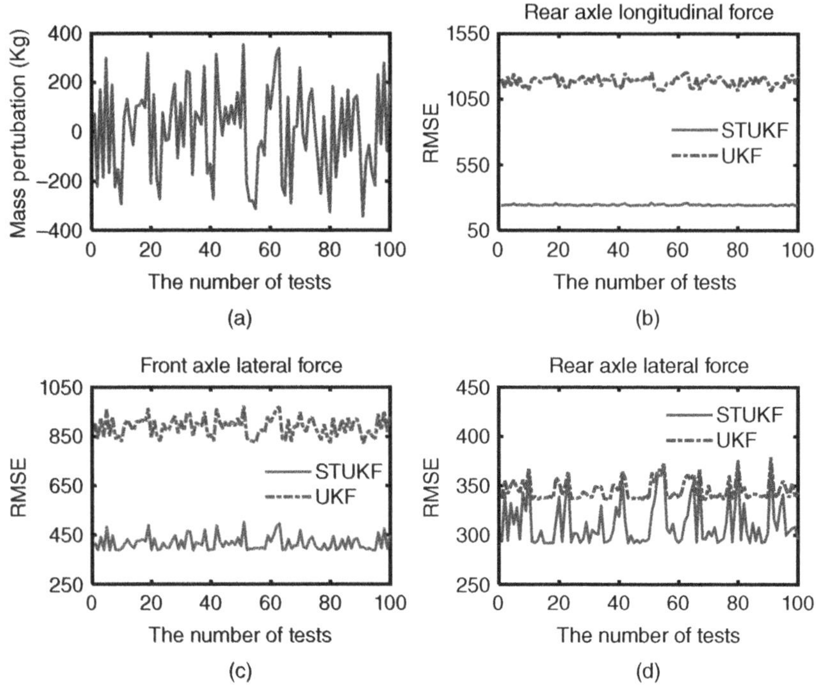

Figure 5.8 The STUKF estimated performance test when the mass parameter is mismatched. (a) Mass parameters perturbation. (b) RMSE of rear axle longitudinal force. (c) RMSE of front axle lateral force. (d) RMSE of rear axle lateral force.

model, while "UKF-lat " relies on a lateral dynamics model. The aquamarine blue and brownish-yellow solid lines represent the maximum and minimum values of TRFC for dry asphalt roads, respectively. Figure 5.10 depicts the variation of the longitudinal and lateral model probabilities in the two-hybrid estimation methods. The legend text "IMM-lon" indicates the probability change of the longitudinal model in the IMMUKF method. Other legends have similar meanings in Fig. 5.10.

In Fig. 5.9, the lateral dynamics model-based UKF can track to the TRFC minimum in about 0.42 seconds. Then there is an overshoot in the interval [0.92, 1], which is related to the sharp change in FWSA. The longitudinal dynamics model-based UKF does not track to the TRFC minimum until 2.22 seconds. This is because the LV changes less in the interval [0, 2.1] before it starts to accelerate. It can be clearly seen that the trend of the MSE estimation curve is consistent with the change in estimation results based on the longitudinal and lateral models. The MSE identification results are obtained by direct weighting based on the probabilities in Fig. 5.10 and the estimated outputs of the longitudinal and lateral

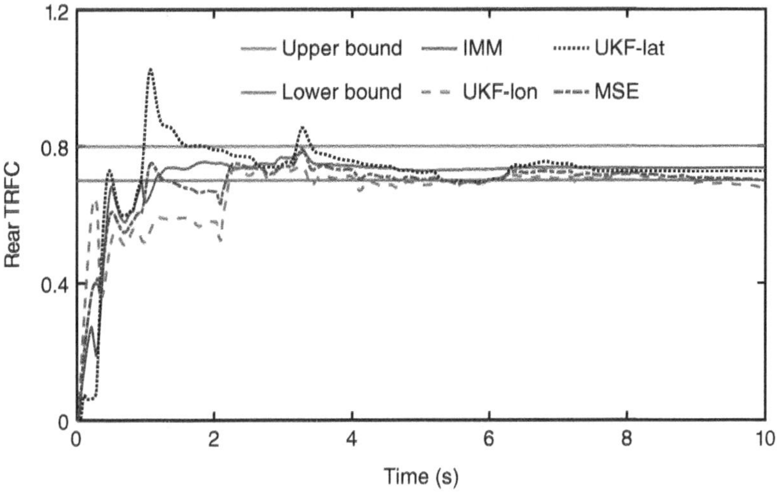

Figure 5.9 The TRFC estimation on dry asphalt road.

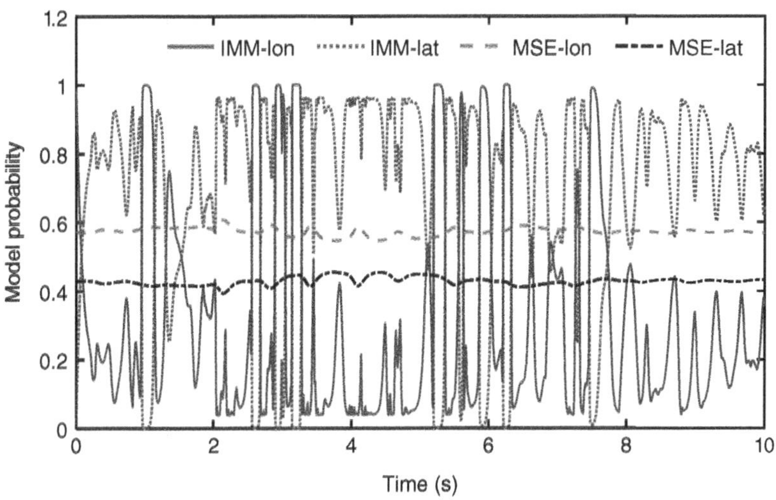

Figure 5.10 The model probability in the two-hybrid estimation methods.

models in Fig. 5.9. The prediction curve based on the IMMUKF method does not fluctuate with the output of each model, which is especially evident in the interval [4–10]. The probability of each model can be dynamically adjusted based on the information of the predicted covariance in the IMM-UKF. The IMM-UKF can integrate the advantages of a single model-based estimation algorithm so that it has the best estimation accuracy among all estimation methods. It is important to note that the change in the IMMUKF prediction curve in Fig. 5.9 cannot be analyzed simply based on the change in model probabilities in the IMMUKF in Fig. 5.10. The input of each model at the next moment in IMMUKF is the fusion of all outputs of all models at the previous moment. The predicted results of each model in IMMUKF differ significantly from the estimates based on longitudinal or lateral dynamics models in the ST.

5.3.3.2 The Test on the Ice-Snow Road

The other test of the experiment involved steering maneuvers while the vehicle speed drops rapidly. For the axial force estimation, the $\mathbf{Q}$ utilized in the STUKF is selected as $\mathbf{Q} = diag[5, 0.5, 0.5]$. The remaining parameters are not listed in STUKF, and the mass parameters are the same as in the dry asphalt road test conditions. In the test conditions on dry asphalt roads, all parameters not specified for adjustment in STUKF remain consistent. These conditions include the FWSA, LA, and LV, which are graphically depicted in Fig. 5.11. Figures 5.12, 5.13, and 5.14 illustrate the detailed estimation results, showing how the STUKF algorithm adjusts and compensates for changes in the vehicle dynamics more effectively than the UKF. The visual representation of the data in these figures provides clear evidence of the STUKF's superior tracking and estimation capabilities. Table 5.2 further supports these findings by listing the RMSE values for each estimation method. The lower RMSE values associated with the STUKF indicate a smaller deviation from the true values, confirming the algorithm's higher accuracy. This quantitative data, combined with the visual

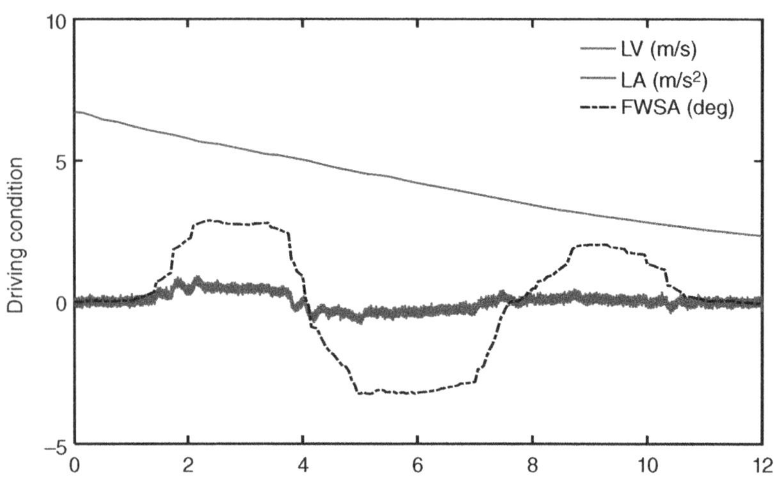

Figure 5.11 The change of FWSA, LA, and LV on ice-snow roads.

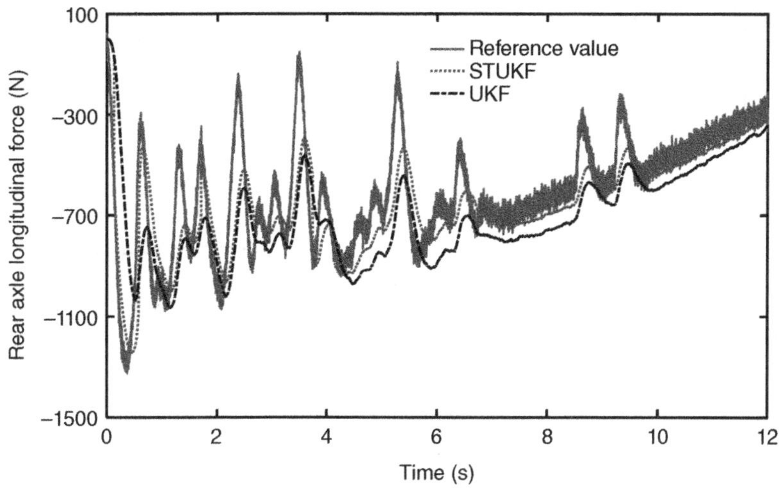

Figure 5.12 The rear axle longitudinal force on ice-snow road.

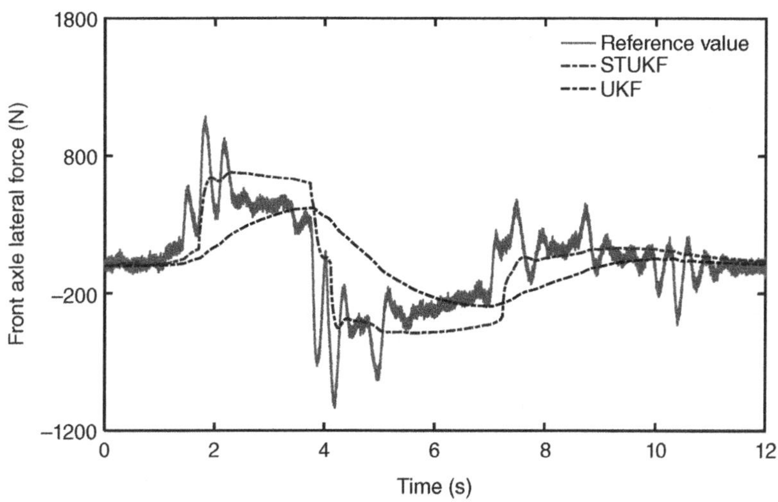

Figure 5.13 The front axle lateral force on ice-snow road.

analysis from the figures, provides a comprehensive validation of the STUKF's performance. The effectiveness of the STUKF in improving estimation precision across dynamic and varied operational scenarios is particularly important for applications in advanced vehicle control systems. Accurate estimation of axle forces is crucial for maintaining vehicle stability, enhancing safety, and improving overall performance. The STUKF's robust performance under varying conditions

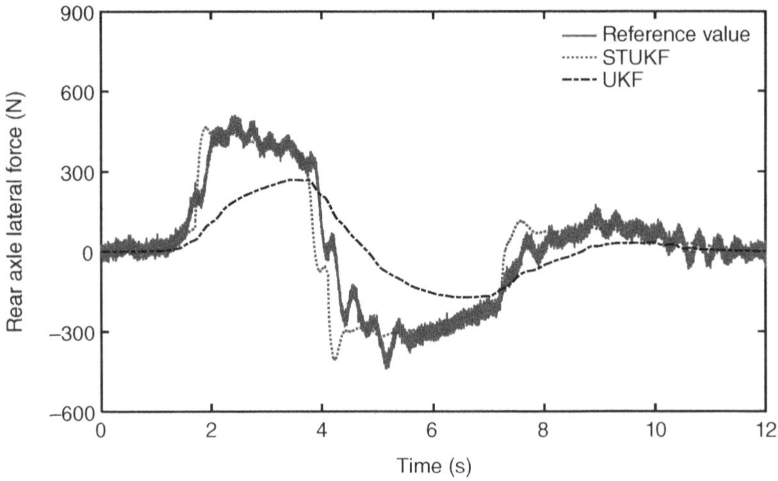

Figure 5.14 The rear axle lateral force on ice-snow road.

Table 5.2 RMSE of vehicle axle forces.

RMSE	F_{xr}	F_{yf}	F_{yr}
UKF	201.9	325.1	136.8
STUKF	142.3	216.3	77.1

makes it an ideal choice for these applications. Moreover, the STUKF's ability to handle parameter mismatches and dynamic changes effectively suggests that it can be reliably used in real-world conditions where such variations are common. This adaptability not only enhances the accuracy of the estimations but also contributes to the overall reliability and robustness of the vehicle's control system.

Similar to the offline tests conducted for the first test condition, 100 offline tests were employed to verify the robustness of the STUKF. The results of these extensive tests are illustrated in Fig. 5.15. In Fig. 5.15(a), the change in mass perturbation values over 100 experiments is presented. These perturbation values were randomly selected for each experiment to ensure a comprehensive evaluation of the filter's robustness under varying conditions. This randomness simulates real-world scenarios where the mass parameter can fluctuate due to various factors, testing the adaptability and reliability of the STUKF. Figure 5.15(b)–(d) depict the variation of the RMSE of the axial force estimated using different methods relative to the reference value across the 100 experiments. These figures compare the performance of the STUKF with that of the UKF and

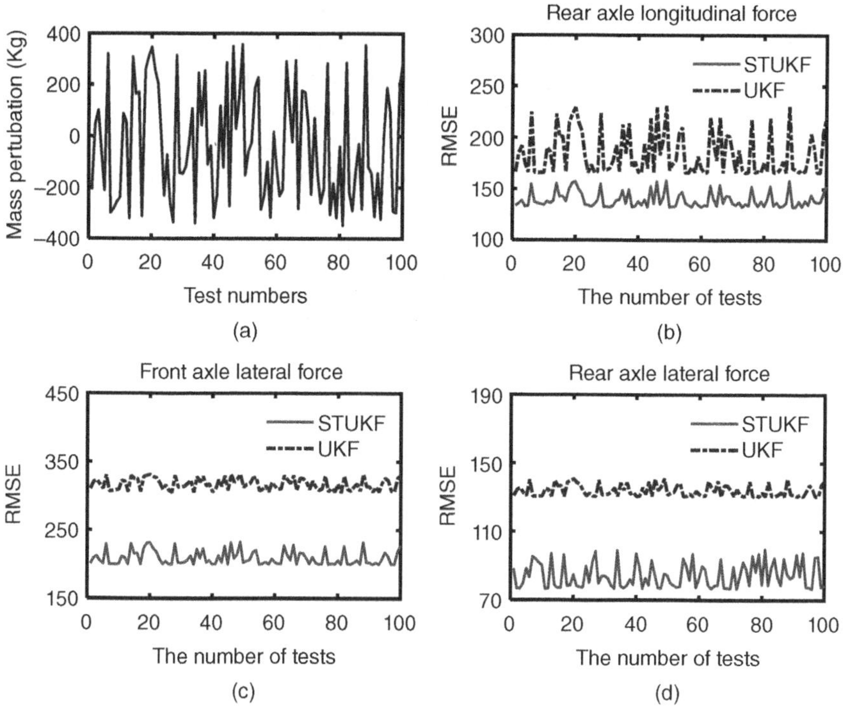

Figure 5.15 The STUKF estimated performance test when the mass parameter is mismatched. (a) Mass parameters perturbation. (b) RMSE of rear axle longitudinal force. (c) RMSE of front axle lateral force. (d) RMSE of rear axle lateral force.

other estimation methods. The RMSE is a critical metric as it quantifies the deviation of the estimated values from the true values, providing a clear measure of estimation accuracy. The results demonstrate that the estimation accuracy of the STUKF consistently surpasses that of the UKF. The STUKF not only provides more precise estimates but also exhibits remarkable robustness to different mass parameter perturbations. This robustness is crucial in practical applications where the mass of the vehicle can change due to varying loads and other dynamic factors. The superior performance of the STUKF is attributed to its strong tracking capabilities, which allow it to adapt to sudden changes and maintain high estimation accuracy. The strong tracking algorithm embedded within the STUKF enhances its ability to handle model parameter perturbations, ensuring that the filter remains reliable under a wide range of conditions. These extensive offline tests underscore the effectiveness of the STUKF in maintaining robust and accurate state estimation even when faced with significant parameter variations. The consistent performance across 100 experiments highlights the filter's capability to adapt to different scenarios, making it a reliable choice for real-world applications in vehicle dynamics and control systems. In conclusion, the results presented in Fig. 5.15 clearly indicate that the STUKF outperforms the UKF and other estimation methods in terms of estimation accuracy and robustness to mass parameter perturbations. The STUKF's ability to provide reliable estimates under varying conditions makes it a valuable tool for enhancing the performance and safety of vehicle state estimation systems. These findings validate the robustness and superiority of the STUKF, confirming its potential for broad application in the automotive industry, where precise and reliable state estimation is critical.

Figure 5.16 shows the results of identifying the TRFC based on the axial force obtained from STUKF estimation. The parameters used in IMM-UKF and MSE are the same as in the first test condition except for $\overline{P} = [0.936, 0.064; 0.064, 0.936]$. Figure 5.17 illustrates the variations in the probabilities of the longitudinal and lateral models within the two hybrid estimation methods being evaluated. Consistent with the first test conditions, the IMM-UKF continues to demonstrate the highest estimation accuracy. A notable deviation from the first test condition, however, is observed in the tracking behavior of the longitudinal dynamics-based model UKF. In Fig. 5.16, this model reaches the TRFC minimum before the lateral dynamics-based model UKF. This early tracking is attributed to the rapid decrease in LV at the beginning of the test, which influences the dynamics and priority of the longitudinal model. The trend observed in the MSE estimation curve is in harmony with the changes in the estimation results derived from both the longitudinal and lateral models. Specifically, the accuracy of the MSE is

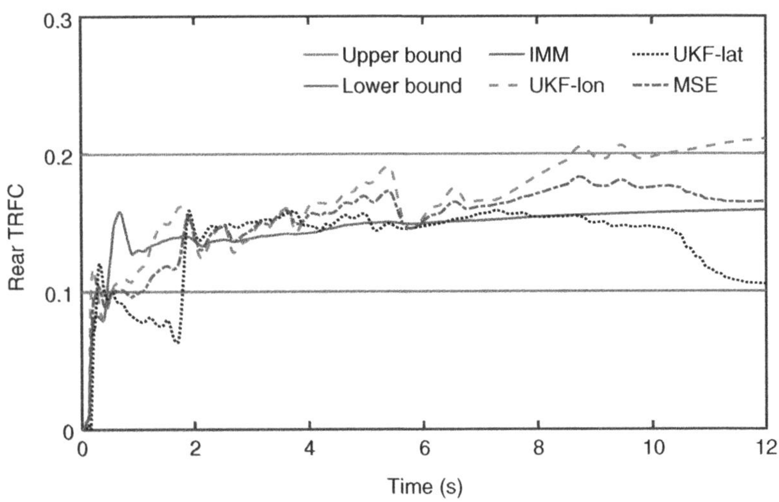

Figure 5.16 The TRFC estimation on ice-snow road.

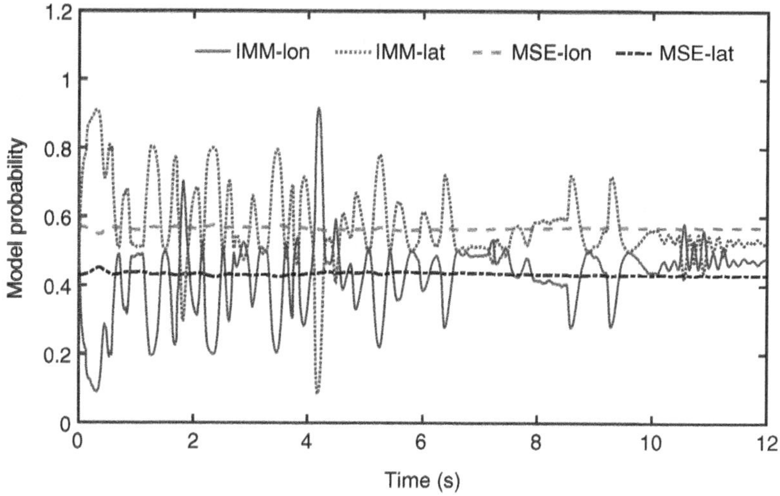

Figure 5.17 The model probability on ice-snow road.

significantly impacted by the estimated output of individual models. When using a single model, any deviations or inaccuracies directly translate to the MSE, affecting its overall performance. In contrast, the prediction curve based on the IMM-UKF method exhibits remarkable stability, particularly in the interval from [1–10]. Unlike the single-model approach, the IMM-UKF method does not show fluctuations corresponding to the output of individual models. This stability is achieved because the IMM-UKF can dynamically adjust the probability of each model based on the predicted covariance information. By leveraging the covariance data, the IMM-UKF effectively balances and integrates the contributions of both longitudinal and lateral models, optimizing the estimation process. The dynamic adjustment capability of the IMM-UKF method allows it to seamlessly integrate the strengths of single model-based estimation algorithms. This integration results in superior estimation accuracy across all tested methods. The ability to adapt to changing model probabilities and adjust based on real-time data ensures that the IMM-UKF provides the most accurate and reliable state estimation. The proposed estimation framework, particularly the IMM-UKF method, demonstrates the best performance across various working conditions. This versatility highlights not only the high estimation accuracy but also the strong adaptability of the method. It confirms that the IMM-UKF can maintain robust performance even when faced with diverse and dynamic operational scenarios. These results underscore the efficacy of the IMM-UKF in handling complex state estimation tasks in vehicle dynamics. The method's robustness is particularly evident in its ability to manage rapid changes in longitudinal velocity and adjust the model probabilities accordingly. This adaptability is crucial for applications in real-world vehicular systems where conditions can change rapidly and unpredictably. The comprehensive evaluation and comparison of the IMM-UKF, UKF, and MSE methods provide clear evidence of the superiority of the IMM-UKF. Its dynamic probability adjustment mechanism and integration of multiple model outputs lead to consistently accurate and stable state estimations. This makes it an ideal choice for advanced vehicular state estimation and control systems. In summary, the IMM-UKF method excels in providing precise and reliable state estimations under varying conditions. Its ability to dynamically integrate multiple models and adjust based on real-time data ensures optimal performance, making it a highly effective tool for modern vehicle state estimation applications. The experimental results confirm the method's high estimation accuracy and adaptability, showcasing its potential for broad application in the automotive industry.

5.4 A Fault-Tolerant Scheme for Multi-model Ensemble Estimation of Tire–Road Friction Coefficient with Missing Measurements

In this section, we introduce a fault-tolerant estimation framework designed to accurately estimate the TRFC. This framework integrates an FTUKF with an event-driven fusion method to enhance estimation performance, especially under challenging conditions such as sensor signal loss. First, we develop two separate FTUKFs, each tailored for specific tasks: one for estimating longitudinal tire forces and the other for lateral tire forces. These FTUKFs are engineered to maintain robust performance even when there is a loss of sensor signals, ensuring that the estimation process remains reliable and accurate despite potential data interruptions. Once the longitudinal and lateral tire forces are estimated using the respective FTUKFs, the next step involves calculating the longitudinal and lateral TRFCs based on this tire force information. This step is crucial as it provides the foundational data required for accurate TRFC estimation, reflecting the tire–road interaction dynamics under varying conditions. To further refine the TRFC estimation, we introduce an event-driven multi-model integrated method. This innovative approach is designed to handle varying degrees of data loss effectively. By dynamically adjusting to the extent of data loss, the method performs a weighted fusion of the longitudinal and lateral TRFCs. The weighted fusion process is critical as it combines the strengths of both longitudinal and lateral estimations, enhancing the overall accuracy and reliability of the TRFC estimation. The event-driven aspect of the fusion method ensures that the system responds promptly to changes in data availability, optimizing the use of available information. This dynamic response mechanism is key to maintaining high estimation performance in real-time applications, where conditions can change rapidly and unpredictably.

The overall framework is illustrated in Fig. 5.18, comprising four main components: sensor signals with data loss, the longitudinal TRFC estimation module, the lateral TRFC estimation module, and the event-driven fusion module. First, in the longitudinal TRFC estimation module, an FTUKF is employed to estimate longitudinal tire forces. This estimator is specifically designed to handle in-vehicle sensor signals that may experience data loss, ensuring robust performance despite potential signal interruptions. Following this, an FTUKF-based longitudinal TRFC estimator is developed, which leverages the relationship between tire forces and the TRFC to provide accurate estimates of the longitudinal TRFC. The lateral TRFC estimation module operates in a similar manner. It begins with an FTUKF-based estimator for lateral tire forces, also built to cope with sensor signal loss. This estimator then feeds into a lateral TRFC estimator, constructed on the same principles as the longitudinal estimator, to derive the lateral TRFC values. Once the longitudinal and lateral TRFCs are obtained, they are processed by the event-driven fusion module. This module performs a weighted fusion of the longitudinal and lateral TRFCs to enhance overall estimation accuracy. Central to this module is an event-triggered mechanism and an RLS-based weight update algorithm. The event-triggered mechanism assesses the degree of data loss to determine whether a weight update is necessary. If the event is triggered, indicating significant data loss, the latest model weights are applied to update the TRFC estimation. Conversely, if the event is not triggered, the module retains the previous weights, maintaining stability in the estimation process. The integration of these components within the framework ensures a resilient and accurate estimation of TRFC, even under conditions of sensor signal loss. The FTUKF estimators for both longitudinal and lateral forces provide robust initial estimates, which are then refined through the event-driven fusion process. This approach not only mitigates the impact of data loss but also dynamically adapts to changing conditions, ensuring the reliability and precision of the TRFC estimates. This comprehensive framework, with its robust handling of sensor data loss and dynamic fusion of TRFC estimates, is designed to enhance vehicle control systems. By ensuring accurate and reliable TRFC estimation, it contributes to improved vehicle stability and safety, particularly in dynamic and variable driving conditions. The advanced use of FTUKF and event-driven mechanisms within this framework demonstrates significant potential for broad application in the automotive industry, offering a sophisticated solution for real-time vehicle dynamics estimation.

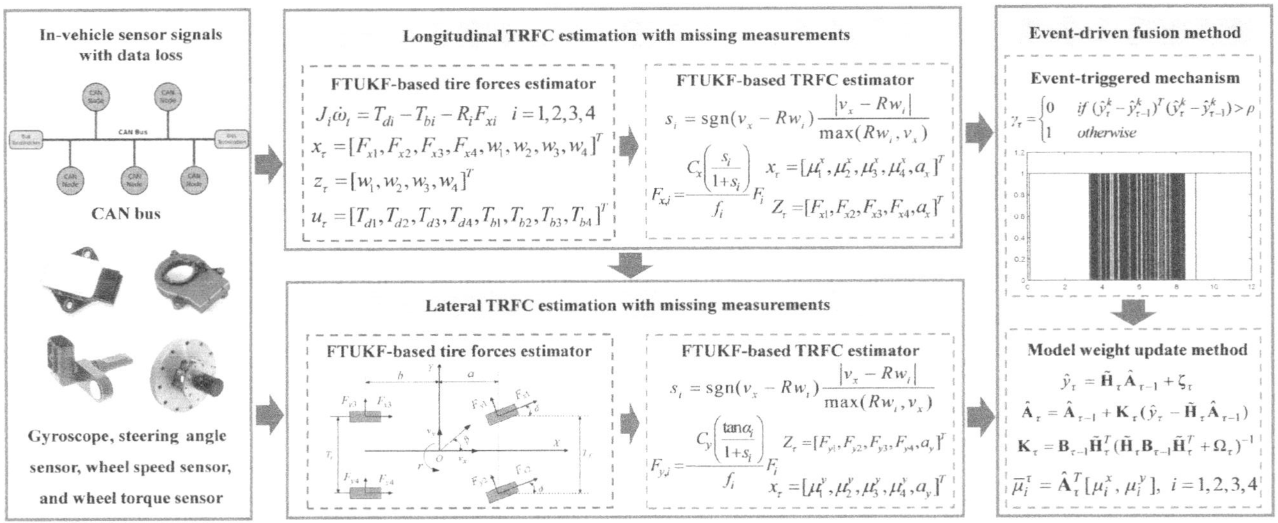

Figure 5.18 The fault-tolerant estimation scheme.

5.4.1 Vehicle Model and Problem Statement

The four-wheel vehicle model (see Fig. 5.19) is employed to depict the dynamic characteristics of the vehicle. The four-wheel vehicle dynamic model is a crucial tool in automotive engineering, designed to predict and analyze vehicle behavior under various driving conditions. Originating from the principles of physics and mechanics, it models the interactions between the vehicle's components and external forces. Key assumptions include a rigid body structure, constant mass, simplified suspension systems, and accurate tire–road interaction models. The theoretical framework is grounded in Newton's laws of motion, kinematic equations, and control systems that simulate driver inputs and vehicle responses. This dynamic model is extensively used in design optimization, allowing engineers to test and refine vehicle parameters such as suspension geometry, weight distribution, and tire characteristics before physical prototyping. It is vital in safety testing, simulating crash scenarios, and vehicle responses to ensure compliance with safety standards. Additionally, it supports the development and testing of advanced driver assistance systems (ADAS), such as adaptive cruise control and lane-keeping assistance, and is integral to autonomous vehicle technology, predicting and controlling vehicle behavior in real-time for safe navigation. The model also facilitates performance analysis under varying conditions, helping to fine-tune vehicles for optimal performance. Furthermore, it enables virtual prototyping, reducing the time and cost associated with building physical prototypes. Overall, the four-wheel vehicle dynamic model plays an essential role in advancing vehicle safety, performance, and innovation in the automotive industry. Some specific equations are as follows.

$$\dot{v}_x = a_x + rv_y \tag{5.51}$$

$$\dot{v}_y = a_y - rv_x \tag{5.52}$$

$$a_x = [(F_{x1} + F_{x2})\cos\delta + F_{x3} + F_{x4} - (F_{y1} + F_{y2})\sin\delta]/m \tag{5.53}$$

$$a_y = [(F_{x1} + F_{x2})\sin\delta + F_{y3} + F_{y4} + (F_{y1} + F_{y2})\cos\delta]/m \tag{5.54}$$

$$\dot{r} = \left\{a[(F_{x1} + F_{x2})\sin\delta + (F_{y1} + F_{y2})\cos\delta] - b(F_{y3} + F_{y4})\right\}/I_z \tag{5.55}$$

$$J_i\dot{\omega}_t = T_{di} - T_{bi} - R_iF_{xi} \quad i = 1, 2, 3, 4 \tag{5.56}$$

where F_{xi} and F_{yi} are longitudinal tire forces and lateral tire forces, $i = 1, 2, 3, 4$, which correspond to the left-front, right-front, left-rear, and right-rear wheels, respectively. a and b are distances from the center of gravity to the front axle and rear axle; r and m are the yaw rate, vehicle mass; T_f and T_r are front track width and rear track width; δ is the front wheel angle; I_z is the inertia moment about the vehicle vertical axis; v_x, v_y, and β are the longitudinal vehicle velocity, lateral vehicle velocity, and sideslip angle; T_{bi}, T_{di}, and J_i are the braking torque, driving torque, and moment of inertia; ω_i is the wheel rotational speed; and

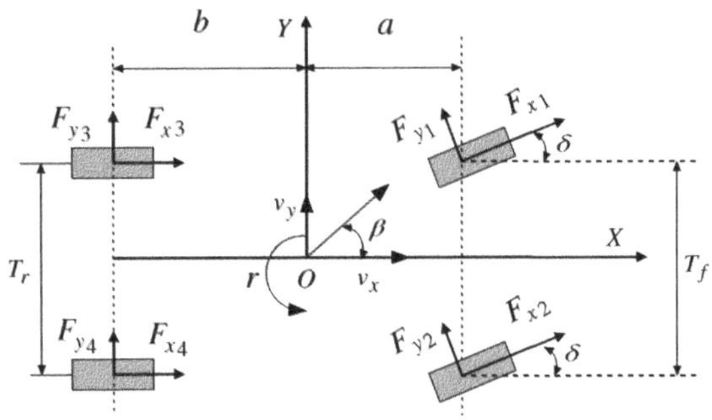

Figure 5.19 The four-wheel model.

R_i is the wheel radius. A combined longitudinal and lateral brushed tire model is used to describe the dynamic characteristics of tires. The brushed tire model is a fundamental concept in vehicle dynamics, used to simulate and analyze the interaction between a tire and the road surface. Originating from simplified physical and mathematical principles, this model assumes that the tire is represented by a series of bristles or brushes that deform under load, capturing the essential characteristics of tire behavior. The primary assumptions include a linear elastic response of the bristles and a constant contact patch length. The theoretical framework is based on the balance of forces and moments within the contact patch, allowing for the calculation of longitudinal and lateral forces as well as aligning moments. These forces and moments are crucial for understanding tire grip, handling, and overall vehicle stability. The brushed tire model is extensively applied in various automotive simulations, from basic handling analysis to advanced vehicle dynamics studies. It is particularly useful in the development and tuning of suspension systems, traction control systems, and stability control algorithms. By providing insights into how tires react under different conditions, the brushed tire model helps engineers optimize tire and vehicle performance, improve safety, and enhance the driving experience. Despite its simplicity, this model offers a valuable approximation for many practical applications, making it a staple in both academic research and industry practice. Some specific equations are as follows:

$$F_{x,i} = \frac{C_x\left(\dfrac{s_i}{1+s_i}\right)}{f_i}F_i \tag{5.57}$$

$$F_{y,i} = -\frac{C_y\left(\dfrac{\tan\alpha_i}{1+s_i}\right)}{f_i}F_i \tag{5.58}$$

$$F_i = \begin{cases} f_i - \frac{1}{3\mu_i^{\vartheta}F_{z,i}}f_i^2 + \frac{1}{27(\mu_i^{\vartheta})^2F_{z,i}^2}f_i^3, & \text{if } f_i \leq 3\mu_i^{\vartheta}F_{z,i} \\ \mu_i^{\vartheta}F_{z,i} & \text{else} \end{cases} \tag{5.59}$$

$$f_i = \sqrt{C_x^2\left(\frac{s_i}{1+s_i}\right)^2 + C_y^2\left(\frac{\tan\alpha_i}{1+s_i}\right)^2} \tag{5.60}$$

$$F_{z,1} = \frac{mgb}{2(a+b)} - \frac{ma_xh}{2(a+b)} - \frac{ma_yh}{T_f}\cdot\frac{b}{a+b} \tag{5.61}$$

$$F_{z,2} = \frac{mgb}{2(a+b)} - \frac{ma_xh}{2(a+b)} + \frac{ma_yh}{T_f}\cdot\frac{b}{a+b} \tag{5.62}$$

$$F_{z,3} = \frac{mga}{2(a+b)} + \frac{ma_xh}{2(a+b)} - \frac{ma_yh}{T_r}\cdot\frac{a}{a+b} \tag{5.63}$$

$$F_{z,4} = \frac{mga}{2(a+b)} + \frac{ma_xh}{2(a+b)} + \frac{ma_yh}{T_r}\cdot\frac{a}{a+b} \tag{5.64}$$

$$s_i = \mathrm{sgn}(v_x - R\omega_i)\frac{|v_x - R\omega_i|}{\max(R\omega_i, v_x)} \tag{5.65}$$

$$\alpha_1 = \delta - \arctan\left(\frac{v_y + ar}{v_x - T_f r/2}\right) \tag{5.66}$$

$$\alpha_2 = \delta - \arctan\left(\frac{v_y + ar}{v_x + T_f r/2}\right) \tag{5.67}$$

$$\alpha_3 = -\arctan\left(\frac{v_y - br}{v_x - T_r r/2}\right) \tag{5.68}$$

$$\alpha_4 = -\arctan\left(\frac{v_y - br}{v_x + T_r r/2}\right) \tag{5.69}$$

where a_x and a_y are longitudinal acceleration and lateral acceleration; h is the height of the center of gravity; μ_i^ϑ is TRFC, for longitudinal tire forces computation: $\mu_i^\vartheta = \mu_i^x$, for lateral tire forces computation: $\mu_i^\vartheta = \mu_i^y$; C_x, C_y, and $F_{z,i}$ are the tire longitudinal, lateral stiffness coefficients, vertical tire forces; s_i and α_i are the longitudinal slip ratio, wheel sideslip angle; $i = 1, 2, 3, 4$, it has the same physical meaning as the vehicle model.

5.4.2 Methodology

1) **Longitudinal Tire Forces Estimation:**
For tire forces estimation in case of missing measurements, based on eq.5.6, the discrete-time recursive model is given by

$$\begin{cases} x_{\tau+1} = f(x_\tau, u_\tau) + \psi_\tau \\ z_\tau = \Pi_\tau \tilde{h}(x_\tau, u_\tau) + \varsigma_\tau \end{cases} \tag{5.70}$$

$$x_\tau = [F_{x1}, F_{x2}, F_{x3}, F_{x4}, \omega_1, \omega_2, \omega_3, \omega_4]^T \quad z_\tau = [\omega_1, \omega_2, \omega_3, \omega_4]^T$$
$$u_\tau = [T_{d1}, T_{d2}, T_{d3}, T_{d4}, T_{b1}, T_{b2}, T_{b3}, T_{b4}]^T$$

where ψ_τ is a process noise and its covariance is Q_τ, ς_τ is a measurement noise and its covariance is Ω_τ, f is the state transition function, τ is the sampling instant, x_τ is the state vector, $\tilde{h}$ is the measurement output function. The process noise and the measurement noise are uncorrelated with each other. z_τ is the measurement vector with data loss, u_τ is the input vector. In this study, we use the Bernoulli distribution to describe the phenomenon of data loss. $\Pi_\tau = diag\left\{\tilde{\varepsilon}_\tau^1, \tilde{\varepsilon}_\tau^2, \ldots \tilde{\varepsilon}_\tau^n\right\}$, where $\tilde{\varepsilon}_\tau^i (i = 1, 2 \cdots n)$ are n independent Bernoulli random variables. They are not related to all noise signals. diag{} is a diagonal matrix.

2) **Lateral Tire Forces Estimation:**
   For estimating lateral tire forces, the recursive model is the same as (5-70). Based on the eqs.(5.51)–(5.55) we have $z_\tau = [a_x, a_y, r]^T$, $u_\tau = [F_{x1}, F_{x2}, F_{x3}, F_{x4}, \delta]^T$, an $x_\tau = [F_{y1}, F_{y2}, F_{y3}, F_{y4}, r, v_x, v_y]^T$. To perform state estimation using discrete measurements, these nonlinear equations need to be expressed in a discrete matrix form.

In the estimation process of longitudinal tire forces, the driving torque, braking torque, and wheel rotational speed signals can be subject to random losses. This means that during data collection and signal processing, these critical inputs might intermittently be unavailable. Similarly, when estimating lateral tire forces, signals such as longitudinal acceleration, front wheel angle, lateral acceleration, and yaw rate can also experience random data loss. To accurately model and address this data loss phenomenon, we employ the Bernoulli distribution, which provides a probabilistic framework for predicting the occurrence of signal dropouts. To mitigate the impact of these missing measurements on tire force estimation, we design an FTUKF. The FTUKF is specifically tailored to handle the random loss of measurement data. In this approach, the data loss only affects the measurement update step of the FTUKF, where the filter adjusts its estimates based on the new incoming data. When measurement data is unavailable, the filter relies on its prediction step, which remains unchanged from the standard UKF. The UKF is a powerful estimation algorithm that extends the capabilities of the KF to handle nonlinear systems with non-Gaussian noise distributions. It addresses limitations of the KF, which assumes linear dynamics and Gaussian noise, by employing a deterministic sampling approach to approximate the state distribution more accurately. The UKF was first introduced by Jeffrey Uhlmann in 1997 as an alternative to the EKF, which linearizes nonlinear functions using Taylor series expansion. Unlike the EKF, which can suffer from inaccuracies in highly nonlinear systems, the UKF operates by computing a set of carefully chosen sample points (sigma points) around the current mean and propagating these through the nonlinear functions of the system. This deterministic sampling approach captures the true distribution of the state variables and their uncertainties more accurately, even under conditions where the system dynamics are significantly nonlinear. The UKF excels in estimating states of nonlinear systems where traditional linearization methods like the EKF may fail due to inaccuracies. By capturing the true distribution of state variables through sigma points, the UKF provides more accurate state estimates compared to linearized filters, especially in scenarios with significant nonlinearities. It can handle non-Gaussian noise distributions better than linearized filters, making it suitable for real-world applications where noise characteristics are complex and dynamic. Compared to particle filters that use Monte Carlo sampling, the UKF achieves comparable accuracy with fewer computational resources, making it feasible for real-time applications. Its deterministic sampling approach and straightforward implementation make the UKF adaptable across various domains, from automotive systems to aerospace and robotics, enhancing the reliability and performance of estimation tasks in complex and dynamic environments. In automotive applications, UKF finds extensive use in sensor fusion: Combining data from multiple sensors (e.g., GPS, IMU, radar) to estimate vehicle position, velocity, and orientation accurately. Vehicle Dynamics: Tracking and predicting vehicle states such as acceleration, steering angle, and tire forces under varying road conditions. ADAS: Enhancing the performance of systems that monitor surroundings, predict collisions, and assist in adaptive cruise control. In summary, the UKF represents a significant advancement in estimation theory, offering robust solutions for nonlinear state estimation problems across diverse application domains. Its ability to accurately model and propagate uncertainties through deterministic sampling sets it apart as a valuable tool in modern control and estimation engineering practices.

In the time update step, where the filter projects the current state estimate into the future, it continues to operate normally, unaffected by the data loss. This design ensures that the FTUKF maintains robust performance even in the presence of intermittent signal losses, providing reliable tire force estimations. By accommodating missing measurements during the update step and maintaining a consistent time update step, the FTUKF effectively balances the need for accuracy and resilience, making it a powerful tool in automotive dynamics and control systems. This approach enhances the reliability of tire force estimations, crucial for vehicle stability and safety systems, especially in environments where sensor data might be prone to interruptions.

The iterative steps of the FTUKF can be expressed as

1) Initialization

   The initial mean x and covariance matrix P are given by

$$\hat{x}_0 = E(x_0) \tag{5.71}$$

$$P_0 = E\left[(x_0 - \hat{x}_0)(x_0 - \hat{x}_0)^T\right] \tag{5.72}$$

2) Time Update

The weight ϕ_c^j, ϕ_m^j, and sigma sampling points $\sigma_{\tau-1}^j$ can be expressed as

$$\begin{cases} \phi_m^0 = \lambda/(n+\lambda), \phi_c^0 = \lambda/(n+\lambda) + 1 + \theta - \alpha^2 \\ \phi_m^j = \phi_c^j = \lambda/(2(n+\lambda)), \quad j = 1, 2, \cdots 2n \end{cases} \tag{5.73}$$

$$\begin{cases} \sigma_{\tau-1}^0 = \hat{x}_{\tau-1} \\ \sigma_{\tau-1}^j = \hat{x}_{\tau-1} + \sqrt{n+\lambda}\left(\sqrt{P_{\tau-1}}\right)_j \quad j = 1, 2, \cdots n \\ \sigma_{\tau-1}^j = \hat{x}_{\tau-1} - \sqrt{n+\lambda}\left(\sqrt{P_{\tau-1}}\right)_j \quad j = n+1, \cdots 2n \end{cases} \tag{5.74}$$

where n is the dimension of system states, $\lambda = \alpha^2(n + \ell) - n$ is a scaling parameter, ℓ is a secondary scaling parameter, α is a small positive value, θ is used to incorporate prior knowledge of the distribution of state.

The propagated sigma points are given by

$$\aleph_{\tau/\tau-1}^{*(j)} = f\left(\sigma_{\tau-1}^{(j)}, u_{\tau-1}\right) \tag{5.75}$$

The prior state $\hat{x}_{\tau/\tau-1}$ and covariance matrix $P_{\tau/\tau-1}$ are updated by

$$\hat{x}_{\tau/\tau-1} = \sum_{j=0}^{2n} \phi_m^j \aleph_{\tau/\tau-1}^{*(j)} \tag{5.76}$$

$$P_{\tau/\tau-1} = \sum_{j=0}^{2n} \phi_c^j \left(\aleph_{\tau/\tau-1}^{*(j)} - \hat{x}_{\tau/\tau-1}\right)\left(\aleph_{\tau/\tau-1}^{*(j)} - \hat{x}_{\tau/\tau-1}\right)^T + Q_{\tau-1} \tag{5.77}$$

In our experiments, the process noise is assumed to obey a Gaussian distribution, reflecting the random variability inherent in the system dynamics. This assumption allows us to model the process noise with a normal distribution, characterized by its mean and variance, which provides a realistic representation of the uncertainties and disturbances that can affect the vehicle's performance. To simulate the uncertainty in the model, we artificially set certain parameters to fluctuate within predefined ranges. These parameters include critical aspects of the vehicle's dynamics, such as the position of the center of gravity and the rotational inertia. By allowing these parameters to vary, we can account for the natural variations and potential inaccuracies in the vehicle's physical properties, which might arise due to manufacturing tolerances, changes in load distribution, or other factors. Furthermore, we assume that the effects of these fluctuating parameters are independent of each other. This assumption simplifies the modeling process and allows us to independently analyze the impact of each parameter on the vehicle's behavior.

3) Measurement Update
The new sigma points σ_τ^j are as follows.

$$\begin{cases} \sigma_\tau^0 = \hat{x}_{\tau/\tau-1} \\ \sigma_\tau^j = \hat{x}_{\tau/\tau-1} + \sqrt{n+\lambda}\left(\sqrt{P_{\tau/\tau-1}}\right)_j \quad j = 1, 2, \cdots n \\ \sigma_\tau^j = \hat{x}_{\tau/\tau-1} - \sqrt{n+\lambda}\left(\sqrt{P_{\tau/\tau-1}}\right)_j \quad j = n+1, \cdots 2n \end{cases} \tag{5.78}$$

The overall iterative steps in the time updating process of the FTUKF are the same as in the traditional UKF. The difference is that due to data loss, we have the extra matrix Π_τ when embedding individual sigma points into Eq. (5.70) to predict the measured variables, then we have the following equation

$$Z_{\tau/\tau-1}^{*(j)} = \Pi_\tau h\left(\sigma_\tau^{(j)}, u_\tau\right) \tag{5.79}$$

The estimated measurement $\hat{Z}_{\tau/\tau-1}$ is obtained by a weighted average of the predicted values of $2n+1$ sigma points.

$$\hat{Z}_{\tau/\tau-1} = \sum_{j=0}^{2n} \phi_m^j Z_{\tau/\tau-1}^{*(j)} = \overline{\Pi}_\tau \sum_{j=0}^{2n} \phi_m^j h\left(\sigma_\tau^{(j)}, u_\tau\right) \tag{5.80}$$

where $\overline{\Pi}_\tau$ is an expectation of Π_τ.
For the covariance matrix $P_{z,\tau}$, the following formula can be derived

$$P_{z,\tau} = \overline{\Pi}_\tau \sum_{j=0}^{2n} \phi_c^j h\left(\sigma_\tau^{(j)}, u_\tau\right) h^T\left(\sigma_\tau^{(j)}, u_\tau\right) - \overline{\Pi}_\tau (I - \overline{\Pi}_\tau) \hat{Z}_{\tau/\tau-1} \hat{Z}_{\tau/\tau-1}^T + \Omega_\tau \tag{5.81}$$

where I *is* a unit matrix.

The cross-covariance matrix $P_{xz,\tau}$ can be expressed as

$$P_{xz,\tau} = \Pi_\tau \sum_{i=0}^{2n} \phi_c^j \left(\aleph_{\tau/\tau-1}^{*(j)} - \hat{x}_{\tau/\tau-1}\right)\left(h\left(\sigma_\tau^{(j)}, u_\tau\right) - \hat{Z}_{\tau/\tau-1}\right)^T \tag{5.82}$$

The gain matrix K_τ, covariance matrix P_τ, and the posterior state $\hat{x}_\tau$ can be calculated by

$$K_\tau = P_{xz,\tau} P_{z,\tau}^{-1} \tag{5.83}$$

$$\hat{x}_\tau = \hat{x}_{\tau/\tau-1} + K_\tau (Z_\tau - \hat{Z}_{\tau/\tau-1}) \tag{5.84}$$

$$P_\tau = P_{\tau/\tau-1} - K_\tau P_{z,\tau} K_\tau^T \tag{5.85}$$

5.4.2.1 TRFC Estimation

After obtaining tire forces, we can use the brushed tire model to identify the TRFC. The state-space model is the same as Eq. (5.70). Although much of the input information during TRFC estimation can be obtained from the tire force estimation module, some of the input information comes directly from the onboard sensor. Specifically, the information of v_x, v_y, r, w_i in Eqs. (5.66)–(5.69) has been directly obtained in tire force estimation. However, a_x, a_y, δ are still from on-board sensors and there is a possibility of data loss. Therefore, we continue to use FTUFK to estimate TRFC considering the effect of data loss.

1) **Longitudinal TRFC Estimation:**
   Based on the longitudinal tire model, we have

$$x_\tau = \left[\mu_1^x, \mu_2^x, \mu_3^x, \mu_4^x, a_x\right]^T, \ Z_\tau = [F_{x1}, F_{x2}, F_{x3}, F_{x4}, a_x]^T.$$

2) **Lateral TRFC Estimation:**
   Based on the lateral tire model, we have

$$x_\tau = \left[\mu_1^y, \mu_2^y, \mu_3^y, \mu_4^y, a_y\right]^T, \ Z_\tau = [F_{y1}, F_{y2}, F_{y3}, F_{y4}, a_y]^T.$$

For the specific iterative steps, it is the same as eqs. (5.71)–(5.85).

5.4.2.2 Event-Driven Multi-model Fusion Method

The single-vehicle model often struggles to adapt to all driving conditions, and the inherent non-linearity of the vehicle system further reduces the accuracy of these models. To address this issue, the multi-model fusion method has been developed, which fuses the outputs of individual models to enhance estimation accuracy. However, determining the optimal weight or validity function for each model in multi-model fusion methods remains a significant challenge. A commonly used approach to calculate the validity function involves deploying a Gaussian function based on bias, which helps in adjusting the influence of each model according to its performance. In this study, we propose a weight-updating strategy based on the RLS principle. RLS is a widely used adaptive algorithm in signal processing and estimation theory, particularly effective in scenarios where data is processed sequentially over time. The RLS algorithm is designed to estimate the parameters of a linear system in real-time, based on incoming data samples. It belongs to the family of adaptive filtering techniques, which continuously update parameter estimates as new data becomes available. The core principle of RLS is to minimize the weighted sum of squared errors between observed data and the predicted values using a recursive formulation. RLS employs a weighted least squares criterion to update parameter estimates iteratively. It calculates the optimal parameter values by minimizing the sum of squared prediction errors, weighted by the covariance matrix of the input data. Unlike batch methods that process all data at once, RLS updates parameters sequentially as new data points arrive. It maintains and updates estimates of the system parameters and their covariance matrices efficiently over time. The efficiency of RLS is enhanced by the Matrix

Inversion Lemma, which simplifies the computation of inverse matrices during each update step, reducing computational complexity compared to direct matrix inversion methods. RLS adapts dynamically to changes in the system parameters, making it suitable for applications where parameters vary over time. By updating parameters incrementally, RLS requires less memory and computational resources compared to batch estimation methods. RLS typically converges faster than other adaptive algorithms, providing accurate parameter estimates quickly after initialization. In summary, RLS is a powerful algorithm in signal processing and estimation, offering real-time adaptability, efficiency, and fast convergence. Its applications span across diverse fields, from system identification to adaptive control and signal processing, where dynamic parameter estimation is crucial for enhancing performance and accuracy in complex systems.

The proposed method dynamically adjusts the weights assigned to each model in the fusion process, ensuring that the most accurate models have the greatest influence on the final output. The RLS approach continuously refines the weight estimates by minimizing the sum of the squares of the errors, thereby improving the reliability and precision of the multi-model system. For a multi-model system, the optimal output is obtained by combining the individual model outputs in a weighted manner, where the weights are updated recursively according to the RLS principle. This strategy ensures that the system can adapt to changing driving conditions and maintain high accuracy by leveraging the strengths of multiple models. By continuously refining the weights based on real-time data, the proposed method enhances the overall performance of the vehicle model, making it more robust and reliable across various scenarios. This approach not only addresses the limitations of single-vehicle models but also provides a systematic and effective solution for multi-model fusion, paving the way for more advanced and accurate automotive systems.

For a multi-model system, the optimal output $\hat{y}_\tau$ is given by

$$\hat{y}_\tau = \frac{\sum_{k=1}^{N} \omega_k \hat{y}_\tau^k}{\sum_{k=1}^{N} \omega_k} \tag{5.86}$$

where $\hat{y}_\tau^k$ represents k^{th} model output at $t = \tau$, N is the number of submodules, and ω_k is the weight of kth model. Furthermore, the normalized validity formulas are given by

$$A_\tau^k = \frac{\omega_\tau^k}{\sum_{k=1}^{N} \omega_\tau^k} \tag{5.87}$$

The following matrix is constructed

$$\mathbf{H} = \begin{bmatrix} \hat{y}_1^1 & \hat{y}_1^2 & \cdots & \hat{y}_1^N \\ \vdots & \vdots & \cdots & \vdots \\ \hat{y}_\tau^1 & \hat{y}_\tau^2 & \cdots & \hat{y}_\tau^N \end{bmatrix}_{\tau \times N} \tag{5.88}$$

$$\mathbf{y} = [\hat{y}_1 \ \hat{y}_2 \cdots \hat{y}_\tau]^T \tag{5.89}$$

$$\mathbf{A} = \left[A_\tau^1 \ A_\tau^2 \cdots A_\tau^N\right]^T \tag{5.90}$$

If we have $\mathbf{y} = \mathbf{HA}$, then each model has the optimal validity function. To obtain the optimal validity function, we define the following cost function

$$J = (\mathbf{y} - \mathbf{HA})^T(\mathbf{y} - \mathbf{HA}) \tag{5.91}$$

We perform a partial derivative operation on the trace of J, then we have

$$\frac{\partial J}{\partial \mathbf{A}} = -\mathbf{y}^T\mathbf{H} - \mathbf{y}^T\mathbf{H} + 2\mathbf{A}^T\mathbf{H}^T\mathbf{H} \tag{5.92}$$

Then let the Eq. (5.92) equal zero, the optimal validity function $\hat{\mathbf{A}}$ is given by

$$\hat{\mathbf{A}} = (\mathbf{H}^T\mathbf{H})^{-1}\mathbf{H}^T\mathbf{y} = \mathbf{Ly} \tag{5.93}$$

$\mathbf{L}$ is the pseudo-inverse of $\mathbf{H}$. The $\mathbf{L}$ exists as long as $\tau \geq N$ and $\mathbf{H}$ is a full rank matrix.

For nonlinear vehicle systems, their input and output are time-series signals. We write the validity function in recursive form. Meanwhile, real on-board measurement signals usually contain measurement noise, thus $\hat{y}_\tau$ can be expressed by

$$\hat{y}_\tau = \tilde{\mathbf{H}}_\tau \hat{\mathbf{A}}_{\tau-1} + \zeta_\tau \tag{5.94}$$

$$\hat{\mathbf{A}}_\tau = \hat{\mathbf{A}}_{\tau-1} + \mathbf{K}_\tau(\hat{y}_\tau - \tilde{\mathbf{H}}_\tau \hat{\mathbf{A}}_{\tau-1}) \tag{5.95}$$

where $\tilde{\mathbf{H}}_\tau = \left[\hat{y}_\tau^1 \; \hat{y}_\tau^2 \cdots \hat{y}_\tau^N\right]^T$ and $\mathbf{K}_\tau$ represents the optimal gain matrix.

To compute $\mathbf{K}_\tau$, we define the update error $\boldsymbol{\varepsilon}_\tau$, then

$$\begin{aligned}\boldsymbol{\varepsilon}_\tau &= \mathbf{A} - \hat{\mathbf{A}}_\tau = \mathbf{A} - \hat{\mathbf{A}}_{\tau-1} - \mathbf{K}_\tau(\hat{y}_\tau - \tilde{\mathbf{H}}_\tau \hat{\mathbf{A}}_{\tau-1}) \\ &= \boldsymbol{\varepsilon}_{\tau-1} - \mathbf{K}_\tau \tilde{\mathbf{H}}_\tau(\mathbf{A} - \hat{\mathbf{A}}_{\tau-1}) - \mathbf{K}_\tau \zeta_\tau \\ &= (\mathbf{I} - \mathbf{K}_\tau \tilde{\mathbf{H}}_\tau)\boldsymbol{\varepsilon}_{\tau-1} - \mathbf{K}_\tau \zeta_\tau \end{aligned} \tag{5.96}$$

Since $\boldsymbol{\varepsilon}_\tau$ and ζ_τ are uncorrelated with each other, based on (Eq. 5.94), we have the error covariance matrix $\mathbf{B}_\tau$

$$\begin{aligned}\mathbf{B}_\tau &= E\left(\boldsymbol{\varepsilon}_\tau \boldsymbol{\varepsilon}_\tau^T\right) \\ &= E\left\{[(\mathbf{I} - \mathbf{K}_\tau \tilde{\mathbf{H}}_\tau)\boldsymbol{\varepsilon}_{\tau-1} - \mathbf{K}_\tau \zeta_\tau][(\mathbf{I} - \mathbf{K}_\tau \tilde{\mathbf{H}}_\tau)\boldsymbol{\varepsilon}_{\tau-1} - \mathbf{K}_\tau \zeta_\tau]^T\right\} \\ &= (\mathbf{I} - \mathbf{K}_\tau \tilde{\mathbf{H}}_\tau)\mathbf{B}_{\tau-1}(\mathbf{I} - \mathbf{K}_\tau \tilde{\mathbf{H}}_\tau)^T + \mathbf{K}_\tau \Omega_\tau \mathbf{K}_\tau^T \end{aligned} \tag{5.97}$$

where E represents math exception.

Take the partial derivative of $\mathbf{B}_\tau$ with respect to $\mathbf{K}_\tau$, we have

$$\frac{\partial \mathbf{B}_\tau}{\partial \mathbf{K}_\tau} = 2(\mathbf{I} - \mathbf{K}_\tau \tilde{\mathbf{H}}_\tau)\mathbf{B}_{\tau-1}\left(-\tilde{\mathbf{H}}_\tau^T\right) + 2\mathbf{K}_\tau \Omega_\tau \tag{5.98}$$

Let the partial derivative be zero, we have

$$\mathbf{K}_\tau = \mathbf{B}_{\tau-1}\tilde{\mathbf{H}}_\tau^T\left(\tilde{\mathbf{H}}_\tau \mathbf{B}_{\tau-1}\tilde{\mathbf{H}}_\tau^T + \Omega_\tau\right)^{-1} \tag{5.99}$$

The fusion weights for longitudinal TRFC and lateral TRFC are updated by longitudinal and lateral acceleration. Since there is data loss in the acceleration signal, we design an event-triggered mechanism to control whether the fusion weights are updated or not. The equation of γ_τ is as follows

$$\gamma_\tau = \begin{cases} 0 & \textit{if} \left(\hat{y}_\tau^k - \hat{y}_{\tau-1}^k\right)^T \left(\hat{y}_\tau^k - \hat{y}_{\tau-1}^k\right) > \rho \\ 1 & \textit{otherwise} \end{cases} \tag{5.100}$$

where ρ represents the event-triggered threshold, $\hat{y}_{\tau-1}^k$ is received acceleration signal at the last moment. $\hat{y}_\tau^k = [a_{x,\tau}, a_{y,\tau}]^T$ indicates acceleration signal at the current moment. γ_τ is an indicator of whether the fusion weights are updated or not. If $\gamma_\tau = 1$, it means that the updated weights of each sub-model are fused with the output of each current sub-model to obtain TRFC. Otherwise, the weights of each sub-model from the last time will be used. Therefore, the fused TRFC $\overline{\mu}_i$ of each wheel can be expressed as

$$\overline{\mu}_i^\tau = \hat{\mathbf{A}}_\tau^T \left[\mu_i^x, \mu_i^y\right], \quad i = 1, 2, 3, 4 \tag{5.101}$$

Since there are only two submodules, $\hat{\mathbf{A}}$ is a (1×2) matrix. The iteration process of the event-driven fusion approach is shown in Algorithm 5.1.

Algorithm 5.1 Pseudocode of the fusion method

Initialize: Set the initial value of $\hat{\mathbf{A}}_0$, $\mathbf{B}_0$, $\bar{\mu}_i^0$

for $\tau \leftarrow 0$ to T **do**
  If $\gamma_\tau = 1$
  for the validity function update process **do**
    Calculate the measurement value of $\hat{y}_\tau$ via (5.94)
    Compute optimal gain $\mathbf{K}_\tau$ by (5.99)
    Update the validity function $\hat{\mathbf{A}}_\tau$ using (5.95)
    Update estimation error covariance $\mathbf{B}_\tau$ by (5.97)
  end for
  Else $\gamma_\tau = 0$
  for for the validity function update process **do**
    $\hat{\mathbf{A}}_\tau = \hat{\mathbf{A}}_{\tau-1}$
  end for
  Update the fused TRFC of each wheel by (5.101)
end for

5.4.3 Simulation and Experiment Tests

To validate the effectiveness of the fault-tolerant estimation framework, a complex joint simulation platform based on CarSim and Simulink is constructed (see Fig. 5.20).

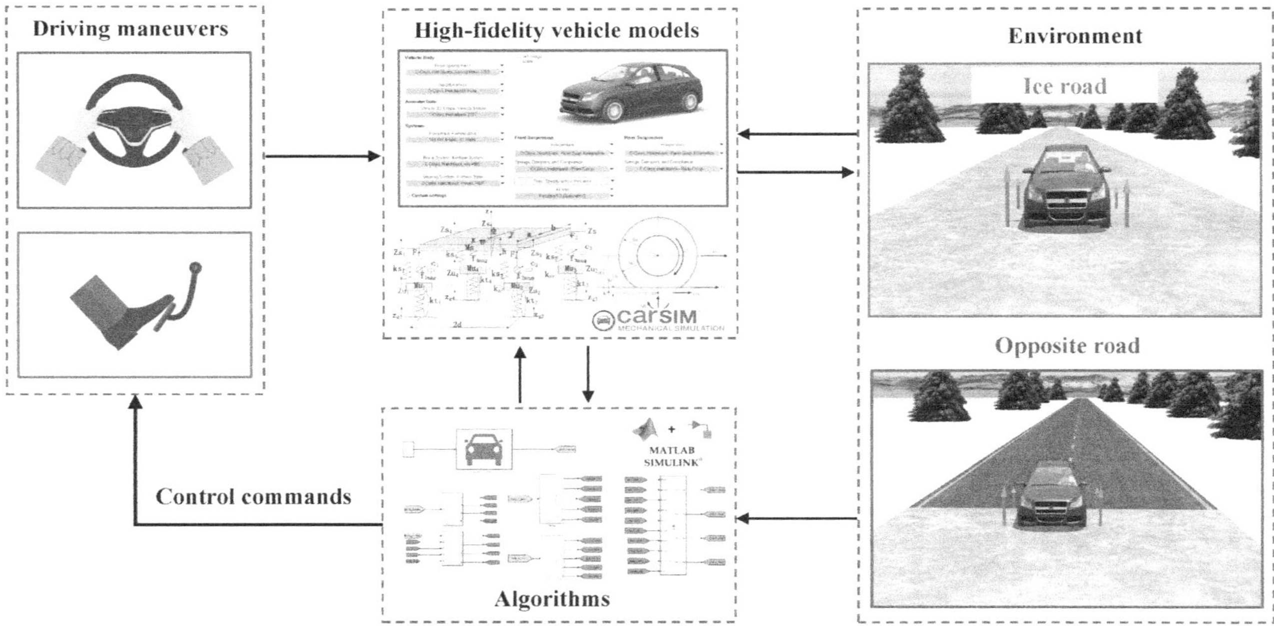

Figure 5.20 The joint simulation platform.

Table 5.3 Parameters of the vehicle model and the value of event-triggered method.

Symbol	Value	Symbol	Value
m	1410 kg	I_z	1536.7 kg·m^2
a	1.015m	b	1.895 m
h	0.54 m	T_f	1.675 m
T_r	1.675m	ρ	1e-6

The CarSim software integrates a high-fidelity vehicle model renowned for its realistic dynamic response, making it an ideal source for reference values (RV) of tire forces and TRFC. In our experimental setup, acceleration, deceleration, and steering commands for the vehicle are generated using Simulink, which interfaces seamlessly with CarSim. Within Simulink, we implement algorithms for estimating tire forces and TRFC using an FTUKF and an event-driven fusion method. To simulate real-world conditions accurately, we introduce data loss into the measurement signals. Initially, Gaussian noise is added to the sensor signals retrieved from CarSim. These signals are subsequently subjected to a Bernoulli distribution module, which introduces random data loss, mimicking the intermittent nature of sensor failures in practical scenarios. To validate the robustness and applicability of our estimation framework, rigorous verification tests are conducted on challenging terrains such as ice roads and opposite roads. Although the vehicle dynamics model in CarSim is comprehensive and includes a wide array of parameters, not all are detailed in Table 5.3. Notably, parameters related to the suspension system are omitted since they do not directly impact the calculation of vertical tire forces. This approach ensures that our estimation algorithms perform reliably under varying conditions, providing accurate assessments of tire forces and TRFC despite potential sensor data disruptions. By leveraging CarSim's advanced vehicle model and integrating sophisticated estimation techniques in Simulink, we aim to enhance the reliability and effectiveness of automotive control systems across diverse operational environments.

5.4.3.1 The Simulation Test

The Test on the Ice Road A steering maneuver with deceleration on ice roads is carried out. The TRFC is 0.2. For the longitudinal tire force estimation, the parameters utilized in the UKF/FTUKF are selected as follows: $\mathbf{Q} = diag[20, 20, 20, 20, 1, 1, 1, 1]$, $\alpha = 0.5$, $\theta = 2$, $\ell = 1$, $\mathbf{R} = diag[1, 1, 1, 1] \cdot 100$. For the lateral tire force estimation, the parameters utilized in the UKF/FTUKF are selected as follows: $Q = diag\,[500,\ 500,\ 500\ 500,\ 1,\ 1,\ 1]$, $\mathbf{R} = diag\,[20,\ 20,\ 20]$.

The rest of the algorithm parameters are the same as the longitudinal tire force estimation algorithm. The front wheel angle, longitudinal acceleration, and lateral acceleration are depicted in Figs. 5.21–5.23. The sampling frequency of the relevant measurement data is 100 Hz. This frequency determines how often data points are recorded from the physical system under observation. To simulate potential data loss, the original data is multiplied by the output of a binomial

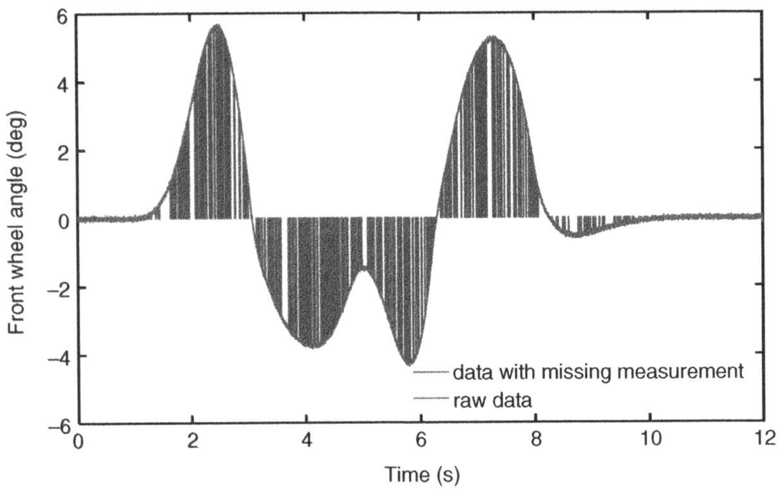

Figure 5.21 The front wheel angle on the ice road.

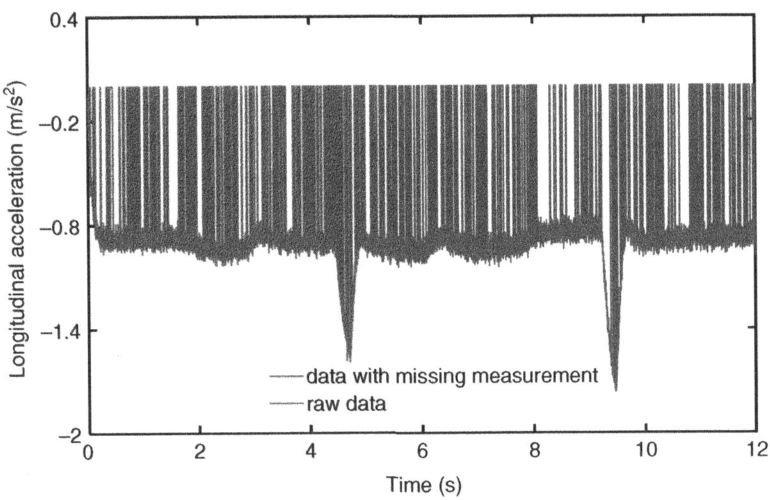

Figure 5.22 The longitudinal acceleration on the ice road.

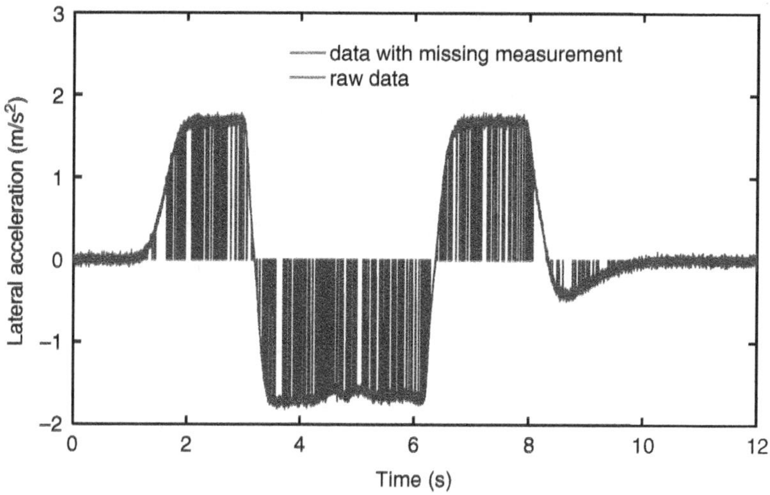

Figure 5.23 The lateral acceleration on the ice road.

Bernoulli distribution. This approach mimics scenarios where data points may not be captured due to various factors in data acquisition processes. In the current setup, the probability of experiencing data loss stands at 30%. Visualizing the impact of this simulated data loss, the raw measurement signal is represented by a red solid line, while the blue solid line depicts the measurement signal with simulated data loss. Notably, the blue line intermittently drops to zero at specific instances, indicating periods where the measurement signal is absent due to simulated data loss. In practical applications, such as evaluating the TRFC on icy roads, the reliability of measurement signals becomes crucial. Ice roads typically offer limited traction, making vehicle maneuvers like acceleration, deceleration, or steering prone to pushing tires into nonlinear operating regions. Despite potential variations in vertical tire force due to load transfers during these maneuvers, the effect on estimating the TRFC remains relatively minor. Understanding these dynamics is essential for refining measurement techniques and ensuring accurate assessments of parameters like TRFC under variable road conditions. By addressing simulated data loss and its implications, researchers can enhance the robustness of their analytical methods in challenging environments, such as those posed by icy road surfaces.

Figure 5.24(a)–(d) describe the estimated longitudinal tire forces from different algorithms. The blue dashed curve indicates the RV. The green solid curve represents the estimated value from the UKF. The red dash-dot curve is the estimation results of the FTUKF. We can see that FTUKF fits more closely with the RV. The estimated curve of UKF is far from the RV because it lacks the ability to deal with

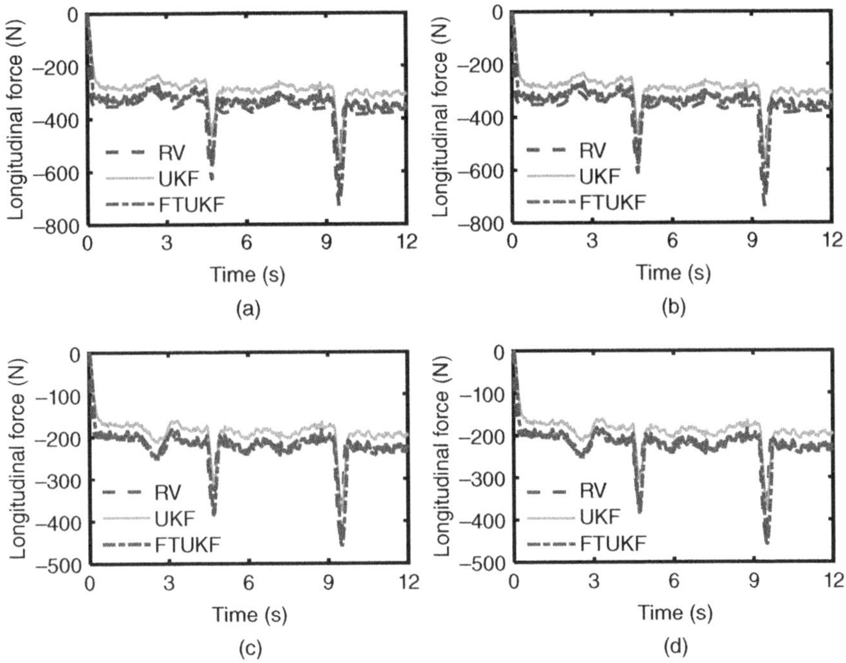

Figure 5.24 The longitudinal tire forces of four wheels on the ice road.

Table 5.4 RMSE of longitudinal tire forces estimation on the ice road.

Symbol	F_{x1}	F_{x2}	F_{x3}	F_{x4}
UKF	81.84	81.43	37.85	37.52
FTUKF	42.87	42.66	22.21	21.89

data loss. Table 5.4 displays the estimated RMSE obtained from various methods. It can be seen that the RMSEs of FTUKF are smaller than those of UKF. This means the FTUKF has higher estimation accuracy than the UKF. The FTUKF has better estimation performance due to data loss being considered in the modeling process and the related covariance formula being updated. Figure 5.24(a)–(d) present comparisons of estimated longitudinal tire forces generated by different algorithms, each offering insights into their respective performance characteristics. The blue dashed curve signifies the RV against which the other estimates are evaluated. Notably, the solid green curve depicts estimates derived from the UKF, while the dash-dot red curve illustrates results obtained using the

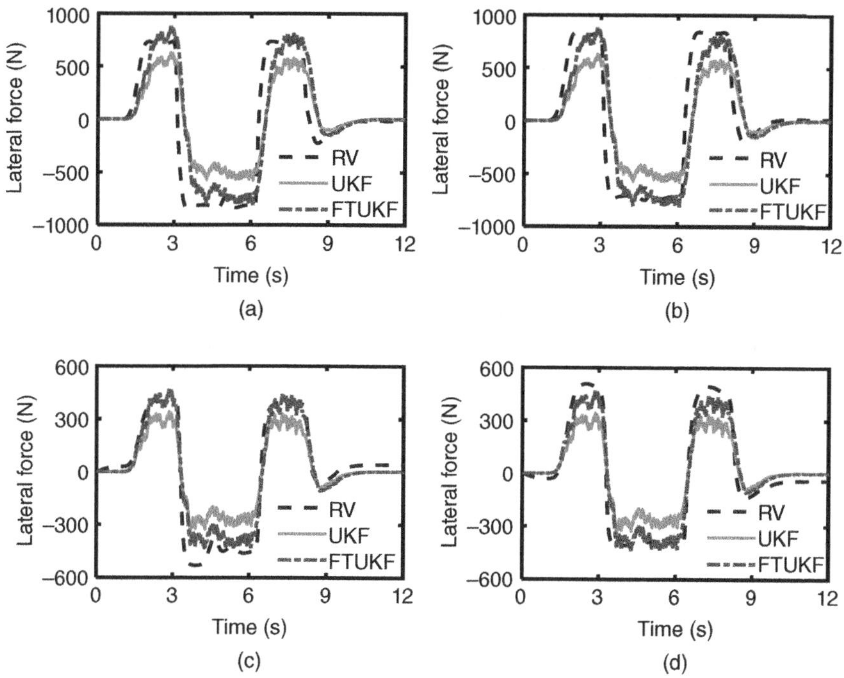

Figure 5.25 The lateral tire forces of four wheels on the ice road.

FTUKF. A discernible observation is that the FTUKF aligns more closely with the RV compared to the UKF. The discrepancy between the UKF and the RV can be attributed to the UKF's inherent limitation in handling data loss scenarios effectively. This deficiency impacts its ability to maintain accuracy in estimation tasks under conditions where data points are intermittently missing or corrupted. In contrast, the FTUKF incorporates mechanisms specifically designed to account for data loss during the modeling process. This includes updates to the covariance formula associated with the estimation, thereby enhancing its robustness and alignment with actual RV values. Figure 5.25(a)–(d) illustrate the estimated lateral tire forces from different approaches. We can see that FTUKF fits more closely to the RV. The estimated curve of UKF is far from the RV. These results show that it is effective to enhance estimation accuracy using FTUKF in the case of data loss. Table 5.5 displays the RMSEs obtained from various methods. The RMSEs of FTUKF are lower than that of UKF. This means the FTUKF has higher estimation accuracy than the UKF.

Further insights into the comparative performance of these algorithms are provided in Table 5.4, which summarizes the RMSE derived from their respective estimations. The RMSE values serve as quantitative indicators of estimation

Table 5.5 RMSE of lateral tire forces estimation on the ice road.

Symbol	F_{y1}	F_{y2}	F_{y3}	F_{y4}
UKF	284.15	282.56	122.63	122.97
FTUKF	251.63	249.71	72.10	68.41

accuracy, with smaller values indicating closer alignment with the true RV. As observed, the RMSEs associated with the FTUKF are consistently lower than those of the UKF. This disparity underscores the superior estimation accuracy achieved by the FTUKF methodology compared to its UKF counterpart. The enhanced performance of the FTUKF can be attributed to its comprehensive approach to modeling, which integrates considerations for data loss into the estimation framework. By dynamically adjusting the covariance formula in response to potential data disruptions, the FTUKF mitigates the adverse effects of missing or corrupted data on estimation outcomes. This proactive adaptation ensures that the estimated longitudinal tire forces closely match actual values, thereby bolstering the reliability and applicability of the algorithm in practical scenarios.

Figure 5.26(a)–(d) depict the estimated TRFC from different algorithms. The blue dashed curve indicates the RV. For the longitudinal and lateral TRFCs estimation, the parameters utilized in the FTUKF are selected as follows: $\mathbf{Q} = diag[0.001, 0.001, 0.001, 0.001, 1]$, $\mathbf{R} = \text{diag}[5,\ 5,\ 5,\ 5,\ 0.2]$. The remaining algorithm parameters remain consistent with those used in the tire force estimation algorithm. Employing the FTUKF, the green solid curve illustrates the estimated longitudinal tire force values. Conversely, the pink dashed curve represents estimates derived from the FTUKF using lateral tire force information. Additionally, the red dash-dot curve portrays results obtained through the event-driven MMEM. In analyzing these estimations, it becomes evident that the longitudinal tire force-based FTUKF method promptly begins tracking the RV from the outset of measurement. In contrast, the FTUKF approach based on lateral tire force information exhibits a delay, only commencing its alignment with the RV after approximately 1.1 seconds of observation. Meanwhile, the MMEM estimation curves demonstrate a convergence toward the RV over time, showcasing the method's ability to refine its estimates effectively. Table 5.6 provides a quantitative summary of estimation performance across these methods, detailing their respective RMSEs. Notably, the MMEM stands out with the highest estimation accuracy, characterized by RMSE values of 0.025, 0.025, 0.023, and 0.024 across different evaluation metrics. The effectiveness of these methods hinges on their ability to adaptively track and predict tire forces under varying

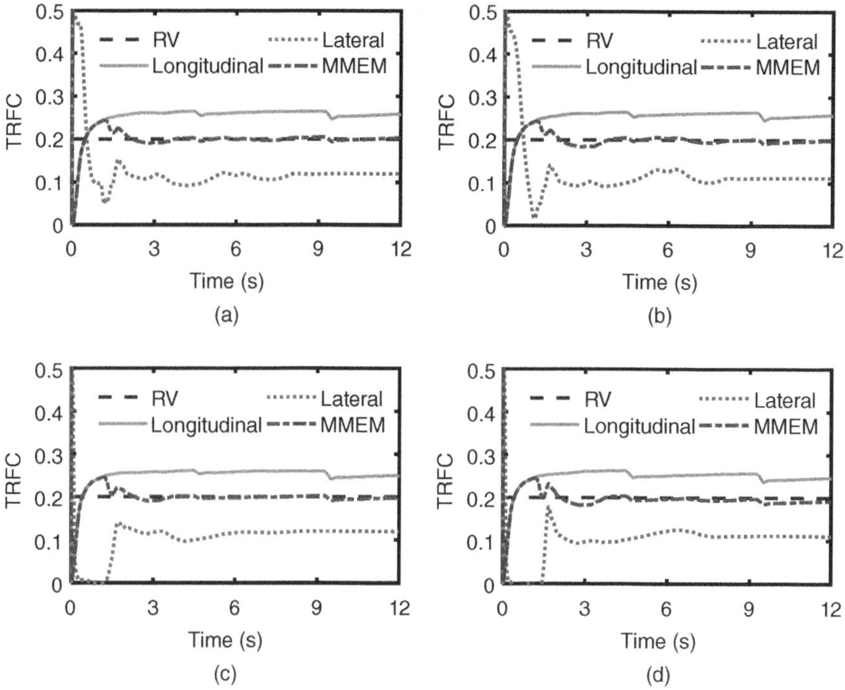

Figure 5.26 The TRFCs of four wheels on the ice road.

Table 5.6 RMSE of TRFC estimation on the ice road.

Symbol	$\overline{\mu}_1^\tau$	$\overline{\mu}_2^\tau$	$\overline{\mu}_3^\tau$	$\overline{\mu}_4^\tau$
Longitudinal	0.061	0.060	0.056	0.254
Lateral	0.099	0.105	0.104	0.109
MMEM	0.025	0.025	0.023	0.024

conditions. The FTUKF excels in promptly aligning its longitudinal tire force estimates with the RV, leveraging its tailored modeling approach. In contrast, while the FTUKF, based on lateral tire force information, introduces a delay in convergence, it ultimately achieves alignment with the RV, demonstrating its versatility across different input modalities. The MMEM, characterized by its event-driven multi-model ensemble approach, exhibits robust performance by consistently minimizing estimation errors and achieving close alignment with the RV. This methodological strength is underscored by its ability to integrate diverse model outputs dynamically, enhancing overall estimation accuracy.

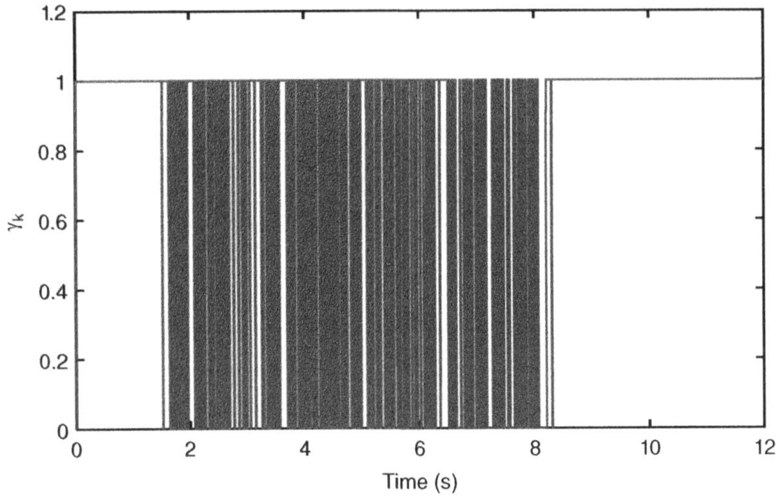

Figure 5.27 The event-triggered times on the ice road.

Figure 5.27 depicts the event-triggered times of the MMEM. We can see that the frequency of $\gamma_k = 0$ increases between [1.5 s, 8.3 s]. At other times, γ_k is almost always equal to one. Therefore, the proposed event-triggered mechanism can respond to the driving conditions well. When the data loss is serious, the weights of the last moment are used to update TRFC. Otherwise, the updated weights will be used. This dynamic update strategy can significantly reduce the impact of data loss.

The Test on the Opposite Road A continuous steering maneuver is initiated on opposite roads, prompting a significant shift in the TRFC from 0.85 to 0.2. This transition is pivotal in understanding the dynamic changes in vehicle behavior under varying road conditions. Figures 5.28–5.30 provide detailed visualizations of the front wheel angle, longitudinal acceleration, and lateral acceleration throughout this maneuver. Initially, as the vehicle operates on a high friction coefficient road during the first 4 seconds, observable changes in both front wheel angle and lateral acceleration exhibit a synchronous pattern. These dynamics reflect the vehicle's stable interaction with the road surface, characterized by predictable responses to steering inputs and changes in lateral dynamics. Subsequently, the vehicle transitions to a low-friction-coefficient road in the subsequent 4 seconds, marking its entry into a nonlinear operating zone. Here, the reduced road friction challenges the vehicle's traction capabilities, influencing its stability and handling characteristics. This shift introduces unpredictable variations in vehicle behavior, necessitating adaptive responses to maintain control and stability during steering maneuvers. These observations underscore

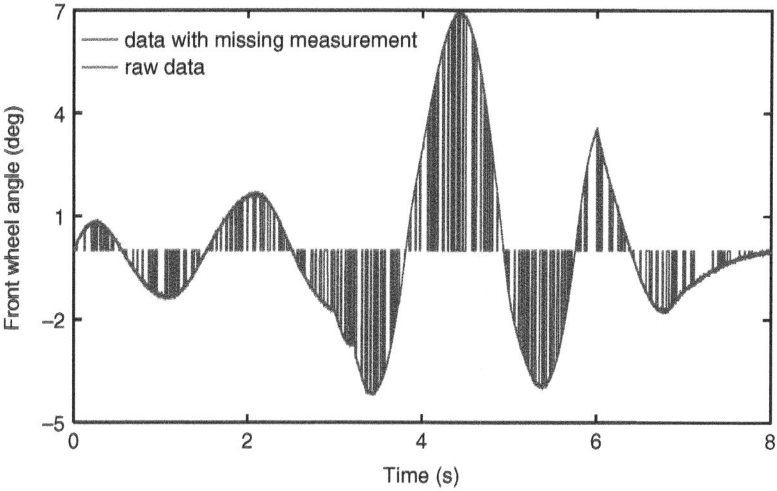

Figure 5.28 The front wheel angle on the opposite road.

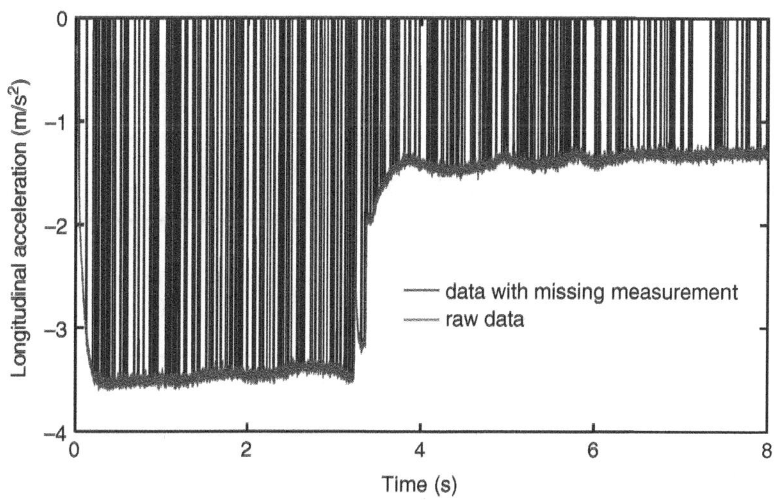

Figure 5.29 The longitudinal acceleration on the opposite road.

the critical role of TRFC in governing vehicle dynamics, particularly in diverse road conditions where friction coefficients vary significantly.

Figure 5.31(a)–(d) describe the estimated longitudinal tire force from different methods. For the longitudinal tire force estimation, the parameters utilized in the UKF/FTUKF are $Q = \text{diag}[500,500,500\ 500,1,\ 1,\ 1,\ 1]$ and the other parameters, and they are the same as on ice roads. Upon comparison, it is evident that the

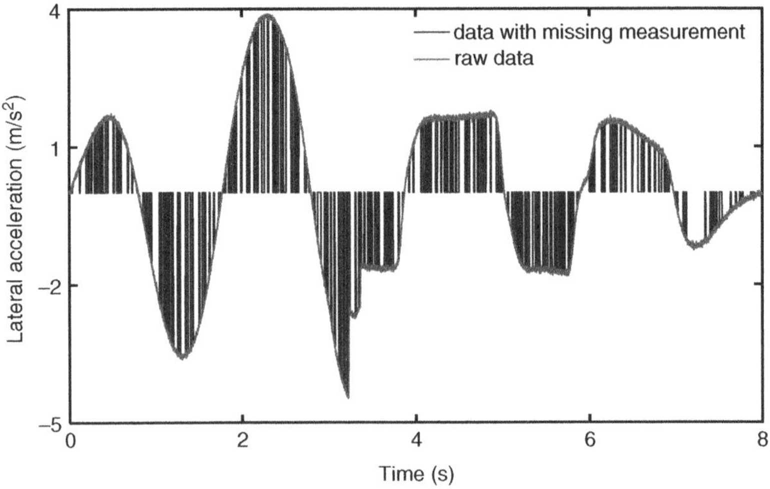

Figure 5.30 The lateral acceleration on the opposite road.

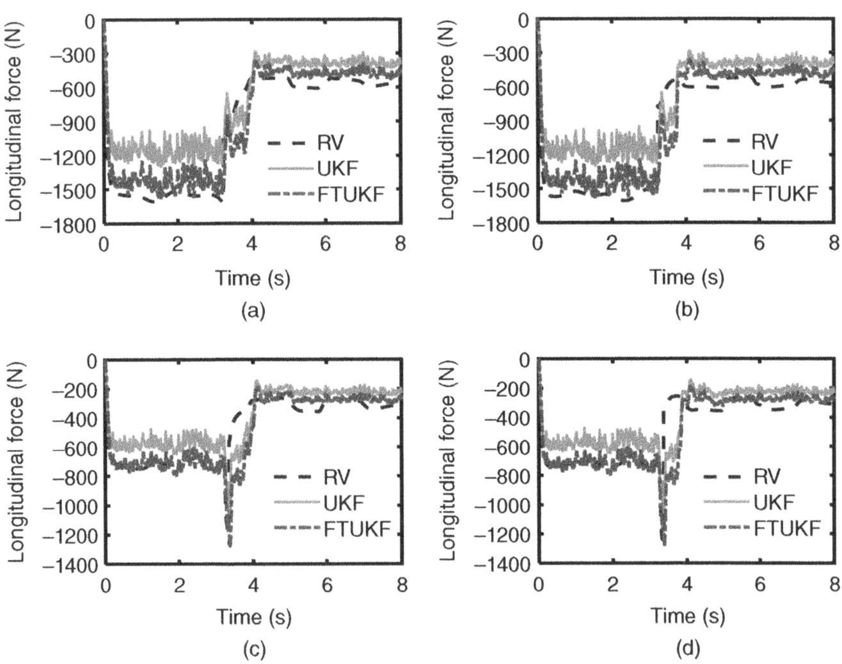

Figure 5.31 The longitudinal tire forces of four wheels on the opposite road.

Table 5.7 RMSE of longitudinal tire forces estimation on the opposite road.

Symbol	F_{x1}	F_{x2}	F_{x3}	F_{x4}
UKF	291.73	299.06	137.54	154.58
FTUKF	155.44	176.21	122.32	136.26

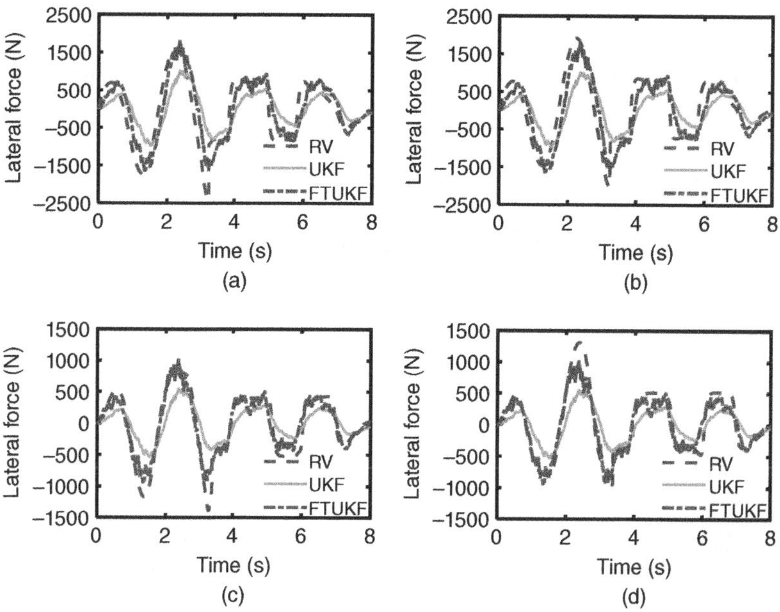

Figure 5.32 The lateral tire forces of four wheels on the opposite road.

FTUKF closely aligns with the reference value (RV), indicating superior accuracy in estimation. Conversely, the UKF exhibits considerable deviation from the RV in its estimated curve. In scenarios where significant longitudinal deceleration occurs, adjustments are made by setting a larger Q matrix. This modification aims to facilitate rapid tracking of the RV by the estimated curve. However, the consequence of increasing Q is an elevated level of fluctuation in the estimated curve. Hence, striking a balance and selecting an appropriate Q matrix becomes crucial to achieving accurate and stable estimation outcomes. Table 5.7 presents a comprehensive overview by detailing the RMSE obtained from various estimation methods. Comparing these metrics reveals that the RMSE values associated with FTUKF are consistently smaller than those of UKF. This disparity underscores FTUKF's enhanced capability for precise estimation, demonstrating its superiority in accuracy over UKF. The effectiveness of FTUKF in minimizing estimation

Table 5.8 RMSE of lateral tire forces estimation on the opposite road.

Symbol	F_{y1}	F_{y2}	F_{y3}	F_{y4}
UKF	636.14	626.63	323.75	325.57
FTUKF	397.90	389.95	162.97	172.10

errors can be attributed to its refined tuning and adaptive filtering mechanisms. By optimizing parameter settings and adapting dynamically to changing conditions, FTUKF mitigates inaccuracies associated with data fluctuations and variations in system dynamics. This robust performance makes FTUKF a preferred choice for applications demanding reliable and precise estimation of critical parameters such as longitudinal deceleration. Figure 5.32(a)–(d) illustrate the estimated lateral tire force from different approaches. The estimated curve of UKF is far from the RV. Table 5.8 displays the RMSE obtained from various methods. The results indicated that the RMSE value of FTUKF is lower than that of UKF.

Figure 5.33(a)–(d) depict the estimated TRFCs from different algorithms. Different from the first working condition, the lateral tire force-based estimation approach also quickly tracks the RV at first. This is because the steering maneuver is applied immediately at the beginning of the simulation. Table 5.9 displays the RMSEs obtained from various methods. The MMEM has a higher estimation accuracy than that of other methods. Figure 5.34 depicts the event-triggered times of the MMEM. Different from the first condition, we can see that the alternating occurrence is equal to 0 or 1. This is because a continuous steering maneuver is performed on the vehicle in this test. In summary, these findings emphasize the importance of advanced estimation techniques, such as FTUKF and MMEM, in enhancing the accuracy and responsiveness of tire force and TRFC estimations. By leveraging adaptive filtering strategies and multi-model approaches, these methods mitigate the limitations observed in traditional UKF-based approaches, thereby improving overall performance and reliability in dynamic operational scenarios.

5.4.3.2 The Hardware-in-the-Loop Test

To further verify the reliability of the algorithm, we performed hardware-in-the-loop testing, and the whole system is shown in Fig. 5.35. The vehicle dynamics model and the proposed algorithms are run in the dSPACE device. The dSPACE device is a sophisticated hardware platform widely utilized in automotive and aerospace industries for rapid control prototyping, hardware-in-the-loop (HIL) simulation, and testing of embedded systems. This versatile tool plays a crucial

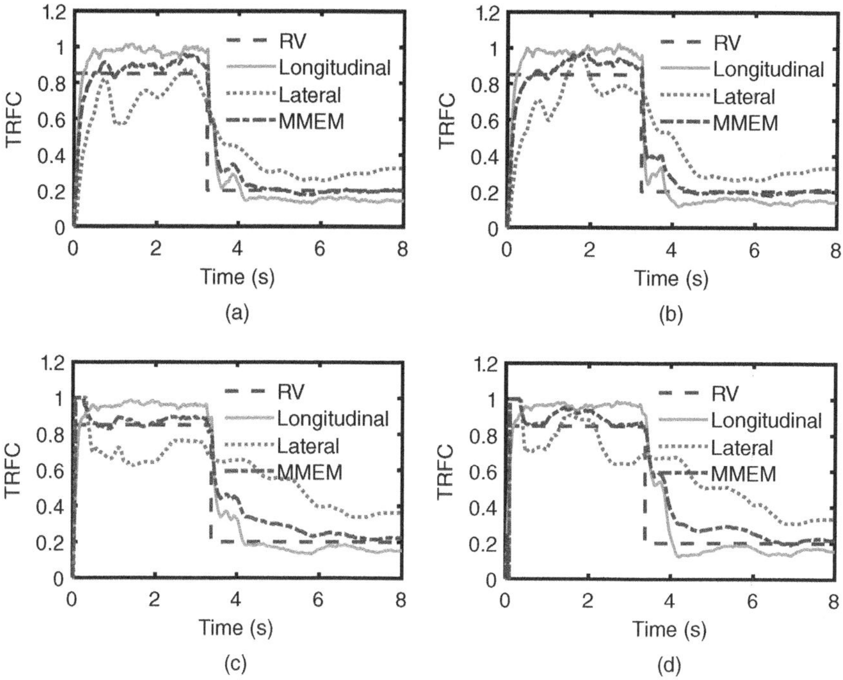

Figure 5.33 The TRFCs of four wheels on the opposite road.

Table 5.9 RMSE of TRFC estimation on the opposite road.

Symbol	$\overline{\mu}_1^\tau$	$\overline{\mu}_2^\tau$	$\overline{\mu}_3^\tau$	$\overline{\mu}_4^\tau$
Longitudinal	0.135	0.128	0.120	0.144
Lateral	0.197	0.213	0.261	0.261
MMEM	0.119	0.117	0.111	0.133

role in developing and validating complex electronic control units (ECUs) and software algorithms that govern vehicle dynamics, powertrain performance, and flight control systems. At its core, the dSPACE device integrates powerful real-time processors, input/output (I/O) modules, and comprehensive software tools tailored to facilitate efficient development and testing cycles. It serves as a bridge between virtual simulation environments and physical prototypes, enabling engineers to validate ECU functionalities under realistic operating conditions before deployment. Central to the dSPACE platform is its high-performance processor

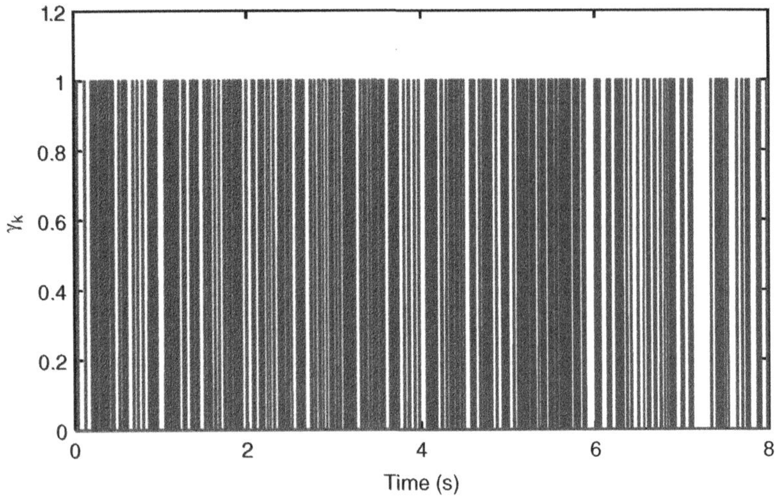

Figure 5.34 The event-triggered times on the opposite road.

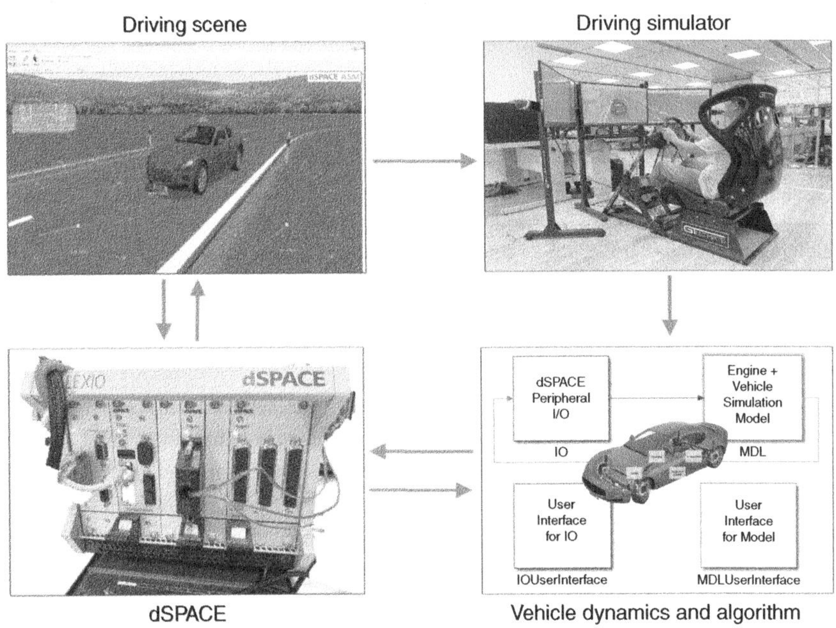

Figure 5.35 The hardware-in-the-loop system.

capable of executing complex control algorithms in real-time. This capability ensures that simulations closely emulate actual system behaviors, facilitating accurate performance evaluation and optimization. The dSPACE devices are equipped with versatile I/O modules that interface with sensors, actuators, and vehicle networks. These modules facilitate bidirectional communication between the ECU under test and the simulated environment or physical test setup, enabling seamless integration and data exchange. The dSPACE software environment provides a comprehensive suite of tools for model-based development, simulation setup, and test automation. Engineers can design control algorithms using industry-standard modeling languages like MATLAB/Simulink and seamlessly deploy them onto the dSPACE platform for real-time execution. One of the primary uses of dSPACE devices is in rapid control prototyping, where engineers iteratively develop and refine control strategies in a simulated environment. HIL simulation extends this capability by interfacing the ECU with virtual models of the vehicle or aircraft systems, allowing for comprehensive validation of ECU responses to various operational scenarios. Testing and Validation: Beyond development, dSPACE devices are instrumental in the validation phase of ECU development. Engineers can conduct systematic tests to verify the functionality, performance, and safety of embedded software and control algorithms before deployment in production vehicles or aircraft. The versatility of dSPACE devices spans across automotive applications such as powertrain control, chassis dynamics, and ADAS. In aerospace, they are used for flight control development, avionics testing, and simulation of mission-critical systems. The dSPACE devices support seamless integration with existing development workflows and tools, offering compatibility with industry standards and protocols. This facilitates collaborative development efforts and enhances interoperability across different stages of the product lifecycle. In conclusion, the dSPACE device stands as a cornerstone in modern engineering practices, enabling efficient development, validation, and testing of embedded systems in the automotive and aerospace industries. Its robust hardware capabilities, coupled with advanced software tools, empower engineers to innovate and deliver safe, reliable, and high-performance electronic control solutions for next-generation vehicles and aircraft.

In the conducted experiment, the driver actively engages in real-time vehicle control using a driving simulator equipped with dynamic driving scenario animations. This setup allows for immediate validation of the output curves generated by the estimation algorithm, assessing its operational efficacy under varying conditions. Specifically, human maneuvers are employed to simulate vehicle movements on asphalt roads, providing a practical test bed for evaluating the performance of a fault-tolerant estimation framework. During these tests, the TRFC is maintained at 0.85, mirroring parameters previously established for ice road conditions. This consistency ensures a comparative evaluation across different road surfaces. The experimental setup captures and analyzes key dynamics, including front wheel angle, longitudinal acceleration, and lateral acceleration, presented in Figs. 5.36–5.38. These figures serve to visually depict how the vehicle responds to driver inputs and varying road conditions. They provide insights into the stability, handling characteristics, and

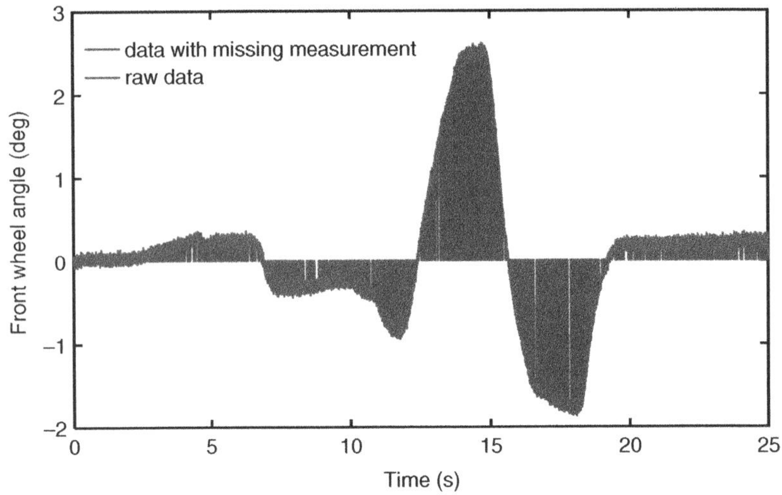

Figure 5.36 The front wheel angle on the asphalt road.

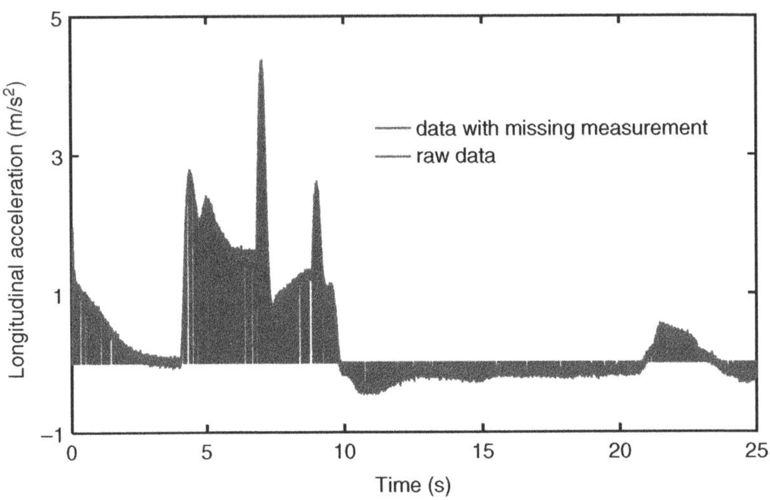

Figure 5.37 The longitudinal acceleration on the asphalt road.

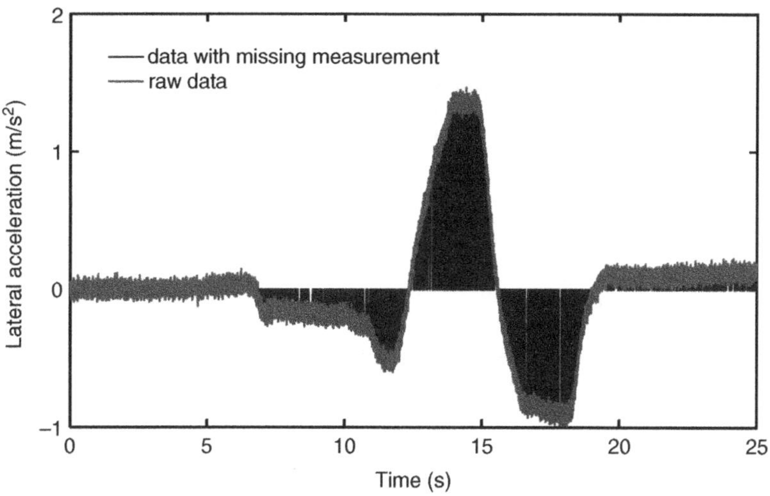

Figure 5.38 The lateral acceleration on the asphalt road.

overall performance of the fault-tolerant estimation framework under realistic driving scenarios. By correlating observed data with algorithm outputs in real time, engineers can effectively gauge the accuracy and reliability of the estimation framework in dynamically adapting to changes in TRFC and other driving parameters. Such experiments are crucial for validating the robustness of algorithms designed to enhance vehicle control systems, ensuring they can effectively operate under diverse and challenging real-world conditions. The integration of driving simulators and advanced estimation algorithms represents a significant step toward developing safer, more responsive automotive technologies capable of adapting to unpredictable road environments.

Due to the inherent variability in human operation compared to machine control, significant fluctuations can occur in acceleration and steering wheel angle curves during the driving simulation. These fluctuations reflect the dynamic nature of human driving behaviors and underscore the challenges in maintaining consistent vehicle control inputs. Figure 5.39(a)–(d) provide detailed representations of estimated longitudinal tire forces derived from different estimation methods. Notably, the FTUKF exhibits a closer alignment with the reference value (RV) compared to other methods. In contrast, the UKF shows noticeable deviation from the RV, primarily attributed to its limitations in handling data loss effectively. A quantitative assessment of estimation accuracy is presented in Table 5.10, where RMSEs obtained from various methods are detailed. These metrics demonstrate that the RMSE values associated with FTUKF are consistently lower than those of UKF, indicating superior estimation accuracy. This superiority

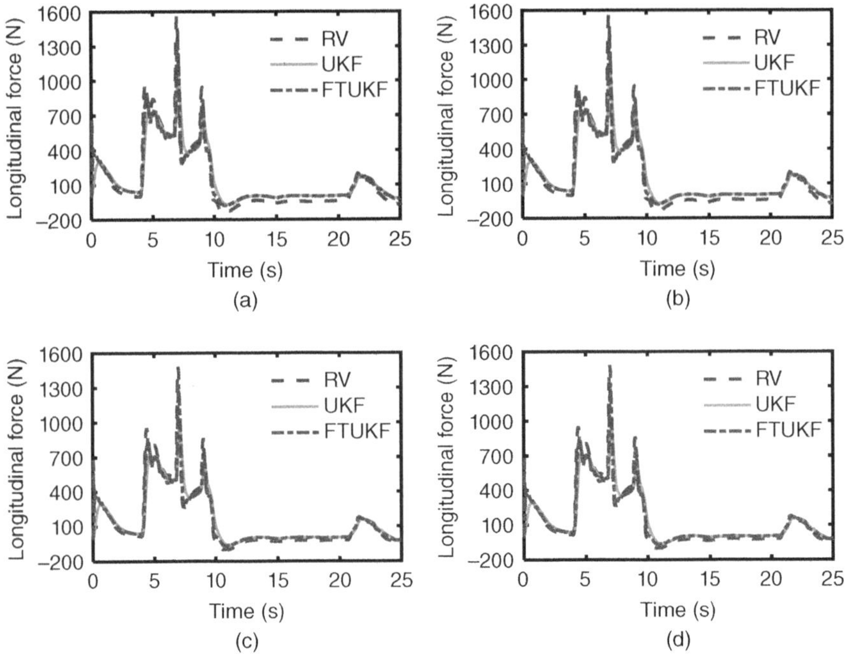

Figure 5.39 The longitudinal tire forces of four wheels on the asphalt road.

Table 5.10 RMSE of longitudinal tire forces estimation on the asphalt road.

Symbol	F_{x1}	F_{x2}	F_{x3}	F_{x4}
UKF	117.53	117.50	107.56	107.47
FTUKF	68.21	68.23	57.64	57.62

underscores FTUKF's capability to mitigate the impact of data fluctuations and uncertainties inherent in human-driven scenarios, thereby enhancing reliability in estimating longitudinal tire forces. Such findings highlight the critical role of advanced estimation algorithms, like FTUKF, in improving the performance and robustness of vehicle control systems under real-world driving conditions. By integrating adaptive filtering strategies and optimizing parameter settings, FTUKF ensures more precise and responsive estimation of key vehicle dynamics, contributing to safer and more efficient automotive technologies.

Figure 5.40(a)–(d) provide graphical representations of estimated lateral tire forces generated by different methodologies. Notably, the UKF shows significant

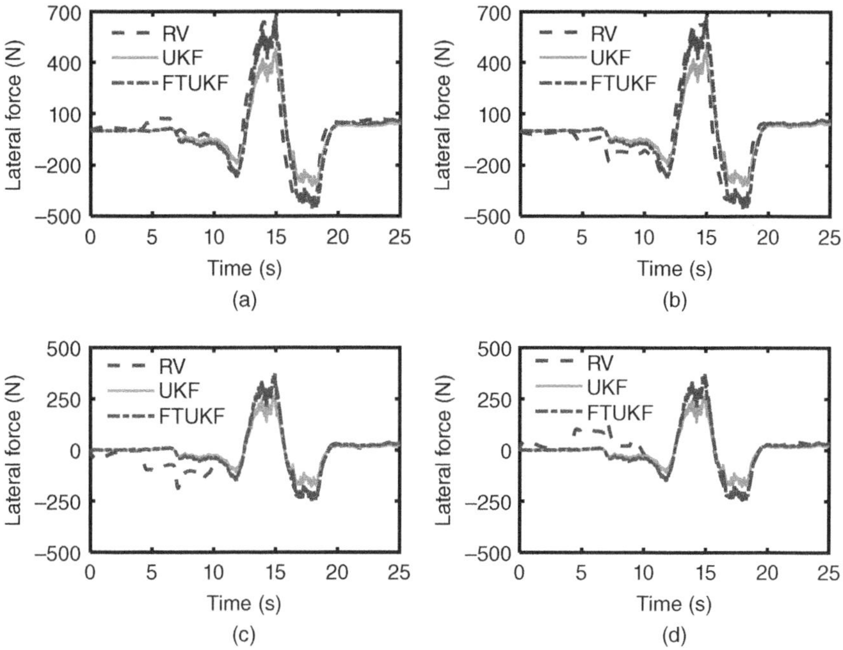

Figure 5.40 The lateral tire forces of four wheels on the asphalt road.

Table 5.11 RMSE of lateral tire forces estimation on the asphalt road.

Symbol	F_{y1}	F_{y2}	F_{y3}	F_{y4}
UKF	95.28	99.67	58.35	58.73
FTUKF	73.89	75.38	45.48	46.50

deviation from the RV, indicating challenges in accurately estimating lateral forces under varying conditions. This disparity underscores the need for advanced estimation techniques to improve alignment with actual measurements. Table 5.11 presents a quantitative analysis of estimation accuracy, detailing the RMSE obtained from each method. Comparisons reveal that the RMSE values associated with the FTUKF are consistently lower than those of the UKF, highlighting FTUKF's superior performance in accurately estimating lateral tire forces. Moving to Fig. 5.41(a)–(d), these graphs illustrate the estimated TRFC derived from different algorithms across varied driving conditions. Unlike previous conditions, where both longitudinal and lateral force-based estimations quickly tracked

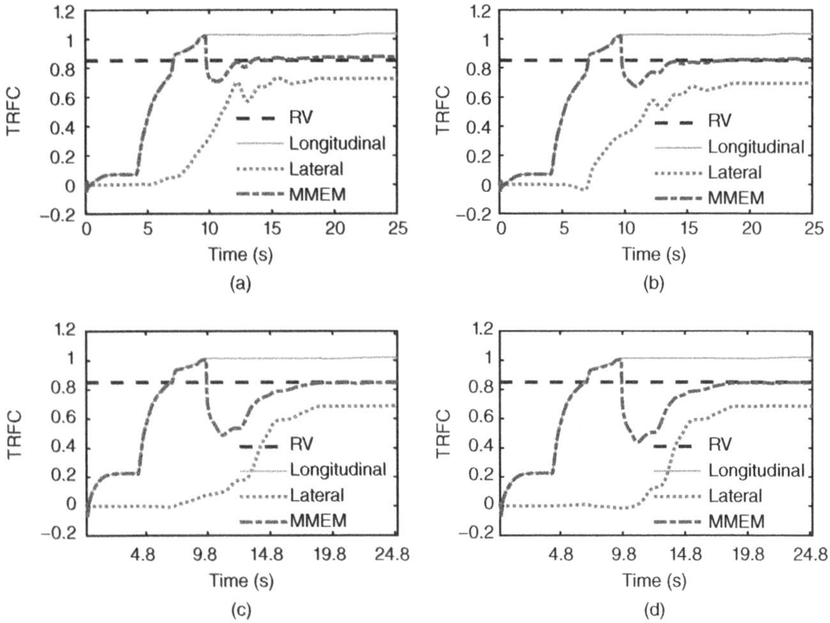

Figure 5.41 The TRFCs of four wheels on the asphalt road.

the RV, the initial 4 seconds show minimal longitudinal and lateral excitations. Consequently, the estimation curves for TRFC based on these forces do not immediately converge to the RV during this period. Subsequently, the longitudinal tire force-based estimation method begins to closely track the RV after 4 seconds, reflecting the initiation of significant vehicle dynamics. In contrast, the lateral tire force-based method shows delayed tracking, aligning with the RV only after 12 seconds due to the commencement of larger steering maneuvers. The MMEM emerges as a dynamic approach capable of adjusting weights between longitudinal and lateral models based on corresponding errors, thereby refining TRFC estimation accuracy. This adaptive strategy enables MMEM's estimation curves to converge effectively toward the RV over time, demonstrating its robust performance in integrating multiple model outputs for enhanced accuracy.

Table 5.12 provides a comprehensive overview of the RMSE obtained from various estimation methods, highlighting the superior accuracy of the MMEM compared to alternative approaches. The data underscores MMEM's capability to achieve more precise estimations across different operational conditions. Figure 5.42 illustrates the event-triggered occurrences documented by MMEM during testing. These events, marked by increased frequency during vehicle acceleration, steering maneuvers, or deceleration, demonstrate the efficacy of the proposed event-driven mechanism in responding dynamically to changing

Table 5.12 RMSE of TRFC estimation on the asphalt road.

Symbol	$\overline{\mu}_1^T$	$\overline{\mu}_2^T$	$\overline{\mu}_3^T$	$\overline{\mu}_4^T$
Longitudinal	0.3505	0.3519	0.3102	0.3156
Lateral	0.5292	0.5372	0.6144	0.6153
MMEM	0.3255	0.3215	0.3042	0.3021

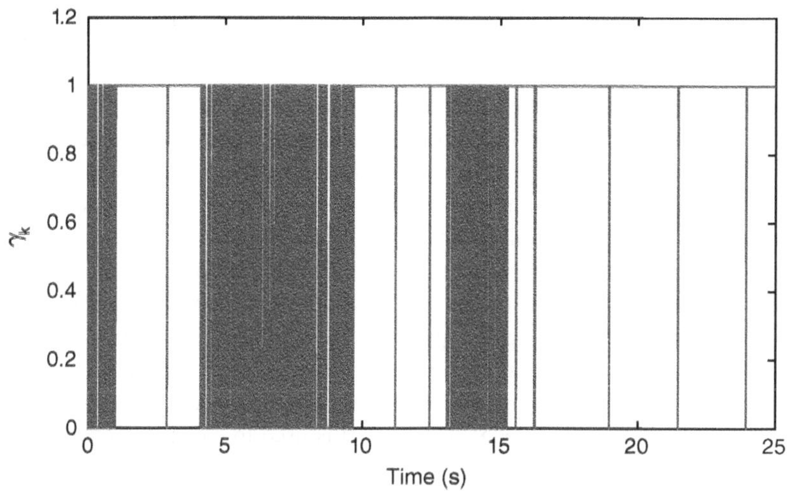

Figure 5.42 The event-triggered times on the asphalt road.

driving conditions. This responsiveness ensures that the estimation framework can adjust promptly to fluctuations in vehicle dynamics, thereby minimizing the impact of data loss and enhancing overall reliability. The dynamic update strategy employed by MMEM proves pivotal in reducing the potential effects of data loss on estimation accuracy. By adapting weights and model outputs based on real-time conditions, MMEM optimizes its estimation process, ensuring robust performance across diverse road surfaces and driving scenarios. The successful tracking of the RV across all three distinct road types further underscores the adaptability and effectiveness of the proposed estimation method. This capability highlights MMEM's robustness in accommodating varying environmental factors and vehicle behaviors, reinforcing its suitability for applications requiring accurate and reliable estimation of critical parameters like TRFC. In conclusion, MMEM stands out as a robust solution for enhancing estimation accuracy in dynamic automotive environments. Its event-driven approach, coupled with adaptive modeling techniques, ensures that the estimation framework remains responsive and effective under challenging conditions, ultimately contributing to safer and more efficient vehicle operations.

5.5 Fundamental Estimation for Tire–Road Friction Coefficient: A Model-Based Learning Framework

In this section, we propose a novel estimation scheme for TRFC that combines model-based and data-driven approaches, specifically designed to handle scenarios with unknown measurement noise and data loss. The proposed scheme involves several key components and steps to ensure accurate and robust TRFC estimation. First, an event-triggered mechanism is developed to detect and assess instances of measurement data loss. This mechanism continuously monitors the incoming sensor data and identifies moments when data is either missing or unreliable. Once a data loss event is detected, the system responds accordingly to mitigate its impact on the overall estimation process. To process the sensor data, the event-triggered mechanism is integrated with the CKF, resulting in an ETCKF. The ETCKF leverages the robustness of the CKF in handling nonlinearities and combines it with the event-triggered approach to enhance its resilience against data loss. This combination ensures that the sensor data is processed accurately and reliably, even under challenging conditions. Next, the processed sensor data is input into a nonlinear tire model. This model is used to compute normalized tire forces, which are essential for understanding the interaction between the tires and the road surface. The nonlinear tire model accounts for various dynamic factors affecting tire performance, providing a more precise representation of the actual forces at play. The final step involves the use of an EKFNet to estimate the TRFC. The EKFNet is an advanced hybrid system composed of an EKF and a multi-layer neural network. The EKF component leverages a vehicle model to track the system's state and dynamics accurately. Meanwhile, the neural network component is trained to learn complex patterns and relationships from the tire force data and other relevant inputs. By combining the strengths of the EKF and the neural network, the EKFNet effectively estimates the TRFC. The EKF provides a robust framework for state estimation and noise reduction, while the neural network enhances the system's ability to capture nonlinearities and intricate dependencies that might be challenging to model explicitly. Together, they offer a comprehensive solution that improves estimation accuracy and robustness. In summary, this section outlines a sophisticated TRFC estimation scheme that integrates model-based and data-driven methodologies to address the challenges of unknown measurement noise and data loss. The proposed approach employs an event-triggered mechanism for data loss detection, an

ETCKF for sensor data processing, a nonlinear tire model for force computation, and an EKFNet for final TRFC estimation. This innovative combination ensures that the estimation process is both accurate and resilient, providing reliable TRFC estimates under diverse and challenging conditions.

The overall estimation architecture is comprehensively illustrated in Fig. 5.43. This architecture is designed to handle the complexities of obtaining and processing signals from vehicles traveling on the road, where measurement noise and data loss are common challenges. These signals are derived from various vehicle subsystems, such as the steering and braking systems, and are transmitted via the vehicle's communication bus. However, these signals often suffer from noise interference and incomplete data due to factors like electromagnetic interference. The first step in this architecture involves normalizing the acquired measurement signals, which typically have different units. Normalization ensures that the data is standardized and comparable across different signal types. Once the data is normalized, an event-triggered rule is applied to assess the extent of data loss. This rule helps determine whether the measurement signals are complete or if some data points are missing. In the event of data loss, the architecture employs only the time update step from the CKF to perform the necessary computations. This approach allows the system to continue processing and updating estimates even when complete data is not available. Conversely, when no data loss occurs, the standard CKF is utilized for data filtering, ensuring that the most accurate and reliable estimates are obtained from the available signals. The integration of the CKF with the event-triggered rule forms what is known as the ETCKF. The data processed by the ETCKF is then input into the Dugoff tire model. This model is used to compute the normalized longitudinal and lateral tire forces, which are critical for understanding the interaction between the vehicle tires and the road surface. The Dugoff tire model accounts for various dynamic factors and provides a precise representation of tire forces under different driving conditions. Following the computation of tire forces, the EKFNet comes into play. The EKFNet is a sophisticated hybrid system that combines the strengths of an EKF and a multi-layer neural network. It performs real-time estimation of the TRFC based on the tire force information and the vehicle model. The EKF component tracks the system's state and dynamics, reducing noise and improving estimation accuracy. Meanwhile, the neural network component captures complex, nonlinear relationships within the data, enhancing the overall robustness of the estimation process.

5.5.1 Vehicle Model and Problem Statement

As shown in Fig. 5.44, the four-wheel vehicle model is employed. The impact of air resistance and the suspension system is disregarded. The front wheels share identical steering angles, while the rear wheels do not have steering capabilities.

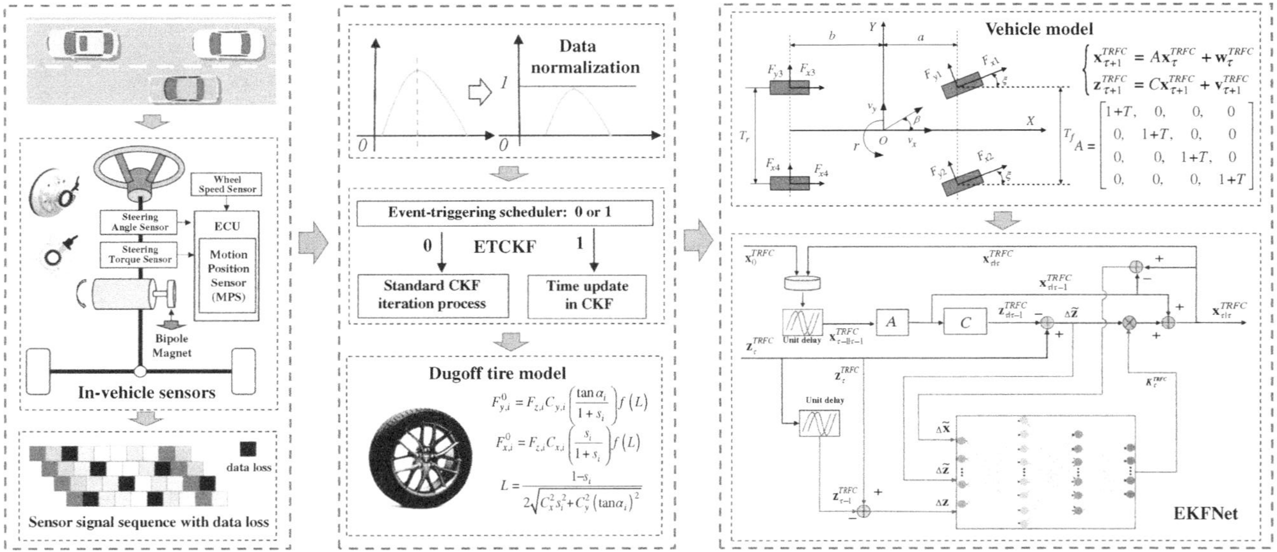

Figure 5.43 The fundamental estimation scheme.

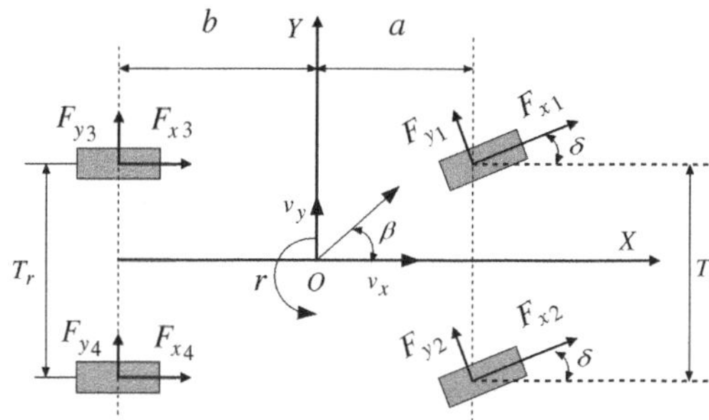

Figure 5.44 The four-wheel model.

Additionally, the center of vehicle gravity is presumed to coincide with the origin of the coordinate system. The relevant equations for the four-wheel vehicle model are as follows.

$$a_x = [(F_{x1} + F_{x2})\cos\delta + F_{x3} + F_{x4} - (F_{y1} + F_{y2})\sin\delta]/m \tag{5.102}$$

$$a_y = [(F_{x1} + F_{x2})\sin\delta + F_{y3} + F_{y4} + (F_{y1} + F_{y2})\cos\delta]/m \tag{5.103}$$

$$\begin{aligned} \dot{r}I_z &= a[(F_{x1} + F_{x2})\sin\delta + (F_{y1} + F_{y2})\cos\delta] - b(F_{y3} + F_{y4}) \\ &\quad - \frac{T_f}{2}(F_{x1} - F_{x2})\cos\delta + \frac{T_f}{2}(F_{y1} - F_{y2})\sin\delta - \frac{T_r}{2}(F_{x3} - F_{x4}) \end{aligned} \tag{5.104}$$

where β, v_x, v_y denote the sideslip angle, longitudinal vehicle velocity, and lateral vehicle velocity, m is the vehicle mass, the parameters a and b denote the distance from the center of gravity to the front axle and rear axle, r is the yaw rate, δ is the front wheel angle, T_f andT_r represent front track width and rear track width, ω_i is the wheel rotational speed, $i = 1, 2, 3, 4$, which correspond to the left-front, right-front, left-rear, and right-rear wheels, respectively. R_i is the wheel radius, F_{xi}, F_{yi} are longitudinal tire forces, and lateral tire forces, I_z is the inertia moment about the vehicle vertical axis.

The correlation between tire motion and forces is delineated by the Dugoff tire model. The Dugoff tire model is a widely used analytical tool in vehicle dynamics and tire behavior analysis. It provides a comprehensive representation of the relationship between tire forces and slip ratios under varying load conditions. Developed by Harry Dugoff in the 1960s, this model offers a significant improvement over simpler linear models by incorporating the nonlinear behavior of tires, which is critical for accurate simulations of vehicle performance, especially under extreme driving conditions. In the Dugoff tire model, the tire–road interaction is characterized by two primary components: longitudinal and lateral forces. These forces are influenced by factors such as slip ratios, slip angles, normal load, and road-tire friction coefficients. The model uses a set of equations to describe how these forces change as a function of the slip ratios and angles, providing a more realistic depiction of tire behavior compared to basic linear models. One of the key features of the Dugoff model is its ability to account for the effects of combined slip. This means it can simultaneously consider both longitudinal slip (such as during acceleration or braking) and lateral slip (such as during cornering), which is essential for accurately modeling scenarios where these conditions occur together. The model does this by introducing a dimensionless parameter that scales the forces based on the combined slip conditions, effectively capturing the complex interactions between the tire and the road surface. The equations of the Dugoff model start with the calculation of the slip ratios and angles, which are then used to determine the combined slip parameter. The longitudinal and lateral forces are then computed using these parameters, normal load, and the friction coefficient. The model typically assumes a parabolic relationship between these forces and the slip conditions, reflecting the nonlinear nature of tire behavior. This approach allows for a more precise estimation of the forces that the tires can generate, which is crucial for accurate vehicle dynamic simulations. Applications of the Dugoff tire model are extensive in the fields of automotive engineering and research. It is particularly valuable in the design and testing of vehicle control systems such as ABS, ESC, and TCS. By providing a detailed and realistic representation of tire behavior, the Dugoff model helps engineers predict how vehicles will respond to various driving inputs and conditions, leading to the development of safer and more reliable automotive systems. Moreover, the Dugoff model is often used in combination with other vehicle models in simulation environments to enhance the accuracy of the overall vehicle dynamics simulation. This integration is critical for tasks such as vehicle design optimization, performance testing, and safety analysis. The model's ability to handle nonlinearities and combined slip conditions makes it an indispensable tool in the toolkit of automotive engineers and researchers. In conclusion, the Dugoff tire model stands out for its detailed and accurate representation of tire–road interactions, particularly under combined slip conditions. Its application in vehicle dynamics modeling and control system design is essential for developing advanced automotive technologies that ensure better handling, safety, and performance of vehicles under a wide range of driving scenarios. By capturing the nonlinear nature of tire forces, the Dugoff model contributes significantly to the field of automotive engineering, driving advancements in vehicle safety and performance.

The specific equations are as follows:

$$F_{x,i} = \mu_i F_{x,i}^0 = \mu_i F_{z,i} C_{x,i} \left(\frac{s_i}{1+s_i} \right) f(L) \tag{5.105}$$

$$F_{y,i} = \mu_i F_{y,i}^0 = \mu_i F_{z,i} C_{y,i} \left(\frac{\tan \alpha_i}{1+s_i} \right) f(L) \tag{5.106}$$

$$f(L) = \begin{cases} L(2-L), & L < 1 \\ 1 & \mathrm{L} \geq 1 \end{cases} \tag{5.107}$$

$$L = \frac{1 - s_i}{2\sqrt{C_x^2 s_i^2 + C_y^2(\tan \alpha_i)^2}} \tag{5.108}$$

$$F_{z,1} = \frac{mgb}{2(a+b)} - \frac{ma_x h}{2(a+b)} - \frac{ma_y h}{T_f} \cdot \frac{b}{a+b} \tag{5.109}$$

$$F_{z,2} = \frac{mgb}{2(a+b)} - \frac{ma_x h}{2(a+b)} + \frac{ma_y h}{T_f} \cdot \frac{b}{a+b} \tag{5.110}$$

$$F_{z,3} = \frac{mga}{2(a+b)} + \frac{ma_x h}{2(a+b)} - \frac{ma_y h}{T_r} \cdot \frac{a}{a+b} \tag{5.111}$$

$$F_{z,4} = \frac{mga}{2(a+b)} + \frac{ma_x h}{2(a+b)} + \frac{ma_y h}{T_r} \cdot \frac{a}{a+b} \tag{5.112}$$

$$s_i = \mathrm{sgn}(v_x - R\omega_i)\frac{|v_x - R\omega_i|}{\max(R\omega_i, v_x)} \tag{5.113}$$

$$\alpha_1 = \delta - \arctan\left(\frac{v_y + ar}{v_x - T_f r/2}\right) \tag{5.114}$$

$$\alpha_2 = \delta - \arctan\left(\frac{v_y + ar}{v_x + T_f r/2}\right) \tag{5.115}$$

$$\alpha_3 = -\arctan\left(\frac{v_y - br}{v_x - T_r r/2}\right) \tag{5.116}$$

$$\alpha_4 = -\arctan\left(\frac{v_y - br}{v_x + T_r r/2}\right) \tag{5.117}$$

5.5.2 Methodology

The overall scheme, as illustrated in Fig. 5.45, begins with the calculation of normalized longitudinal and lateral forces, which are essential for accurate vehicle dynamics analysis. To achieve this, eqs. (5.105)–(5.117) require various types of onboard sensor information, such as acceleration, yaw rate, and other relevant data. However, it is important to note that these sensor data are typically subject to measurement noise and may also be prone to data loss due to factors like electromagnetic interference or sensor malfunctions. Directly inputting noisy and potentially incomplete data into the filtering process can significantly degrade the performance of the filter, potentially leading to erroneous estimations or even divergence of the filtering process. To address these challenges, a robust switching rule is devised within the scheme. This switching rule operates as follows: In the event of data loss, only the time update equation of the filtering process is employed. The time update equation allows the filter to continue updating its state estimates based on the model's predictions without relying

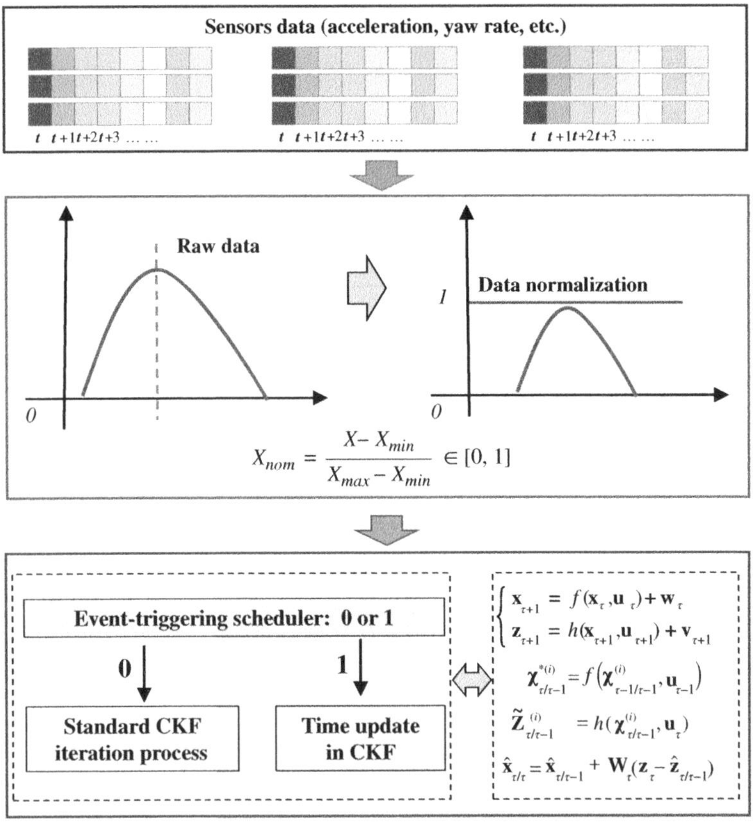

Figure 5.45 The ETCKF scheme.

on the incomplete measurement data. This helps maintain the stability and continuity of the estimation process. When no data loss is detected, the complete filtering computation process is executed, utilizing both the time update and measurement update equations. This ensures that the filter fully leverages the available measurement data to correct and refine its state estimates. For the foundational filter within this scheme, the CKF is chosen due to its superior ability to accurately capture nonlinear relationships. The CKF employs a deterministic sampling approach, which involves selecting a set of sigma points around the mean state. These sigma points are then propagated through the nonlinear system equations, capturing the effects of nonlinearity more effectively than traditional linearization methods like the EKF. The CKF's deterministic sampling approach ensures that the nonlinearities inherent in vehicle dynamics are more accurately represented, leading to more reliable state estimations. This is particularly crucial for applications involving complex vehicle maneuvers and varying road conditions, where linear approximations may fall short.

5.5.2.1 Event-Triggering Scheduler

To process the data, we first establish the state-space model as follows.

$$\begin{cases} \mathbf{x}_{\tau+1} = f(\mathbf{x}_\tau, \mathbf{u}_\tau) + \mathbf{w}_\tau \\ \mathbf{z}_{\tau+1} = h(\mathbf{x}_{\tau+1}, \mathbf{u}_{\tau+1}) + \mathbf{v}_{\tau+1} \end{cases} \tag{5.118}$$

$$\mathbf{x} = [v_x, v_y, r, \dot{r}, a_x, a_y, \omega_1, \omega_2, \omega_3, \omega_4, \delta]^T$$

where $f(\cdot)$ denotes state transition function, $\mathbf{z}$ is the measurement vector, $\mathbf{u}$ is the input vector, $h(\cdot)$ denotes measurement transition function, $\mathbf{x}$ represents the state vector. As direct data processing is conducted here, the state transition function, when discretized, is represented by an identity matrix. $\mathbf{w}$ process noise, and $\mathbf{v}$ measurement noise.

We establish the following event-triggering rules.

$$\gamma_\tau = \begin{cases} 1 & \textit{if}\ (\mathbf{z}_\tau - \mathbf{z}_{\tau-1})^T(\mathbf{z}_\tau - \mathbf{z}_{\tau-1}) > \psi \\ 0 & \textit{otherwise} \end{cases} \tag{5.119}$$

Based on the description of the state-space equations, we can infer that the sensor signals in the vector $\mathbf{z}$ are identical to those in the vector $\mathbf{x}$. However, the signals in the vector $\mathbf{z}$ are augmented by the addition of measurement noise. $\mathbf{z}_\tau$ is the normalized measurement vector at time τ, $\mathbf{z}_{\tau-1}$ represents the normalized measurement vector before time τ, ψ is the event-triggered threshold. Let γ_τ be the indicator for event triggering. When $\gamma_\tau = 1$, it indicates a severe data loss situation. During this period, only the CKF's time update process is utilized for data processing, and the measurement update process is excluded from the iterative loop to mitigate the impact of data loss. When $\gamma_\tau = 0$, normal CKF filtering is applied to measurement signals with noise. The CKF is an advanced state estimation algorithm designed to handle nonlinear systems more effectively than traditional filters like the KF and EKF. Unlike the EKF, which linearizes nonlinear systems using first-order Taylor series expansions, the CKF employs a deterministic sampling approach based on cubature integration to accurately approximate the Gaussian weighted integrals in Bayesian filtering. This method ensures a precise representation of state and measurement distributions by selecting symmetrically placed cubature points around the mean state estimate, capturing the second-order moments of the state distribution. The CKF operates through initialization, cubature point selection, time update (prediction), and measurement update (correction) steps, providing more accurate estimates for nonlinear systems by propagating uncertainties through nonlinear functions. Its ability to handle significant nonlinearities and Gaussian noise assumptions makes it particularly useful in aerospace, automotive systems, robotics, and sensor fusion applications. The CKF offers higher accuracy and robustness in state estimation, reducing the risk of divergence under nonlinear conditions, and achieves efficiency comparable to particle filters with fewer computational resources, making it suitable for real-time applications.

The procedural steps of the CKF are given by

1) Initialization:

$$\hat{\mathbf{x}}_0 = E(\mathbf{x}_0) \tag{5.120}$$

$$\mathbf{P}_0 = E\left[(\mathbf{x}_0 - \hat{\mathbf{x}}_0)(\mathbf{x}_0 - \hat{\mathbf{x}}_0)^T\right] \tag{5.121}$$

In the eqs. (5.120, 5.121), E symbolizes the mathematical expectation and $\mathbf{P}$ denotes the covariance matrix.

The calculation of cubature sampling points, denoted as ξ_i, and corresponding weights, denoted as ℓ_i, can be determined through the following procedure:

$$\ell_i = \frac{1}{d}\ \xi_i = \sqrt{\frac{d}{2}}\left[\begin{pmatrix}1\\0\\\vdots\\0\end{pmatrix}\cdots\begin{pmatrix}0\\0\\\vdots\\1\end{pmatrix}\begin{pmatrix}-1\\0\\\vdots\\0\end{pmatrix}\cdots\begin{pmatrix}0\\0\\\vdots\\-1\end{pmatrix}\right] \quad i = 1, 2, \ldots d, d = 2n \tag{5.122}$$

In this context, d signifies the count of cubature points, and the dimension of $\mathbf{x}$ is denoted as n.

2) Time Update:

Singular value decomposition of $\mathbf{P}_{\tau-1/\tau-1}$

$$\mathbf{P}_{\tau-1/\tau-1} = \mathbf{U}\begin{bmatrix}\mathbf{S} & 0\\0 & 0\end{bmatrix}\boldsymbol{\Lambda}^T \tag{5.123}$$

where $\mathbf{S}$ is a diagonal matrix. Subsequently, the covariance matrix, denoted as $\mathbf{P}_{\tau-1/\tau-1}$, can be expressed as

$$\mathbf{P}_{\tau-1/\tau-1} = \mathbf{U}_{\tau-1/\tau-1}\mathbf{S}_{\tau-1/\tau-1}\boldsymbol{\Lambda}^T_{\tau-1/\tau-1} \tag{5.124}$$

Compute the cubature points

$$\boldsymbol{\chi}^{(i)}_{\tau-1/\tau-1} = \mathbf{S}_{\tau-1/\tau-1}\xi_i + \hat{\mathbf{x}}_{\tau-1/\tau-1} \tag{5.125}$$

where $\hat{\mathbf{x}}_{\tau-1/\tau-1}$ is the state prediction. $\boldsymbol{\chi}^{(i)}_{\tau-1/\tau-1}$ is the cubature point of $\hat{\mathbf{x}}_{\tau-1/\tau-1}$.

Update the propagated cubature points

$$\boldsymbol{\chi}^{*(i)}_{\tau/\tau-1} = f\left(\boldsymbol{\chi}^{(i)}_{\tau-1/\tau-1}, \mathbf{u}_{\tau-1}\right) \tag{5.126}$$

Update the $\mathbf{P}_{\tau/\tau-1}$ is as follows

$$\hat{\mathbf{x}}_{\tau/\tau-1} = \sum_{i=1}^{c} \omega_i \boldsymbol{\chi}^{*(i)}_{\tau-1/\tau-1} \tag{5.127}$$

$$\mathbf{P}_{\tau/\tau-1} = \sum_{i=1}^{c} \ell_i \boldsymbol{\chi}^{*(i)}_{\tau/\tau-1} \boldsymbol{\chi}^{*(i)T}_{\tau/\tau-1} - \hat{\mathbf{x}}_{\tau/\tau-1}\hat{\mathbf{x}}^{T}_{\tau/\tau-1} + \mathbf{Q}_{\tau-1} \tag{5.128}$$

where $\mathbf{Q}$ represents the covariance matrix of $\mathbf{w}$.

3) Measurement Update
Singular value decomposition of $\mathbf{P}_{\tau/\tau-1}$

$$\mathbf{P}_{\tau/\tau-1} = \mathbf{U}_{\tau/\tau-1}\mathbf{S}_{\tau/\tau-1}\boldsymbol{\Lambda}^{T}_{\tau/\tau-1} \tag{5.129}$$

Compute the cubature points

$$\boldsymbol{\chi}^{(i)}_{\tau/\tau-1} = \mathbf{S}_{\tau/\tau-1}\xi_i + \hat{\mathbf{x}}_{\tau/\tau-1} \tag{5.130}$$

Update the propagated cubature points

$$\tilde{\mathbf{Z}}^{(i)}_{\tau/\tau-1} = h\left(\boldsymbol{\chi}^{(i)}_{\tau/\tau-1}, \mathbf{u}_{\tau}\right) \tag{5.131}$$

The predicted measurement vector $\hat{\mathbf{z}}_{\tau/\tau-1}$, the innovation covariance matrix $\mathbf{P}_{zz,\tau/\tau-1}$, and the cross-covariance matrix $\mathbf{P}_{\mathbf{xz},\tau/\tau-1}$ are given by

$$\hat{\mathbf{z}}_{\tau/\tau-1} = \sum_{i=1}^{c} \omega_i \tilde{\mathbf{Z}}^{(i)}_{\tau-1/\tau-1} \tag{5.132}$$

$$\mathbf{P}_{\mathbf{zz},\tau/\tau-1} = \sum_{i=1}^{c} \ell_i \tilde{\mathbf{Z}}^{(i)}_{\tau/\tau-1} \tilde{\mathbf{Z}}^{(i)T}_{\tau/\tau-1} - \hat{\mathbf{z}}_{\tau/\tau-1}\hat{\mathbf{z}}^{T}_{\tau/\tau-1} + \mathbf{R}_{\tau} \tag{5.133}$$

$$\mathbf{P}_{\mathbf{xz},\tau/\tau-1} = \sum_{i=1}^{c} \omega_i \boldsymbol{\chi}^{(i)}_{\tau/\tau-1} \mathbf{Z}^{(i)T}_{\tau/\tau-1} - \hat{\mathbf{x}}_{\tau/\tau-1}\hat{\mathbf{z}}^{T}_{\tau/\tau-1} \tag{5.134}$$

where $\mathbf{R}$ is the covariance matrix of $\mathbf{v}$.
Update the gain matrix $\mathbf{W}_{\tau}$ and the state $\hat{\mathbf{x}}_{\tau/\tau}$

$$\mathbf{W}_{\tau} = \mathbf{P}_{\mathbf{xz},\tau/\tau-1}\mathbf{P}^{-1}_{\mathbf{zz},\tau/\tau-1} \tag{5.135}$$

$$\hat{\mathbf{x}}_{\tau/\tau} = \hat{\mathbf{x}}_{\tau/\tau-1} + \mathbf{W}_{\tau}(\mathbf{z}_{\tau} - \hat{\mathbf{z}}_{\tau/\tau-1}) \tag{5.136}$$

5.5.2.2 TRFC Estimation

The overall framework of TRFC is illustrated in Fig. 5.46. Initially, sensor data from the ETCKF is input into the Dugoff tire model for computation to obtain normalized longitudinal and lateral tire forces. Based on these tire force data, TRFC estimation is performed using EKFNet.

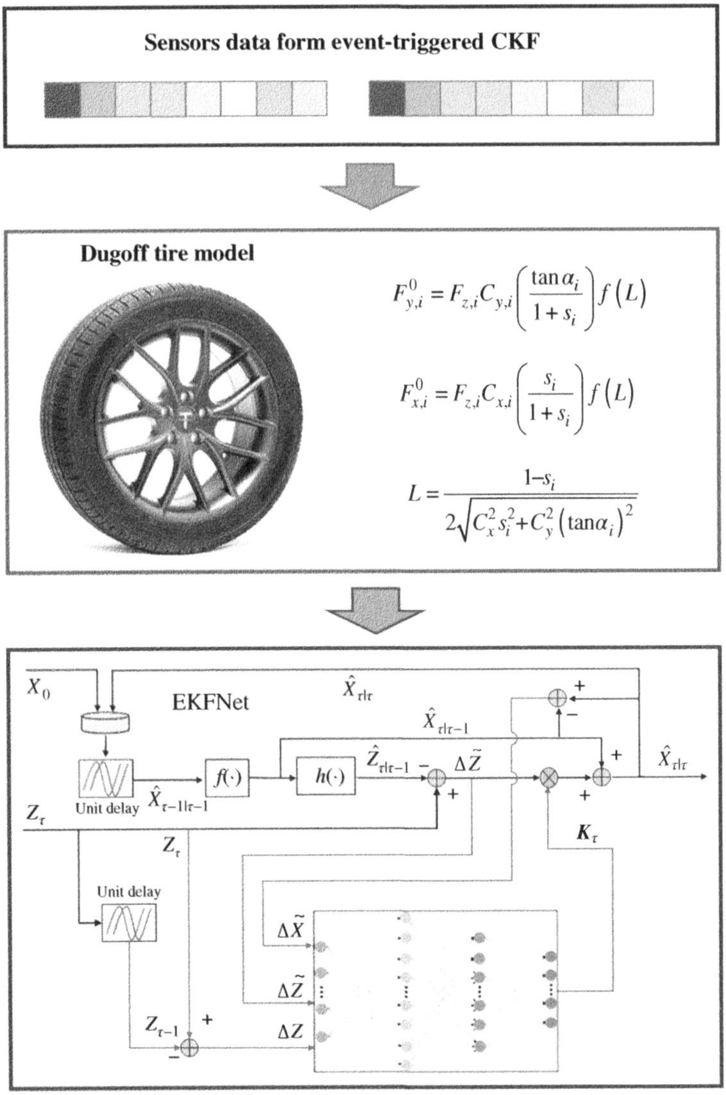

Figure 5.46 The TRFC estimation scheme.

The TRFC can be considered constant over a brief time interval. By utilizing the connection between $F^0_{x,i}$ and $F_{x,i}$, and $F^0_{y,i}$ and $F_{y,i}$ as defined in eqs (5.105, 5.106), the discrete mathematical model can be derived using eqs (5.102–5.104).

$$\begin{cases} \mathbf{x}^{TRFC}_{\tau+1} = A\mathbf{x}^{TRFC}_{\tau} + \mathbf{w}^{TRFC}_{\tau} \\ \mathbf{z}^{TRFC}_{\tau+1} = C\mathbf{x}^{TRFC}_{\tau+1} + \mathbf{v}^{TRFC}_{\tau+1} \end{cases} \tag{5.137}$$

$$\mathbf{x}^{TRFC} = [\mu_1, \mu_2, \mu_3, \mu_4]^T, \quad \mathbf{z}^{TRFC}_{\tau+1} = [a_x, a_y, \dot{r}]^T$$

$$A = \begin{bmatrix} 1+T, & 0, & 0, & 0 \\ 0, & 1+T, & 0, & 0 \\ 0, & 0, & 1+T, & 0 \\ 0, & 0, & 0, & 1+T \end{bmatrix}$$

$$C = \begin{bmatrix} C_{11}, & C_{12}, & C_{13}, & C_{14} \\ C_{21}, & C_{22}, & C_{23}, & C_{24} \\ C_{31}, & C_{32}, & C_{33}, & C_{34} \end{bmatrix}$$

$$\begin{cases} C_{11} = \left(F^0_{x,1}\cos\delta - F^0_{y,1}\sin\delta\right)/m \\ C_{12} = \left(F^0_{x,2}\cos\delta - F^0_{y,2}\sin\delta\right)/m \\ C_{13} = F^0_{x,3}/m \\ C_{14} = F^0_{x,4}/m \end{cases} \tag{5.138}$$

$$\begin{cases} C_{21} = \left(F^0_{x,1}\sin\delta + F^0_{y,1}\cos\delta\right)/m \\ C_{22} = \left(F^0_{x,2}\sin\delta + F^0_{y,2}\cos\delta\right)/m \\ C_{23} = F^0_{y,3}/m \\ C_{24} = F^0_{y,4}/m \end{cases} \tag{5.139}$$

$$\begin{cases} C_{31} = \dfrac{a\left(F^0_{x,1}\sin\delta + F^0_{y,1}\cos\delta\right) - \frac{T_f}{2}\left(F^0_{x,1}\cos\delta - F^0_{y,1}\sin\delta\right)}{I_z} \\ C_{32} = \dfrac{a\left(F^0_{x,2}\sin\delta + F^0_{y,2}\cos\delta\right) + \frac{T_f}{2}\left(F^0_{x,2}\cos\delta - F^0_{y,2}\sin\delta\right)}{I_z} \\ C_{33} = \dfrac{-\frac{T_r}{2}F^0_{x,3} - bF_{y3}}{I_z} \\ C_{34} = \dfrac{-\frac{T_r}{2}F^0_{x,4} - bF_{y4}}{I_z} \end{cases} \tag{5.140}$$

Next, EKFNet will be utilized to estimate the TRFC. EKFNet is an innovative combination of neural networks and the traditional EKF, designed to enhance the estimation process. The neural network component of EKFNet is employed to directly predict the Kalman gain, which significantly mitigates interference from unknown measurement noise. This approach is particularly advantageous because, while the ETCKF is used to process sensor information, traditional CKF relies heavily on prior knowledge of the statistical characteristics of noise. However, the noise characteristics of onboard sensors are not static; they change over time due to factors such as sensor aging and varying operational conditions. Moreover, the diversity of the sensors used in vehicle systems introduces varied noise characteristics, which complicates the estimation process. For instance, the calculation of the slip ratio involves wheel speed data obtained from wheel speed sensors, the yaw rate is derived from a gyroscope, and the front wheel steering angle signal is acquired from the steering wheel angle sensor. Each of these sensors has different noise profiles and levels of accuracy. This variability necessitates a robust method for handling noise uncertainties during TRFC estimation.

By integrating neural networks with the EKF, EKFNet leverages the strength of neural networks in capturing complex, nonlinear relationships and adapting to changing noise characteristics. The neural network's ability to predict the Kalman gain dynamically ensures that the filter can adapt to the current noise environment, providing more accurate and reliable TRFC estimates. This dynamic adaptation is crucial because, as previously mentioned, traditional CKF methods require fixed statistical models for noise, which may not accurately reflect the real-time conditions of onboard sensors. The inclusion of the neural network allows EKFNet to learn from the data and adjust its parameters accordingly, offering a significant improvement over static models. This adaptability makes EKFNet particularly effective in dealing with the diverse and time-varying noise profiles encountered in real-world applications. Consequently, EKFNet can maintain high estimation accuracy even when sensor noise characteristics deviate from the expected norms due to factors like sensor aging or operational stress. The EKF iteration process for TRFC estimation is as follows:

Time Update:

The prior state prediction

$$\mathbf{x}_{\tau|\tau-1}^{TRFC} = A\mathbf{x}_{\tau-1|\tau-1}^{TRFC} \tag{5.141}$$

The error covariance prediction

$$P_{\tau|\tau-1}^{TRFC} = \boldsymbol{A}P_{\tau-1|\tau-1}^{TRFC}\boldsymbol{A}^T + \boldsymbol{Q}^{TRFC} \tag{5.142}$$

Measurement Update:

Compute Kalman gain

$$\boldsymbol{K}_{\tau}^{TRFC} = P_{\tau|\tau-1}^{TRFC}\boldsymbol{C}^T\left(\boldsymbol{C}P_{\tau|\tau-1}^{TRFC}C^T + \boldsymbol{R}^{TRFC}\right)^{-1} \tag{5.143}$$

Update the posterior state

$$\mathbf{x}_{\tau|\tau}^{TRFC} = \mathbf{x}_{\tau|\tau-1}^{TRFC} + \boldsymbol{K}_{\tau}^{TRFC}\left(\mathbf{z}_{\tau}^{TRFC} - C\mathbf{x}_{\tau|\tau-1}^{TRFC}\right) \tag{5.144}$$

Update the error covariance

$$P_{\tau|\tau}^{TRFC} = \left(\boldsymbol{I} - \mathbf{K}_{\tau}^{TRFC}\boldsymbol{C}\right)P_{\tau|\tau-1}^{TRFC} \tag{5.145}$$

Represent the iterative process described above using a flowchart, and replace the Kalman gain calculation part with a neural network prediction module. The framework is shown in Fig. 5.47.

The EKFNet integrates the EKF with deep neural networks to achieve precise estimation of the TRFC, particularly in scenarios where the statistical characteristics of measurement noise are unknown. This innovative approach leverages the strengths of both EKF and neural networks to enhance estimation accuracy and robustness. By examining the EKFNet architecture, it becomes clear that the key to effective system state updates lies in the accurate computation of the Kalman gain. Traditional EKF relies on predefined statistical models to compute this gain, which can be a limitation when dealing with unpredictable or changing noise characteristics. To address this challenge, EKFNet removes the module in EKF responsible for computing the Kalman gain. Instead, it utilizes a deep neural network model to directly predict the optimal Kalman gain by providing it with appropriate inputs. This design choice allows EKFNet to dynamically adjust to varying noise conditions and improve its estimation performance. The deep neural network model is trained to recognize patterns and correlations in the input data that

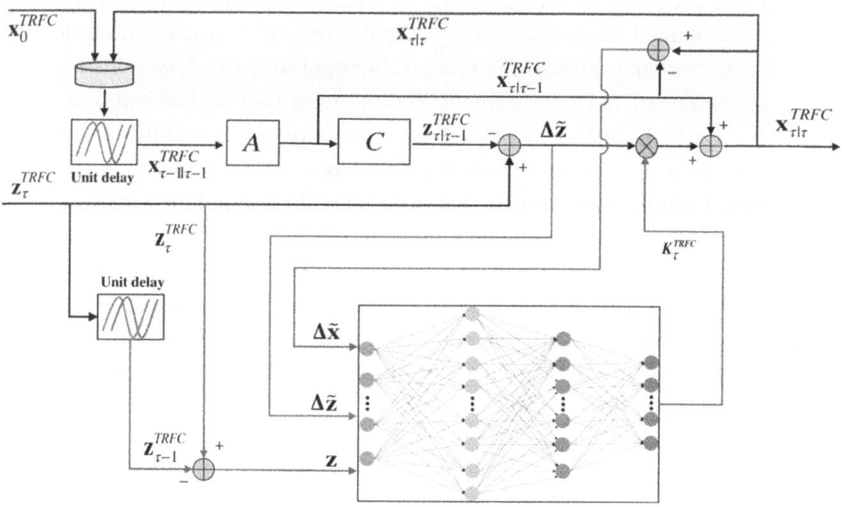

Figure 5.47 The scheme of EKFNet.

affect the Kalman gain, enabling it to make real-time predictions that enhance the filter's accuracy. The entire EKFNet framework consists of two main modules: the system state update and the Kalman gain prediction. The system state update section, depicted in the light green box in the accompanying diagram, follows the traditional EKF methodology to update the state estimates based on the predicted Kalman gain. This involves the propagation of the state and covariance matrices through the system model, ensuring that the state estimates reflect the latest measurements and system dynamics. The Kalman gain prediction section is primarily implemented by a deep neural network model, which replaces the conventional Kalman gain computation process. This neural network is trained using historical data to learn the complex relationships between the measurement inputs and the optimal Kalman gain. By continuously adapting to new data, the neural network can account for changes in measurement noise characteristics that may occur due to sensor aging, environmental factors, or other sources of variability. The integration of these two modules within EKFNet results in a robust and flexible estimation framework that can handle a wide range of operating conditions. The deep neural network's ability to predict the Kalman gain in real-time ensures that the system state updates are always based on the most accurate and current information available. This dynamic approach significantly improves the filter's performance compared to traditional EKF, particularly in environments with unknown or changing noise characteristics.

In the system state update module, the process begins by using the state values from the previous time step, denoted as $\mathbf{x}_{\tau-1|\tau-1}^{TRFC}$, to obtain a prior estimation, $\mathbf{x}_{\tau|\tau-1}^{TRFC}$, through the use of eq. (5.141). Subsequently, the prediction of the measurement variable is obtained using $\boldsymbol{C}$ and $\mathbf{x}_{\tau|\tau-1}^{TRFC}$. Finally, the ultimate estimate is derived by iteratively looping through the process, incorporating the deviations between the actual and predicted measurement vectors, along with the Kalman gain $\boldsymbol{K}_{\tau}^{TRFC}$ and $\mathbf{x}_{\tau|\tau-1}^{TRFC}$. In the Kalman gain prediction section, we opt for a multi-layer neural network to forecast real-time Kalman gains. Based on the traditional EKF gain calculation formula, it is known that accurate gain computation relies on the statistical characteristics of the observed variables and state variables. Therefore, when selecting the input features for the neural network, it is advisable to include information about both the observed variable $\mathbf{z}_{\tau}^{TRFC}$ and the state variable $\mathbf{x}_{\tau-1|\tau-1}^{TRFC}$. Furthermore, taking the discrete difference of these variables over unit time can effectively capture the variations in noise. Therefore, inputs of the network primarily consist of three variables.

a) The forward update error: $\Delta\tilde{\mathbf{x}} = \mathbf{x}_{\tau|\tau}^{TRFC} - \mathbf{x}_{\tau|\tau-1}^{TRFC}$;
b) The observation error: $\Delta\mathbf{z} = \mathbf{z}_{\tau}^{TRFC} - \mathbf{z}_{\tau-1}^{TRFC}$;
c) The innovation error: $\Delta\tilde{\mathbf{z}} = \mathbf{z}_{\tau}^{TRFC} - C\,\mathbf{x}_{\tau|\tau-1}^{TRFC}$.

It can be observed that the multi-layer neural network takes $\Delta\mathbf{z}$, $\Delta\tilde{\mathbf{x}}$, and $\Delta\tilde{\mathbf{z}}$ as inputs. Subsequently, it generates corresponding filtering gains through an activation function. Since the estimated state is four-dimensional and three measurement signals, the Kalman gain is a 4×3 matrix consisting of 12 elements. The performance function consists of the deviation between the predicted value of the multilayer neural network and the ground truth of the Kalman gains, and we use a backpropagation algorithm to optimize this function. In addition, we use the gradient descent method to update the weights and bias. After iterative training with a substantial amount of data, the model is eventually developed to directly predict Kalman gains. This study employs a multi-layer neural network structured with two hidden layers. To ensure a high level of predictive accuracy in the multi-layer neural network, ample training data is imperative. To achieve this, we established a joint simulation platform using CarSim and Simulink. In Simulink, the EKF algorithm is implemented, and noise interference is artificially introduced, with the statistical characteristics of the noise known in advance. This allows us to create a controlled environment where the Kalman gain data collected serves as the ideal value, effectively acting as the true value for the training data. The vehicle operates at different speeds on various surfaces such as asphalt, gravel, and ice and snow. During these simulations, white noise with a mean of zero and a variance within the range of [0.0001, 0.01] is added to emulate realistic measurement noise. By conducting these simulations, we collect extensive input data for the neural network as well as ground truth values for the Kalman gains under a variety of conditions. In total, we amassed 26,000 samples. To ensure a robust model development process, this dataset was strategically divided into three parts: 80% was allocated for training the neural network, 10% was reserved for rigorous testing, and the remaining 10% was designated for validation purposes. This meticulous division ensures that the model is effectively trained, rigorously assessed, and unbiasedly validated. During the training phase, the neural network learns to predict the Kalman gain based on the input data, adjusting its weights and biases to minimize the error between its predictions and the true Kalman gain values. The testing phase involves evaluating the trained model on a separate subset of data to assess its performance and ensure it generalizes well to unseen data. Finally, the validation phase provides an additional layer of evaluation, ensuring that the model's performance is consistent and reliable across different datasets.

By incorporating a diverse range of driving conditions and noise characteristics, the training data encompasses a wide spectrum of scenarios that the neural network might encounter in real-world applications. This comprehensive approach ensures that the model can adapt to various operating environments and maintain high predictive accuracy. Moreover, the joint simulation platform's ability to introduce controlled noise interference allows us to fine-tune the neural network's performance, ensuring that it remains robust even in the presence of unpredictable measurement noise. This aspect is particularly crucial for real-time applications where sensor noise can vary significantly due to factors such as sensor aging, environmental changes, and operational stress.

5.5.3 Simulation Tests

The validity of the model-based learning estimation method is verified by utilizing integrated simulation platforms comprising CarSim and Simulink. In this setup, a high-fidelity vehicle model is incorporated into CarSim, providing a more realistic dynamic response. Consequently, sensor signals and the TRFC from CarSim serve as RV. Vehicle acceleration, deceleration, and steering commands are set via Simulink, where both the ETCKF and EKFNet are executed. To emulate real-world conditions, Gaussian noise is added to the sensor signals, which are then multiplied by a Bernoulli distribution to generate sensor signals with data dropout. This process ensures that the simulation accounts for noise and data loss, providing a robust environment for testing the estimation methods. Verification tests were conducted on various road surfaces, including high-adhesion road surfaces (HARS), middle-adhesion road surfaces (MARS), and opposite road surfaces (ORS), to assess the applicability of the model-based learning method. These tests demonstrate the method's effectiveness across different driving conditions. Additionally, to validate the generalization capability of EKFNet, experiments were performed on both DDEV and FV. In the latest version of CarSim software, models for components such as batteries and motors have been integrated. This integration simplifies the experimental setup for DDEV, as it eliminates the need for additional modeling work. By leveraging these integrated models, the simulations can more accurately reflect the behavior of electric vehicles under various conditions. The use of CarSim and Simulink together creates a comprehensive simulation environment where high-fidelity vehicle dynamics from CarSim provide realistic reference values, while Simulink facilitates the implementation and testing of the estimation algorithms. The incorporation of Gaussian noise and data dropout in the sensor signals ensures that the algorithms are tested under realistic and challenging conditions, further validating their robustness and reliability. In summary, the integration of CarSim and Simulink provides a powerful platform for validating the model-based learning estimation method. The high-fidelity vehicle model in CarSim delivers realistic dynamic responses, while Simulink facilitates the execution of ETCKF and EKFNet. The addition of noise and data dropout in the sensor signals ensures rigorous testing conditions, and the comprehensive verification tests across different road surfaces and vehicle types demonstrate the method's robustness and applicability.

5.5.3.1 The Effectiveness of ETCKF

According to eq. (5.118), it is evident that there are 11 sensor signals involved in the estimation process, each susceptible to noise interference and potential data loss. To manage these challenges effectively, we have designed the ETCKF for processing the sensor data in this study. The ETCKF incorporates a sophisticated mechanism to assess the occurrence of data loss and selectively updates the state estimates based on the reliability of the incoming sensor data. In our methodology, determining the threshold for event triggering is crucial for the ETCKF's operation. Through extensive software-in-the-loop experiments and iterative debugging, we have determined this threshold to be 0.0013. This value represents the criterion at which the ETCKF initiates a decision process regarding the validity of incoming sensor data. When the measured noise or dropout exceeds this threshold, the ETCKF triggers an event to mitigate the impact of unreliable data on the estimation process. By setting a well-calibrated threshold through rigorous experimentation, we ensure that the ETCKF operates optimally in real-world scenarios where sensor signals are prone to variations and uncertainties. This approach enhances the robustness and reliability of the estimation framework, allowing it to adapt dynamically to changing environmental conditions and sensor performance characteristics. The specific matrix parameters for process noise and measurement noise in ETCKF are as follows: $\mathbf{Q} = eye(11) \cdot 0.02$, $\mathbf{R} = diag[1, 1, 1, 1, 1, 1, 1, 1, 1, 1, 1] \cdot 10$.

The longitudinal acceleration data, depicted in Fig. 5.48, provides a comparative analysis of estimation methods under varying conditions. The red solid line represents the RV, which serves as the ideal signal. In contrast, the green solid line illustrates the sensor signals affected by added noise and occasional data loss events. These events are indicated by abrupt transitions to zero in the green solid line, highlighting periods where data reliability is compromised. In this study, two estimation methods are evaluated: the EKF and the ETCKF. The results of these estimations are visually presented in Fig. 5.48. The magenta dashed line represents the output of the EKF, while the blue dashed line with dots corresponds to the estimation results obtained using ETCKF. A noticeable observation is that the EKF estimation curve exhibits minor fluctuations at several instances, indicative of its sensitivity to noisy and intermittently lost data. In contrast, the ETCKF estimation curve aligns more closely with the RV, demonstrating its ability to mitigate the effects of data loss and maintain smoother estimation outputs.

Further insights into vehicle dynamics are provided by Figs. 5.49–5.51, which depict longitudinal velocity, lateral acceleration, and lateral velocity, respectively. Comparing the curves generated by EKF and ETCKF, it becomes evident that the

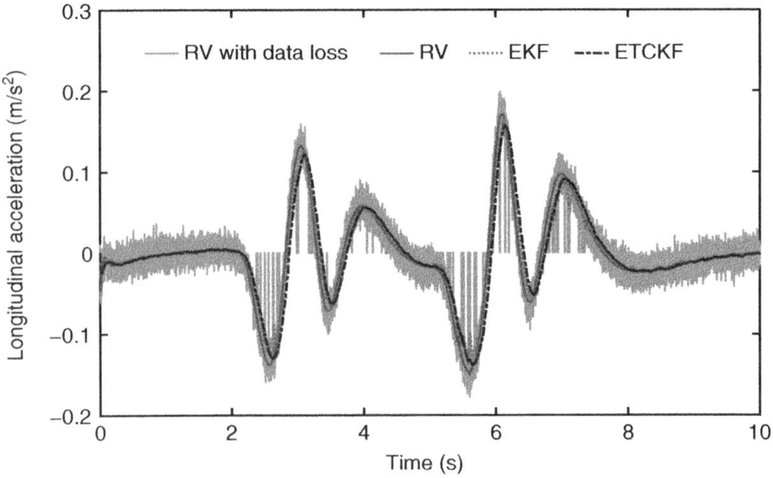

Figure 5.48 The longitudinal acceleration.

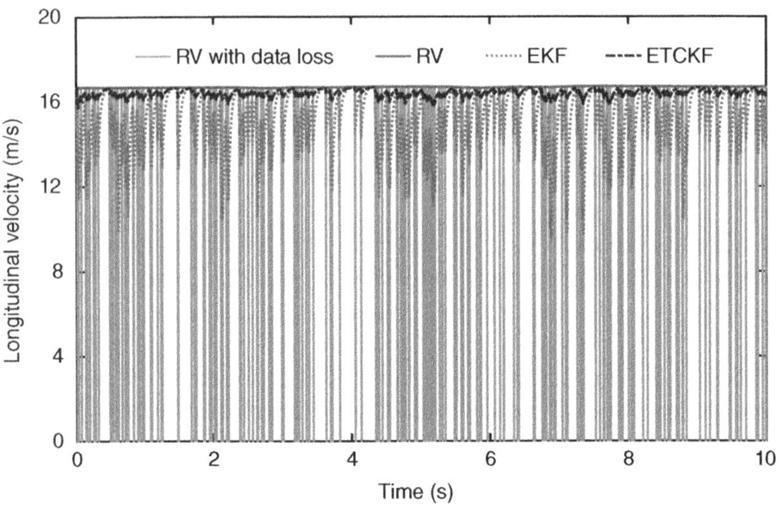

Figure 5.49 The longitudinal velocity.

ETCKF outputs are not only smoother but also closer to the RV. This smoother behavior indicates that ETCKF effectively filters out noise and adapts its estimation process based on the reliability of incoming sensor data, thereby enhancing the accuracy of state estimation. These results underscore the advantages of employing ETCKF in scenarios where sensor signals are prone to noise and occasional data loss. By integrating an event-triggered mechanism, ETCKF

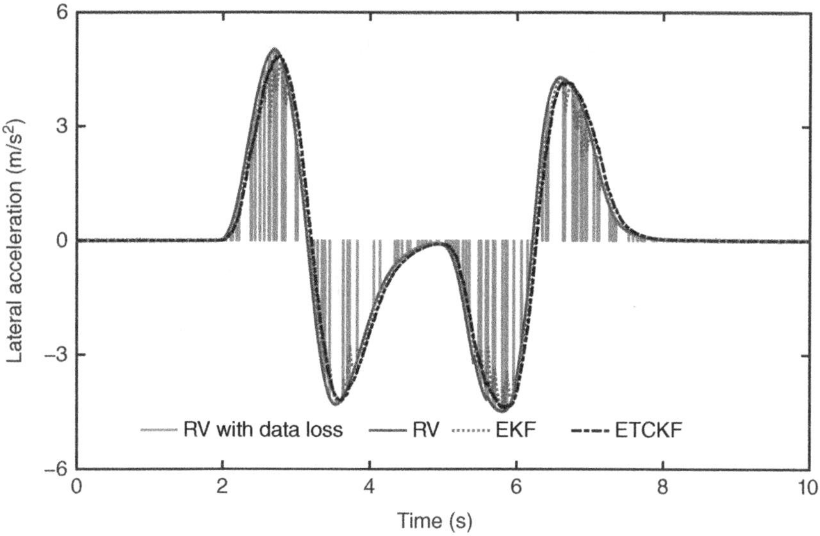

Figure 5.50 The lateral acceleration.

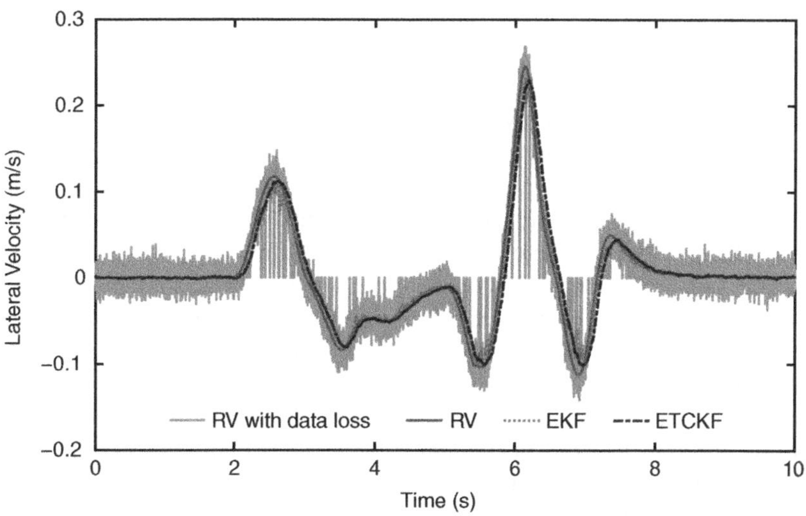

Figure 5.51 The lateral velocity.

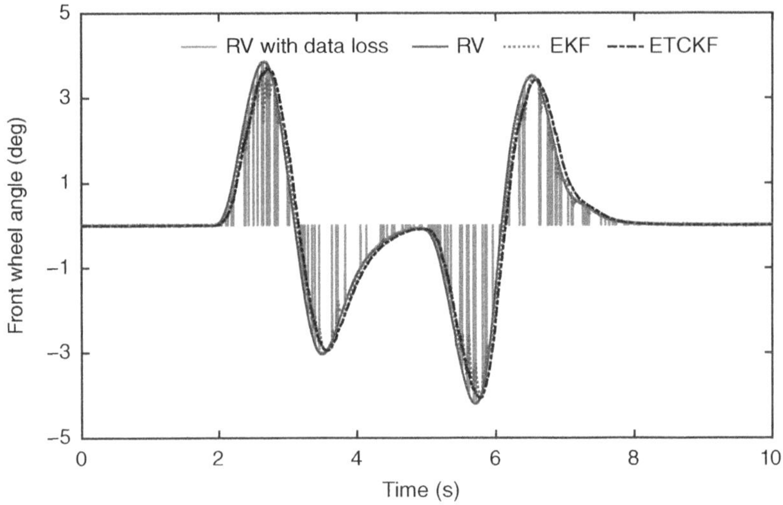

Figure 5.52 The front wheel angle.

optimizes the utilization of available data, ensuring robust and reliable estimation of vehicle dynamics even under challenging operational conditions.

The front wheel steering angle is visually represented in Fig. 5.52, where instances of data loss are evident as abrupt transitions to zero occur intermittently along the curve. This phenomenon underscores the challenges posed by unreliable sensor data in dynamic environments. In Fig. 5.53, the yaw rate curve illustrates how changes in the front wheel steering angle influence vehicle dynamics. The trend in yaw rate corresponds closely to variations in the steering angle, highlighting their interdependent relationship. Figure 5.54 introduces the yaw acceleration curve, where the performance of the ETCKF is prominently displayed. The ETCKF maintains superior performance by accurately capturing the dynamic changes in yaw acceleration despite occasional data loss events. Wheel speed variation is analyzed in Fig. 5.55, focusing on the curve of the left front wheel. Although similar trends are observed across all four wheels, the left front wheel serves as a representative example here. Notably, during periods of data loss, the vehicle speed fluctuation curve exhibits significant variability in the case of EKF. In contrast, the ETCKF curve demonstrates smoother behavior with minor fluctuations, closely tracking the RV throughout the simulation. These observations underscore the robustness and reliability of ETCKF in handling noisy sensor signals and intermittent data loss scenarios. By leveraging an event-triggered mechanism, ETCKF optimizes the utilization of available data, thereby enhancing the accuracy of TRFC estimation and ensuring consistent performance across various dynamic driving conditions.

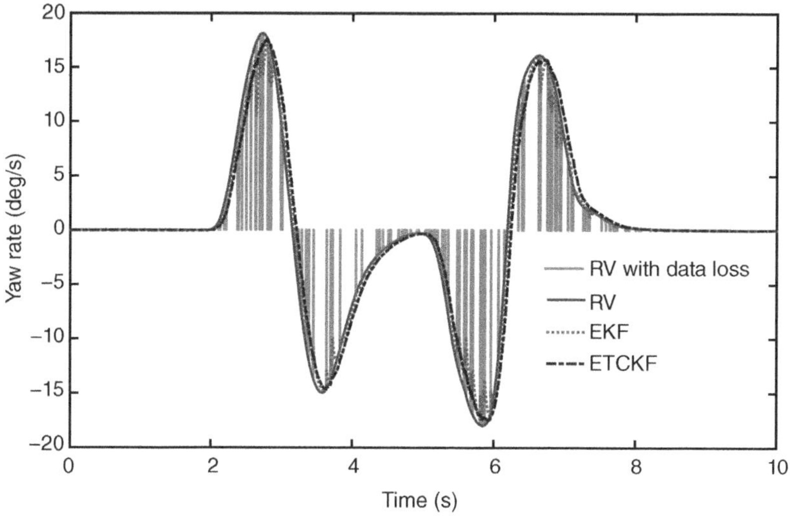

Figure 5.53 The yaw rate.

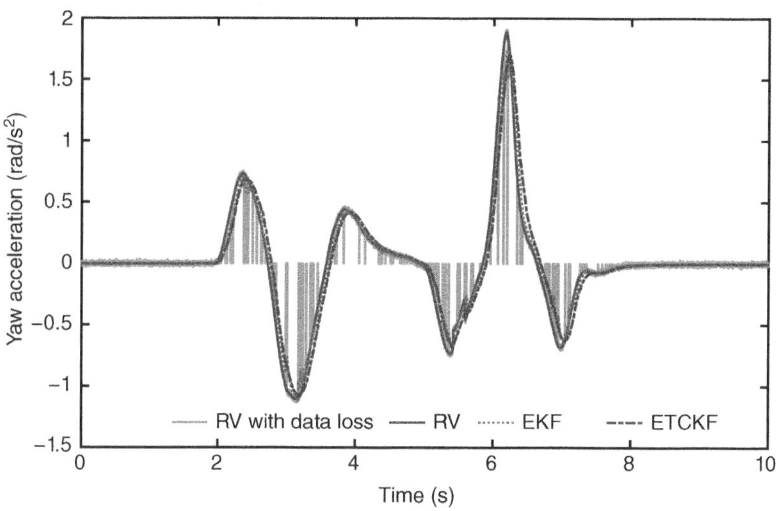

Figure 5.54 The yaw acceleration.

5.5.3.2 The TRFC Estimation Using the DDEV

To validate the effectiveness of EKFNet, we established an initial testing scenario in the CarSim software featuring a dual-lane setup. The vehicle configuration includes four in-wheel motors, each powered by dedicated battery packs, as depicted in Fig. 5.56. This setup aims to simulate realistic driving conditions

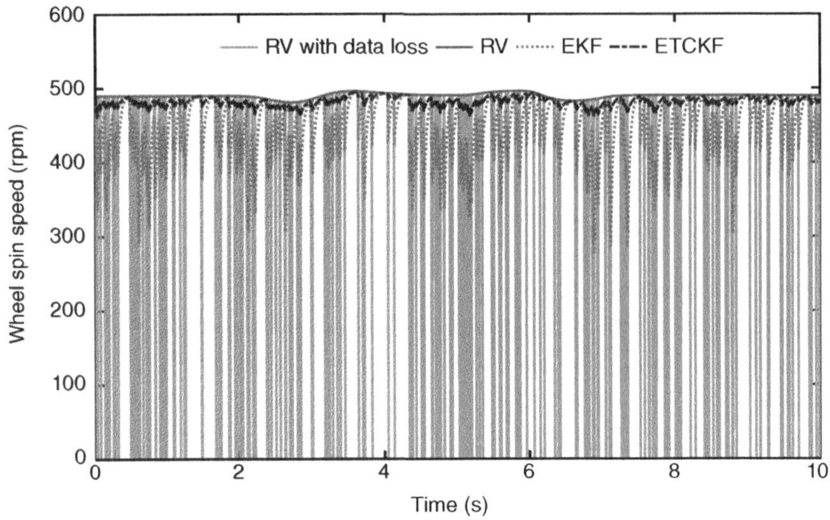

Figure 5.55 The left front wheel spin speed.

Figure 5.56 The double lane change test.

where electric propulsion systems are utilized to drive the vehicle. The dual-lane testing scenario allows us to assess how EKFNet performs under dynamic driving conditions, particularly focusing on its ability to estimate vehicle dynamics and TRFC accurately. By integrating four in-wheel motors and battery packs into the simulation, we replicate the operational environment of electric vehicles, emphasizing the integration and interaction between propulsion systems and vehicle dynamics. This setup is essential for evaluating EKFNet's capability to handle complex dynamics and variations in sensor data while maintaining robust estimation performance. The simulated scenario provides a controlled environment to validate the model's accuracy and reliability under different driving maneuvers and environmental conditions encountered on dual-lane roads. Through this validation process, we aim to demonstrate how EKFNet enhances the estimation accuracy of TRFC and other critical vehicle parameters, leveraging both the vehicle dynamics model in CarSim and the advanced data processing capabilities of EKFNet. This approach ensures that the estimation method is well-suited for real-world applications where electric propulsion systems play a crucial role in vehicle performance and energy efficiency.

The DLC Test on Opposite Roads Conducting consecutive steering maneuvers on opposite roads, the vehicle undergoes transitions from a HARS to a MARS during the estimation process, leading to a change in the TRFC from 0.85 to 0.5. The front wheel angle data is illustrated in Fig. 5.57, depicting the vehicle's path over HARS for the initial 3.91 seconds before transitioning to MARS. Throughout this period, the vehicle maintains a constant speed of 60 km/h. To assess the effectiveness of EKFNet, we conducted comparative analyses with both EKF and UKF. The estimation results generated by these algorithms are presented in Figs. 5.58 and 5.59. To ensure a fair comparison, measurement noise was randomly introduced, with its characteristics deliberately deviating from the true value. This controlled approach allows for a rigorous evaluation of each algorithm's performance under realistic conditions, considering the variability and uncertainty inherent in sensor measurements. By comparing the outputs of EKFNet, EKF, and UKF, we

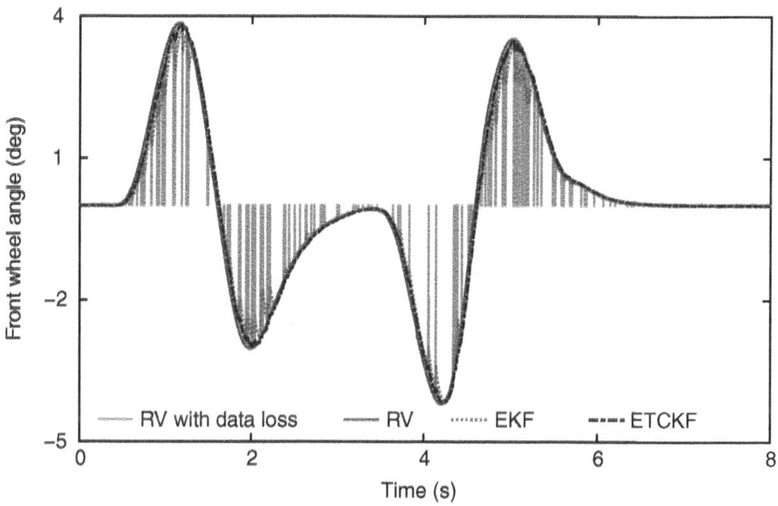

Figure 5.57 The front wheel angle on ORS.

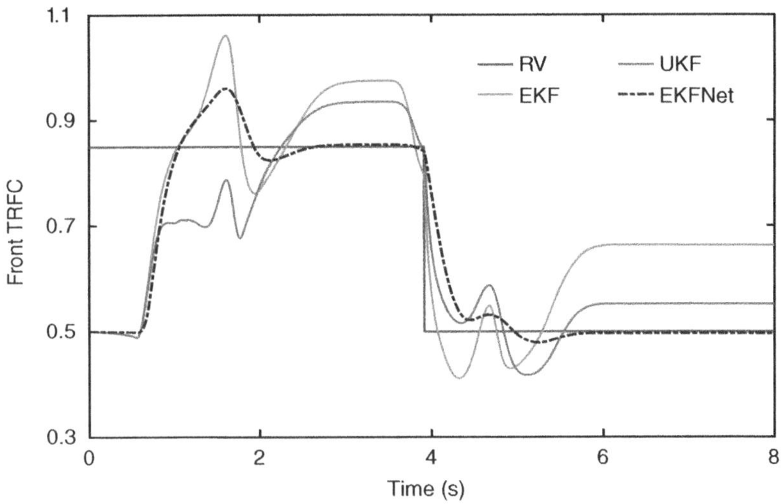

Figure 5.58 The front TRFC on ORS.

aim to demonstrate the advantages of EKFNet in accurately estimating TRFC and other critical vehicle parameters across diverse road surfaces and driving conditions. The comparative analysis highlights EKFNet's ability to leverage advanced data processing techniques and model-based learning to achieve robust estimation performance, ensuring reliable operation in dynamic environments where accurate real-time vehicle dynamics estimation is essential.

In this context, "front TRFC" denotes the average estimation results derived from the left front wheel and the right front wheel, while "rear TRFC" represents the average estimation results from the left rear wheel and the right rear wheel. The true values of TRFC are depicted by the red solid line in the graphs, serving as a RV. The magenta solid line represents the estimation results obtained from the UKF, the green solid line corresponds to the EKF estimation, and the blue dashed line illustrates the results from the EKFNet. Figure 5.58 shows the comparison of these estimation methods. As steering maneuvers commence, all algorithms begin converging toward the RV. Notably, EKFNet exhibits the least amount of fluctuation and eventually aligns closely with the TRFC value of 0.85. However, during the transition of TRFC from 0.8 to 0.5, all estimation curves exhibit overshooting, with EKFNet showing minimized fluctuations compared to the other methods. Eventually, the EKFNet estimation curve stabilizes quickly at 0.5, demonstrating its robust performance under changing TRFC conditions compared to EKF and UKF, which deviate more from the RV. Moreover, UKF demonstrates superior estimation performance over EKF due to its enhanced nonlinear approximation capability, which does not necessitate the computation

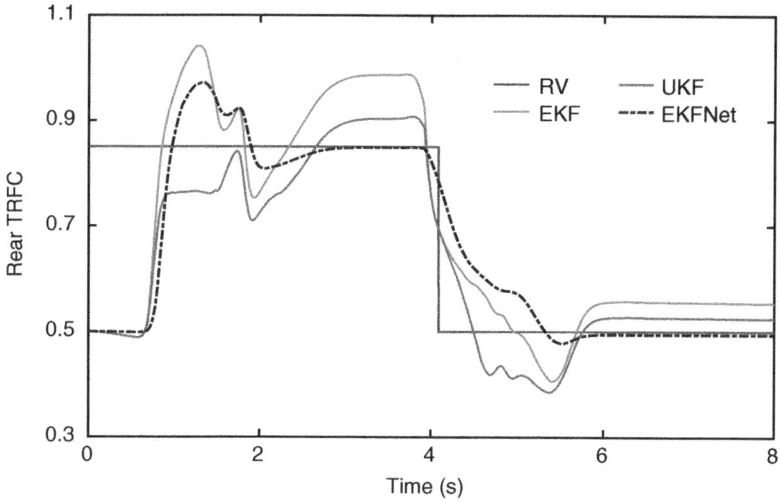

Figure 5.59 The rear TRFC on ORS.

of the Jacobian matrix. EKFNet further improves upon these methods by employing neural networks to predict the Kalman gain directly, thereby avoiding the accuracy reduction associated with Jacobian matrix computations. Extensive training with diverse noise conditions enables the neural network in EKFNet to accurately predict the Kalman gain, contributing to its superior estimation accuracy compared to UKF.

Figure 5.59 illustrates a lag in the TRFC mutation time compared to Fig. 5.58, attributed to the front wheels making contact with the road surface before the rear wheels during steering maneuvers. EKFNet consistently demonstrates the highest estimation accuracy among the different methods evaluated, showcasing its capability to effectively handle dynamic changes in TRFC and provide reliable estimations under varied driving conditions. These findings underscore the effectiveness of EKFNet in enhancing TRFC estimation accuracy, leveraging advanced data processing techniques and model-based learning to achieve reliable performance in real-world scenarios where precise vehicle dynamics estimation is critical.

The Acceleration Maneuver on a High-Adhesion Road In the test scenario, the TRFC is set to 0.7, and the vehicle initiates an acceleration maneuver from an initial speed of 20 km/h. As depicted in Fig. 5.60, the vehicle speed progressively increases throughout the simulation. Notably, instances of data loss cause significant fluctuations in the vehicle speed estimation curves, particularly noticeable in the EKF curve. In contrast, the ETCKF curve demonstrates minimal fluctuations and closely follows the RV.

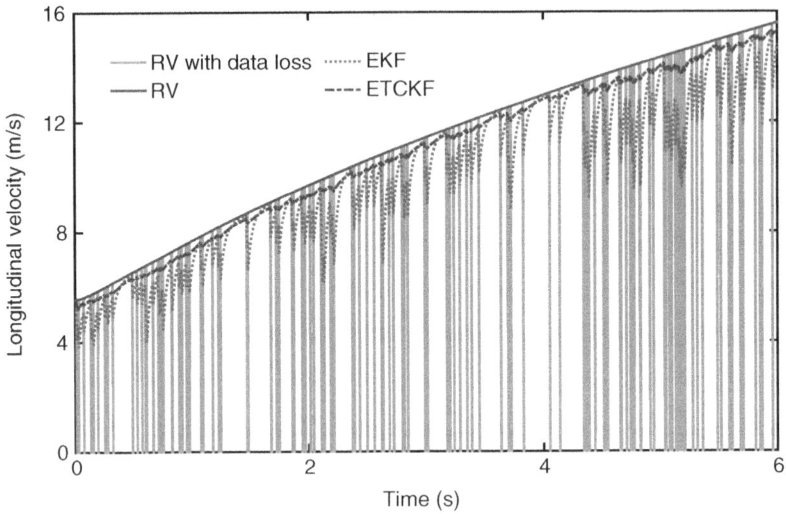

Figure 5.60 The longitudinal velocity on HARS.

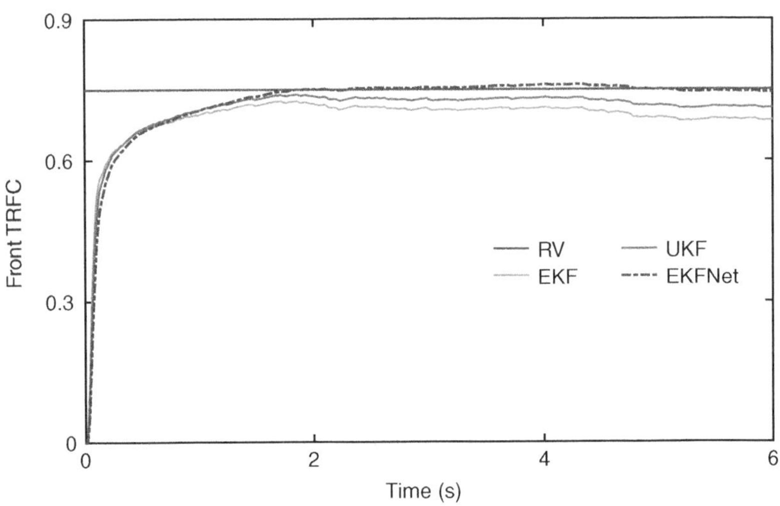

Figure 5.61 The front TRFC on HARS.

Detailed results of various estimation algorithms are presented in Figs. 5.61 and 5.62, showcasing their performance under dynamic conditions and in the presence of data uncertainties. These figures highlight how ETCKF outperforms EKF by maintaining more stable and accurate estimation of vehicle speed despite

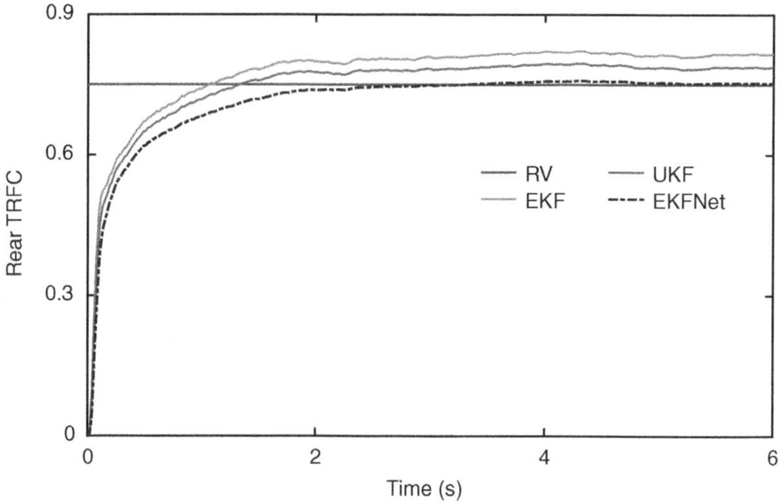

Figure 5.62 The rear TRFC on HARS.

challenges such as noise and data loss. The comparative analysis underscores the robustness of ETCKF in real-time applications where precise vehicle dynamics estimation is crucial for effective control and monitoring. In the test, TRFC is set to 0.7, and the initial speed is 20 km/h with an acceleration maneuver. As shown in Fig. 5.60, the vehicle speed is continuously increasing. It is evident that with data loss, the vehicle speed fluctuation is much more pronounced in the EKF curve, whereas the ETCKF curve exhibits minor fluctuations and closely tracks the RV. The results of different estimation algorithms are shown in Figs. 5.61 and 5.62.

In Figs. 5.61 and 5.62, as acceleration maneuvers commence at the start of the simulation, all three algorithms show rapid initial convergence toward the RV. Initially, their convergence speeds appear comparable, indicating effective adaptation to the changing dynamics of the vehicle. However, a notable divergence occurs after approximately 2 seconds in the estimation curves of the EKF and UKF. This divergence is attributed to inaccuracies stemming from imprecise noise statistics used as inputs, which affect their ability to accurately track the RV. In contrast, EKFNet consistently demonstrates superior estimation performance throughout the simulation period. By leveraging advanced neural network-based techniques, EKFNet mitigates the impact of noise uncertainties, enabling it to maintain closer alignment with the RV compared to traditional filtering methods like EKF and UKF. This resilience is crucial in dynamic environments where precise estimation of parameters such as TRFC is essential for safe and efficient vehicle operation.

The Deceleration Maneuver on a Low-Adhesion Road Verification tests were performed on LARS to evaluate the performance of different estimation algorithms. In this test scenario, the TRFC was set to 0.2, and the vehicle initiated a deceleration maneuver from an initial speed of 60 km/h. As shown in Fig. 5.63, the vehicle speed decreases steadily throughout the simulation. The curve representing the vehicle speed after processing by the ETCKF appears smoother and maintains closer alignment with the RV compared to other estimation algorithms. This indicates ETCKF's ability to effectively mitigate the effects of noise and data loss, thereby providing more accurate estimation of vehicle dynamics under challenging road conditions like LARS. Figure 5.64 and 5.65 present detailed results of different estimation algorithms, offering insights into their respective performances in capturing the dynamic changes in vehicle speed during deceleration on low-adhesion surfaces. These figures highlight ETCKF's robustness and superior estimation accuracy, reinforcing its suitability for real-time applications where reliable vehicle dynamics estimation is crucial for safety and performance. We conducted verification tests on LARS. During the test, TRFC is configured to 0.2, and the initial speed is 60 km/h for a deceleration maneuver. As illustrated in Fig. 5.63, the vehicle speed steadily decreases. The speed curve is smoother and closer to the RV after processing by ETCKF. The outcomes of various estimation algorithms are depicted in Figs. 5.64 and 5.65.

In Figs. 5.64 and 5.65, the initial significant braking input prompts all three algorithms to rapidly converge toward the RV at the start of the simulation. As

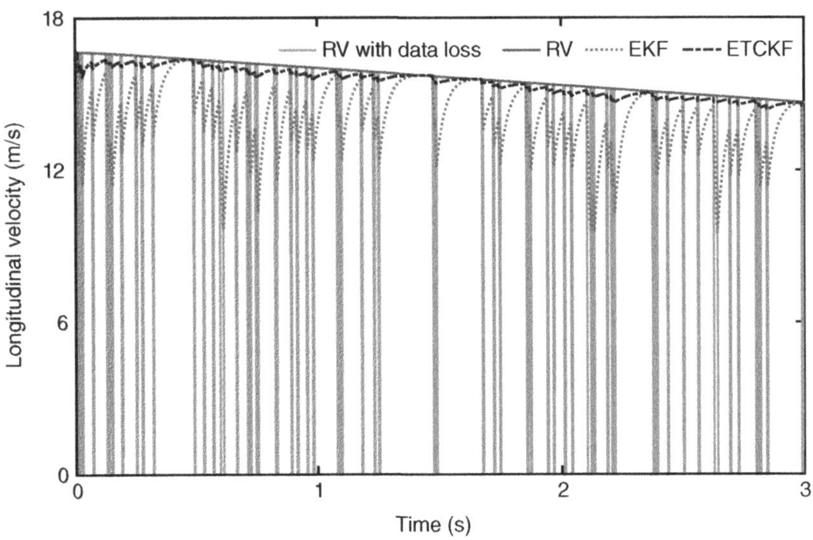

Figure 5.63 The longitudinal velocity on LARS.

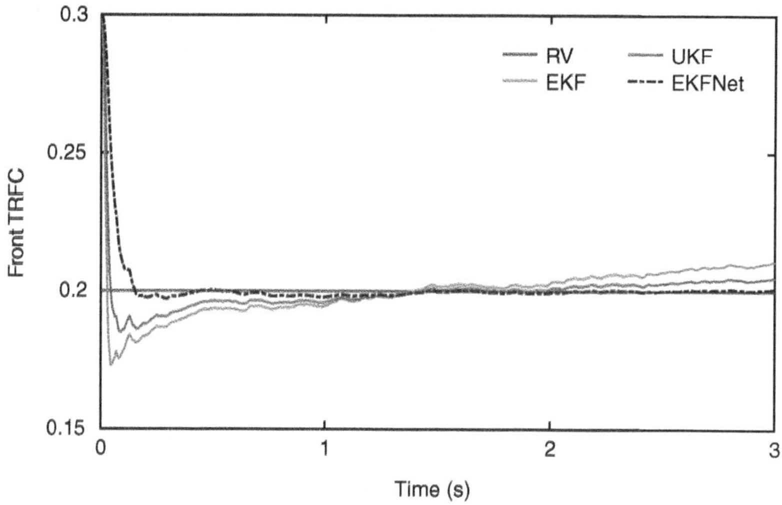

Figure 5.64 The front TRFC on LARS.

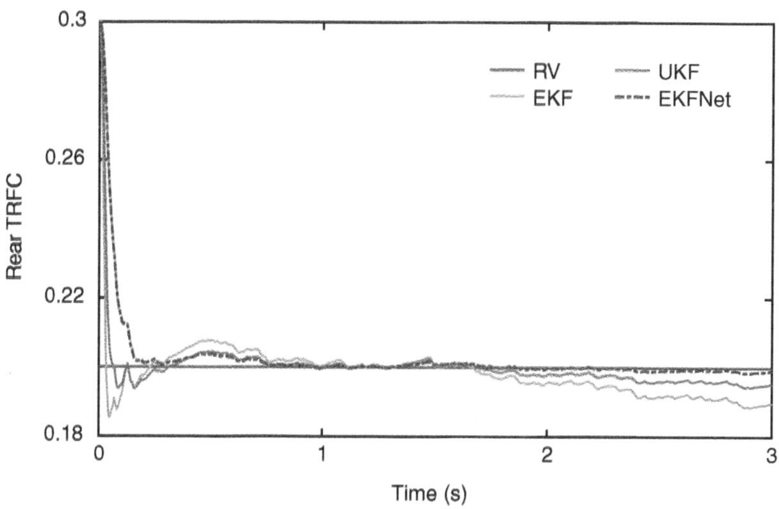

Figure 5.65 The rear TRFC on LARS.

the vehicle begins its descent, overshooting behavior is observed in all three algorithms, with the EKF exhibiting the highest amplitude of deviation and EKFNet showing the least. Following this initial overshoot, all algorithms gradually converge back toward the RV. However, after approximately 1 second into the simulation, noticeable deviations start to appear in the estimated values of EKF and UKF, while the estimation curve of EKFNet remains consistently aligned with the RV. This performance disparity highlights EKFNet's robustness in maintaining accurate estimation of vehicle dynamics, even in the presence of dynamic braking maneuvers and varying road conditions. By leveraging advanced neural network techniques to optimize the estimation process, EKFNet effectively mitigates the impact of noise and data uncertainties, ensuring more reliable and precise results compared to traditional filtering methods like EKF and UKF.

5.5.3.3 The TRFC Estimation Using the FV

To validate the algorithm's generalization further, we transitioned to an FV for testing, as depicted in Fig. 5.66. The test involved consecutive steering maneuvers, during which TRFC transitioned from 0.5 to 0.8. Figure 5.67 illustrates the front wheel angle throughout the test. The curve of the front wheel angle appears smoother and closely aligned with the RV after processing by ETCKF. During the initial 5 seconds of the test, the vehicle operated on a MARS, followed by a transition to a HARS. Throughout this period, the vehicle maintained a constant speed of 54 km/h. This experimental setup aims to assess the algorithm's ability to adapt and accurately estimate TRFC under varying road conditions and dynamic maneuvers.

Estimation results for various algorithms are presented in Figs. 5.68 and 5.69. Randomly generated measurement noise with identical covariance in both EKF and UKF deviates from the true value. In Fig. 5.68, it is evident that as steering maneuvers initiate, all three algorithms quickly converge toward the RV. Notably, EKFNet demonstrates the smallest fluctuation and eventually aligns perfectly with the true value of 0.5. As TRFC transitions from 0.5 to 0.8, the EKFNet estimation curve stabilizes rapidly at 0.8, contrasting with other estimation

Figure 5.66 The test using the FV.

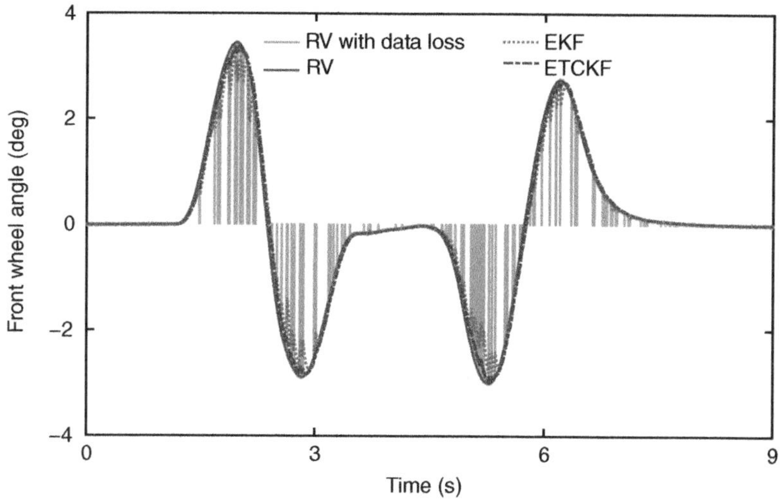

Figure 5.67 The front wheel angle of the FV.

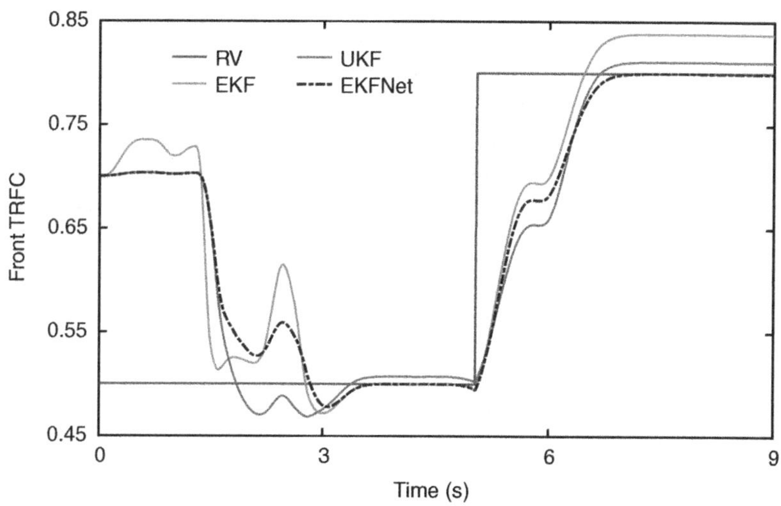

Figure 5.68 The front TRFC estimation using the FV.

algorithms that deviate from the RV. Figure 5.69 shows similar outcomes, with EKFNet exhibiting the highest estimation accuracy throughout the transition period. The performance of EKF and UKF, characterized by fluctuations and deviations from the RV, underscores the superior accuracy and robustness of EKFNet in tracking TRFC under dynamic driving conditions.

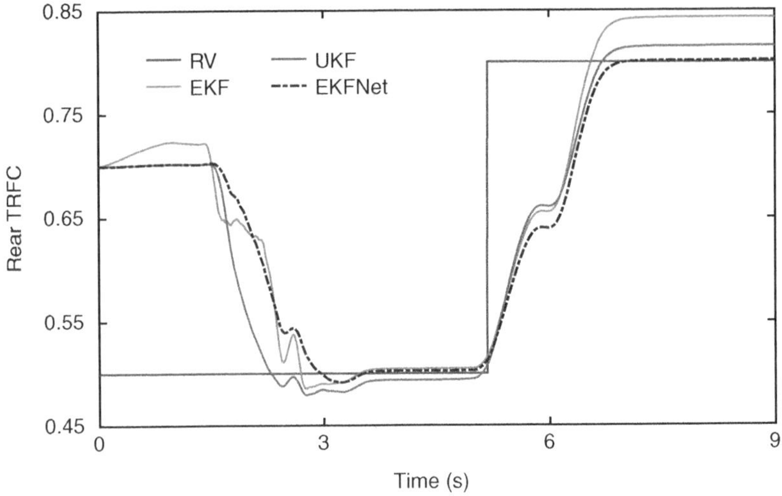

Figure 5.69 The rear TRFC estimation using the FV.

From the simulation results, it is evident that deploying the proposed algorithm on both DDEV and FV enables accurate estimation of TRFC. This success is attributed to several factors. First, employing a generalized vehicle model helps mitigate the influence of specific powertrain configurations on the overall vehicle dynamics modeling. This approach ensures that the estimation framework remains robust and adaptable across different vehicle types and configurations. Second, by collecting data from both DDEV and FV platforms, the applicability of EKFNet is significantly enhanced. This data-driven approach allows the algorithm to learn from a diverse range of driving conditions and scenarios, thereby improving its ability to generalize and perform reliably under various real-world situations. These combined strengths of the model and algorithm underscore the high generalizability and effectiveness of the proposed estimation framework. By addressing vehicle-specific nuances while leveraging comprehensive data integration, EKFNet proves capable of achieving accurate TRFC estimation in diverse operational contexts, enhancing its utility for practical implementation in vehicle dynamics and control systems.

5.6 Summary and Future Research

In this chapter, a novel hybrid estimation approach was proposed by integrating STUKF and IMM-UKF. Real-time experiments have demonstrated that the proposed identification approach exhibits superior estimation accuracy and adaptability compared to existing methods across a variety of driving scenarios. STUKF enhances the adaptive capability of the hybrid estimator to effectively handle significant parameter perturbations encountered in dynamic driving environments. Meanwhile, IMM-UKF leverages the interactive combination of results from different estimators to significantly enhance the overall accuracy of the hybrid estimator. The application potential of this work extends to the dynamics control systems of intelligent vehicles, promising substantial advancements in vehicle safety. Future research directions include exploring the fusion of vehicle dynamics-based approaches with sensing approaches to further improve estimation accuracy. Additionally, while some algorithm parameters are currently set manually, future efforts will focus on implementing adaptive parameter adjustments within the algorithm to enhance its robustness and performance in real-world applications. These advancements aim to continually refine and optimize the proposed hybrid estimation framework for enhanced practical utility and effectiveness in intelligent vehicle systems.

Furthermore, a novel fault-tolerant estimation framework was proposed to estimate TRFC. With the consideration of missing measurements, an FTUKF was employed to predict longitudinal and lateral TRFCs, respectively. To further enhance estimation accuracy, an event-driven multi-model fusion algorithm was designed to dynamically adjust the weight of different models. Different tests are carried out to verify the effectiveness of the fault-tolerant estimation scheme. The results indicate that the designed method has high accuracy and strong adaptability under various roads. Furthermore, the process noise and measurement noise of the system are usually unknown. Therefore, considering the effect of noise uncertainty to improve the estimation performance is a direction for future research. In addition, the current research work has not yet been validated in real vehicles due to expensive equipment and experimental site constraints as well as some safety considerations. In the future, we will perform more validation using real vehicle data where some test equipment is available.

Finally, a model-based learning estimation scheme is proposed to effectively identify TRFC under challenging conditions of unknown measurement noise and data loss. The framework begins with the introduction of an ETCKF to handle missing measurement signals, ensuring robust data processing. Following this, the EKFNet algorithm is deployed to estimate TRFC, leveraging a combination of EKF principles and neural network predictions. Virtual tests conducted with both electric and fuel vehicles validate the efficacy of the proposed approach across diverse driving scenarios, encompassing scenarios involving acceleration, deceleration, and steering maneuvers. These tests underscore the method's adaptability and accuracy in estimating TRFC under dynamic and varied road conditions. It is important to highlight that our estimation framework is based on a two-axle vehicle model, making it particularly suitable for conventional passenger cars.

However, deploying this method directly for TRFC estimation in specialized vehicles like multi-axle trailers presents challenges due to their distinct dynamics and sensor configurations. Furthermore, the absence of an experimental field and complete equipment has precluded real vehicle experiments thus far. Future validation efforts will prioritize conducting experiments in real vehicle settings, pending the availability of appropriate conditions and equipment. This step is crucial to affirming the practical applicability and reliability of the proposed estimation framework in real-world driving environments.

References

1 Wang, Y., Hu, J., Wang, F.A. et al. (2022). Tire road friction coefficient estimation: review and research perspectives. *Chinese Journal of Mechanical Engineering* 335 (6): 1–11.

2 Du, H., Teng, S., Chen, H. et al. (2023). Chat with ChatGPT on intelligent vehicles: An IEEE TIV perspective. *IEEE Transactions on Intelligent Vehicles* 8 (3): 2020–2026.

3 Wang, Y., Hu, F., Tian, C. et al. (2024). FTEKFNet: hybridizing physical and data-driven estimation algorithms for vehicle state. *IEEE Transactions on Intelligent Vehicles*. https://doi.org/10.1109/TIV.2024.3395911.

4 Hu, B.-B., Zhang, H.-T., Yao, W. et al. (2024). Coordinated guiding vector field design for ordering-flexible multi-robot surface navigation. *IEEE Transactions on Automatic Control*. https://doi.org/10.1109/TAC.2024.3355321.

5 Feng, J., Liang, J., Lu, Y. et al. (2024). An integrated control framework for torque vectoring and active suspension system. *Chinese Journal of Mechanical Engineering* 37 (10). https://doi.org/10.1186/s10033-024-00999-6.

6 Zhang, L., Zhang, Z., Wang, Z. et al. (2021). Chassis coordinated control for full X-by-wire vehicles-A review. *Chinese Journal of Mechanical Engineering* 34 (42). https://doi.org/10.1186/s10033-021-00555-6.

7 Liang, C., Xu, X., Wang, F. et al. (2022). CAN-induced asynchronous random delays-considered mode transition system for DM-PHEV based on constrained output feedback robust control strategy. *IEEE Transactions on Vehicular Technology* 71 (6): 5995–6006.

8 Liang, C., Xu, X., Auger, D.J. et al. (2023). Efficient mode transition control for DM-PHEV with mechanical hysteresis based on piecewise affine H∞ strategy. *IEEE Transactions on Transportation Electrification* 9 (3): 4366–4379.

9 Wei, H., Zhang, H., AI-Haddad, K., and Shi, Y. (2024). Ensuring secure platooning of constrained intelligent and connected vehicles against Byzantine attacks: A distributed MPC framework. *Engineering* 33: 35–46.

10 Hu, B.-B., Zhang, H.-T., Yao, W. et al. (2023). Spontaneous-ordering platoon control for multirobot path navigation using guiding vector fields. *IEEE Transactions on Robotics* 39 (4): 2654–2668.

11 Deng, H., Zhao, Y., Nguyen, A.-T., and Huang, C. (2023). Fault-tolerant predictive control with deep-reinforcement-learning-based torque distribution for four in-wheel motor drive electric vehicles. *IEEE/ASME Transactions on Mechatronics* 28 (2): 668–680.

12 Fu, T., Zhou, H., and Liu, Z. (2022). NMPC-based path tracking control strategy for autonomous vehicles with stable limit handling. *IEEE Transactions on Vehicular Technology* 71 (12): 12499–12510.

13 Prathiba, S.B., Raja, G., and Kumar, N. (2022). Intelligent cooperative collision avoidance at overtaking and lane changing maneuver in 6G-V2X communications. *IEEE Transactions on Vehicular Technology* 71 (1): 112–122.

14 Huang, Y. and Chen, Y. (2020). Vehicle lateral stability control based on shiftable stability regions and dynamic margins. *IEEE Transactions on Vehicular Technology* 69 (12): 14727–14738.

15 Du, Y., Liu, C., Song, Y. et al. (2020). Rapid estimation of road friction for anti-skid autonomous driving. *IEEE Transactions on Intelligent Transportation Systems* 21 (6): 2461–2470.

16 Zou, Z., Zhang, X., Zou, Y., and Lenzo, B. (2021). Tire-road friction coefficient estimation method design for intelligent tires equipped with three-axis accelerometer. *SAE International Journal of Vehicle Dynamics, Stability, and NVH* 5 (3): 249–258.

17 Yoon, J.-H., Li, S.E., and Ahn, C. (2016). Estimation of vehicle sideslip angle and tire-road friction coefficient based on magnetometer with GPS. *International Journal of Automotive Technology* 17: 427–435.

18 Leng, B., Jin, D., Xiong, L. et al. (2021). Estimation of tire-road peak adhesion coefficient for intelligent electric vehicles based on camera and tire dynamics information fusion. *Mechanical Systems and Signal Processing* 150: 1–15.

19 Yu, M., Xu, X., Wu, C. et al. (2021). Research on the prediction model of the friction coefficient of asphalt pavement based on tire-pavement coupling. *Advances in Materials Science and Engineering* 2021: 1–10.

20 Gupta, U., Nouri, A., Subramanian, C., and Taheri, S. (2021). Developing an experimental setup for real-time road surface identification using intelligent tires. *SAE International Journal of Vehicle Dynamics, Stability, and NVH* 5 (3): 351–367.

21 Han, K., Lee, E., Choi, M., and Choi, S.B. (2017). Adaptive scheme for the real-time estimation of tire-road friction coefficient and vehicle velocity. *IEEE/ASME Transactions on Mechatronics* 22 (4): 1508–1518.

22 Rajamani, R., Piyabongkarn, N., Lew, J. et al. (2010). Tire-road friction-coefficient estimation. *IEEE Control Systems* 30 (4): 54–69.

23 Xia, X., Xiong, L., Sun, K. et al. (2016). Estimation of maximum road friction coefficient based on Lyapunov method. *International Journal of Automotive Technology* 17 (6): 991–1002.

24 Sharifzadeh, M., Senatore, A., Farnam, A. et al. (2019). A real-time approach to robust identification of tyre-road friction characteristics on mixed-μ roads. *Vehicle System Dynamics* 57 (9): 1338–1362.

25 Enisz, K., Szalay, I., Kohlrusz, G., and Fodor, D. (2015). Tyre-road friction coefficient estimation based on the discrete-time extended Kalman filter. *Proceedings of the Institution of Mechanical Engineers, Part D: Journal of Automobile Engineering* 229 (9): 1158–1168.

26 Castillo, J.J., Cabrera, J.A., Guerra, A.J., and Simón, A. (2016). A novel electro-hydraulic brake system with tire-road friction estimation and continuous brake pressure control. *IEEE Transactions on Industrial Electronics* 63 (3): 1863–1875.

27 Paul, D., Velenis, E., Cao, D.P. et al. (2017). Optimal μ-estimation-based regenerative braking strategy for an AWD HEV. *IEEE Transactions on Transportation Electrification* 3 (1): 249–258.

28 Beal, C.E. (2020). Rapid road friction estimation using independent left/right steering torque measurements. *Vehicle System Dynamics* 58 (3): 377–403.

29 Hu, J., Rakheja, S., and Zhang, Y. (2020). Real-time estimation of tire-road friction coefficient based on lateral vehicle dynamics. *Proceedings of the Institution of Mechanical Engineers, Part D: Journal of Automobile Engineering* 234 (10): 2444–2457.

30 Xu, Y., Chen, B., and Chi, C. (2018). Estimation of road friction coefficient and vehicle states by 3-DOF dynamic model and HSRI model based on information fusion: Estimation of road and vehicle parameters based on information fusion. *Asian Journal of Control* 20 (3): 1067–1076.

31 Liu, Y.-H., Li, T., Yang, Y.-Y. et al. (2017). Estimation of tire-road friction coefficient based on combined APF-IEKF and iteration algorithm. *Mechanical Systems and Signal Processing* 88: 25–35.

32 Zhang, L., Guo, P., Wang, Z., and Ding, X. (2022). An enabling tire-road friction estimation method for four-in-wheel-motor-drive electric vehicles. *IEEE Transactions on Transportation Electrification* early access. https://doi.org/10.1109/TTE.2022.3231707.

33 Biswas, S.K. and Dempster, A.G. (2019). Approximating sample state vectors using the ESPT for computationally efficient particle filtering. *IEEE Transactions on Signal Processing* 67 (7): 1918–1928.

34 Choi, M., Oh, J.J., and Choi, S.B. (2013). Linearized recursive least squares methods for real-time identification of tire–road friction coefficient. *IEEE Transactions on Vehicular Technology* 62 (7): 2906–2918.

35 Ren, H., Chen, S., Shim, T., and Wu, Z. (2014). Effective assessment of tyre–road friction coefficient using a hybrid estimator. *Vehicle System Dynamics* 52 (8): 1047–1065.

36 Li, L., Yang, K., Jia, G. et al. (2015). Comprehensive tire-road friction coefficient estimation based on signal fusion method under complex maneuvering operations. *Mechanical Systems and Signal Processing* 56: 259–276.

37 Chen, L., Bian, M., Luo, Y. et al. (2016). Real-time identification of the tyre-road friction coefficient using an unscented Kalman filter and mean-square-error-weighted fusion. *Proceedings of the Institution of Mechanical Engineers, Part D: Journal of Automobile Engineering* 230 (6): 788–802.

38 Peng, Y., Chen, J., and Ma, Y. (2019). Observer-based estimation of velocity and tire-road friction coefficient for vehicle control systems. *Nonlinear Dynamics* 96 (1): 363–387.

39 Zareian, A., Azadi, S., and Kazemi, R. (2016). Estimation of road friction coefficient using extended Kalman filter, recursive least square, and neural network. *Proceedings of the Institution of Mechanical Engineers, Part K: Journal of Multi-body Dynamics* 230 (1): 52–68.

40 Chen, L., Luo, Y., Bian, M. et al. (2017). Estimation of tire-road friction coefficient based on frequency domain data fusion. *Mechanical Systems and Signal Processing* 85: 177–192.

41 Mirmohammad Sadeghi, S., Mashadi, B., Amirkhani, A. et al. (2022). Maximum tire/road friction coefficient prediction based on vehicle vertical accelerations using wavelet transform and neural network. *Journal of the Brazilian Society of Mechanical Sciences and Engineering* 44: 324.

42 Chen, L., Qin, Z., Hu, M. et al. (2023). Data-enabled tire-road friction estimation based on explainable dynamics mechanism under straight stationary driving maneuvers. *IEEE Transactions on Intelligent Transportation Systems.* https://doi.org/10.1109/TITS.2023.3339333.

43 Yu, Q., Xiong, R., Lin, C. et al. (2017). Lithium-ion battery parameters and state-of-charge joint estimation based on H-infinity and unscented Kalman filters. *IEEE Transactions on Vehicular Technology* 66 (10): 8693–8701.

44 Jwo, D., Yang, C., Chuang, C., and Lee, T. (2013). Performance enhancement for ultra-tight GPS/INS integration using a fuzzy adaptive strong tracking unscented Kalman filter. *Nonlinear Dynamics* 73 (1): 377–395.

45 Xie, G., Gao, H., Qian, L. et al. (2017). Vehicle trajectory prediction by integrating physics-and maneuver-based approaches using interactive multiple models. *IEEE Transactions on Industrial Electronics* 65 (7): 5999–6008.

46 Novikov, A., Novikov, I., and Shevtsova, A. (2018). Study of the impact of type and condition of the road surface on parameters of signalized intersection. *Transportation Research Procedia* 36: 548–555.

6

Conclusions and Recommendations

In the field of intelligent transportation systems, it is crucial to understand and master advanced estimation techniques for MVRIS. Through these techniques, researchers and engineers can effectively estimate the state parameters of both vehicles and roads. By utilizing accurate information, they can make efficient decisions, thereby enhancing vehicle safety and comfort.

This book systematically introduces the latest research work of the authors in this field, which can be summarized as follows:

1) A novel approach named AFTEKF has been developed for the estimation of vehicle states. This inventive method demonstrates its effectiveness in alleviating the impact of partial sensor data loss, parameter perturbations, and state mutations on estimation accuracy. The AFTEKF method dynamically adjusts the filter parameters in response to changes in the system and environment, ensuring robust state estimation even in the presence of uncertainties and disruptions.
2) A novel approach, FARCKF, is introduced for the estimation of sideslip angle and tire cornering stiffness. This method incorporates dynamic updates of the vehicle mass parameter and process noise, aiming to enhance estimation performance throughout the estimation process. By continuously adapting to the changing conditions of the vehicle and road, FARCKF provides more accurate and reliable estimates, which are crucial for maintaining vehicle stability and safety.
3) A hybrid estimation algorithm, FTEKFNet, is introduced for the simultaneous estimation of vehicle velocity and sideslip angle, leveraging a fusion of physical modeling and data-driven approaches. This innovative approach incorporates a neural network to directly predict Kalman gains, thereby enhancing robustness against uncertainties associated with Gaussian noise and improving the algorithm's capability to handle colored noise.

State Estimation of Multi-Agent Vehicle-Road Interaction Systems, First Edition.
Yan Wang, Guodong Yin, and Chao Huang.
© 2026 The Institute of Electrical and Electronics Engineers, Inc. Published 2026 by John Wiley & Sons, Inc.

The integration of neural networks with traditional estimation methods allows for more precise and adaptive state estimation.

4) An innovative event-triggered estimation framework integrates an event-triggered mechanism with an ECKF based on a 3-DOF model. This framework is designed for state estimation of PVs in dynamic environments. This method effectively manages the trade-off between communication rate and estimation performance. By triggering updates only when necessary, the framework reduces communication overhead while maintaining high estimation accuracy, making it suitable for real-time applications in connected and autonomous vehicles.
5) An ETCKF is designed for the host vehicle to estimate the motion state of the preceding vehicle. To further improve the robustness of the algorithm against perturbation of model parameters, a strong tracking algorithm is utilized to optimize the ETCKF to form an STETCKF. The proposed estimation scheme still achieves high accuracy even under low communication rates and model parameter perturbations.
6) A novel hybrid estimation approach for TRFC estimation is proposed by integrating the STUKF and the IMM-UKF. STUKF enhances the adaptive capability of the hybrid estimator to effectively handle significant parameter perturbations encountered in dynamic driving environments. Meanwhile, IMM-UKF leverages the interactive combination of results from different estimators to significantly enhance the overall accuracy of the hybrid estimator. This approach ensures reliable and precise estimation of TRFC, which is critical for vehicle traction control and safety.
7) A novel fault-tolerant estimation framework is proposed to estimate TRFC. Considering the challenges of missing measurements, an FTUKF is employed to predict longitudinal and lateral TRFCs, respectively. To further enhance estimation accuracy, an event-driven multimodel fusion algorithm is designed to dynamically adjust the weight of different models. This comprehensive approach ensures robust and accurate TRFC estimation, essential for maintaining vehicle stability and control under various driving conditions.
8) A model-based learning estimation scheme is proposed to effectively identify TRFC under challenging conditions of unknown measurement noise and data loss. The framework begins with the introduction of an ETCKF to handle missing measurement signals, ensuring robust data processing. Following this, the EKFNet algorithm is deployed to estimate TRFC, leveraging a combination of EKF principles and neural network predictions. This hybrid approach combines the strengths of model-based and data-driven techniques, providing a robust solution for TRFC estimation in real-world conditions.

The contributions summarized in this book highlight the significant advancements made by the authors in the field of nonlinear system state estimation for intelligent vehicles. Through the development of novel algorithms and frameworks such as AFTEKF, FARCKF, FTEKFNet, and various event-triggered and hybrid estimation methods, the authors address critical challenges in vehicle state estimation. These innovative approaches improve the accuracy, reliability, and robustness of state estimation, which are essential for the safety, performance, and comfort of intelligent vehicles. Researchers and engineers in the field will find these contributions invaluable for advancing the state of the art in vehicle estimation technology.

For professionals and researchers engaged in vehicle engineering, autonomous driving technology, and intelligent transportation systems, this book is an essential reference. Here are several reasons why:

Comprehensive Structure: The book is systematically structured, progressing from fundamental concepts to advanced techniques, making it suitable for readers at different levels. Each chapter has clear objectives and content layout, allowing readers to gradually deepen their understanding and mastery of the related knowledge.

High Practicality: Beyond theoretical discussions, the book provides numerous practical case studies and algorithm implementations, facilitating readers' application in real-world work. The examples are based on actual vehicle systems, with detailed explanations of the problem background, model construction, algorithm selection, implementation process, and experimental result analysis.

Cutting-Edge Technology: Covering the latest estimation technologies, the book helps readers stay abreast of industry developments. With the evolution of autonomous driving technology, the demand for vehicle–road interaction system estimation is continually changing and increasing. The book particularly focuses on recent research findings and technological advancements in the field, such as the application of machine learning in estimation and sensor fusion technology.

Interdisciplinary Application: The book is suitable for research and study in multiple disciplines, including vehicle engineering, control theory, computer science, and more. Vehicle–road interaction system estimation is a multidisciplinary field that involves control theory, signal processing, computer science, and vehicle engineering. The book provides detailed explanations of each technique's theoretical basis and application methods, benefiting readers from diverse backgrounds.

In conclusion, vehicle–road interaction system estimation is a complex and crucial discipline, with increasingly widespread and significant applications in modern vehicle engineering and autonomous driving technology. This book provides readers with a high-quality reference through comprehensive theoretical introductions, rich and practical case analyses, cutting-edge technological discussions, and multidisciplinary applicability. Whether you are a beginner or an experienced professional, you will find this book immensely beneficial. By reading and studying this book, readers can not only enhance their theoretical understanding of vehicle–road interaction system estimation but also gain valuable practical experience, laying a solid foundation for future research and application in the field.

Index

State Estimation of Multi-Agent Vehicle-Road Interaction Systems, First Edition.
Yan Wang, Guodong Yin, and Chao Huang.
© 2026 The Institute of Electrical and Electronics Engineers, Inc. Published 2026 by John Wiley & Sons, Inc.

Printed and bound by CPI Group (UK) Ltd, Croydon, CR0 4YY

17/09/2026

14964333-0003